Diet Composition, Eating Habits and Their Impact on Metabolic Diseases

Diet Composition, Eating Habits and Their Impact on Metabolic Diseases

Guest Editors

Silvia V. Conde
Fatima O. Martins

MDPI

Basel • Beijing • Wuhan • Barcelona • Belgrade • Novi Sad • Cluj • Manchester

Guest Editors
Silvia V. Conde
Faculdade de Ciências Médicas
Universidade NOVA de Lisboa
Lisboa
Portugal

Fatima O. Martins
Faculdade de Ciências Médicas
Universidade NOVA de Lisboa
Lisboa
Portugal

Editorial Office
MDPI AG
Grosspeteranlage 5
4052 Basel, Switzerland

This is a reprint of the Special Issue, published open access by the journal *Nutrients* (ISSN 2072-6643), freely accessible at: https://www.mdpi.com/journal/nutrients/special_issues/diet_eating_metabolic.

For citation purposes, cite each article independently as indicated on the article page online and as indicated below:

Lastname, A.A.; Lastname, B.B. Article Title. *Journal Name* **Year**, *Volume Number*, Page Range.

ISBN 978-3-7258-6127-9 (Hbk)
ISBN 978-3-7258-6128-6 (PDF)
https://doi.org/10.3390/books978-3-7258-6128-6

Contents

About the Editors

Silvia V. Conde

Silvia V. Conde is a Professor of Pharmacology and Neurosciences and Principal Investigator at NOVA Medical School (NMS), as well as a Distinguished Beatriz Galindo Senior Professor at the Faculty of Medicine of the University of Valladolid in Spain. She obtained her PhD in Pharmacology from NOVA University Lisbon (Portugal) and in Biotechnology from the University of Valladolid (Spain) in 2007. Her main research interests focus on inter-organ communication and brain–body physiology and pathophysiology, particularly the link carotid body–brain–body in the context of metabolic diseases.

Fatima O. Martins

Fatima O. Martins is a Biomedical Researcher and Affiliated Professor at NOVA Medical School, NOVA University Lisbon, Portugal. She holds a PhD in Experimental Biology and Biomedicine and has extensive experience in metabolic disease research, focusing on glucose homeostasis, insulin sensitivity, and adipose tissue biology. Her work bridges basic and translational research, combining molecular and physiological approaches to explore metabolic regulation under physiological and pathological conditions. Beyond her academic activity, she has been involved in biotechnology entrepreneurship and actively supervises undergraduate and graduate students, fostering scientific innovation and collaboration across disciplines. Moreover, her commitment to science goes beyond the laboratory; she is deeply engaged in science communication and outreach activities, aiming to make scientific knowledge accessible to the broader community.

Preface

This Reprint presents a collection of studies examining the interplay between diet, eating behaviors, and metabolic health. The contributions explore how nutrient composition, specific foods, and environmental factors influence obesity, insulin resistance, and related metabolic disorders. Topics range from proteomic profiling of extracellular vesicles to the effects of golden berry, rare sugars, and cannabis sativa, providing both mechanistic insights and translational implications.

The Reprint highlights how dietary patterns affect peripheral metabolism, central regulation, and the gut–liver–muscle axis, including developmental and sex-dependent effects. By integrating experimental, clinical, and behavioral research, it offers a comprehensive perspective on the factors shaping metabolic outcomes.

This Reprint is intended for researchers, clinicians, and nutrition professionals seeking to understand the biological and behavioral determinants of metabolic health. It underscores the value of interdisciplinary approaches in developing nutrition-based strategies to prevent and manage metabolic diseases.

We sincerely thank all authors and reviewers whose contributions made this Reprint possible, and whose expertise and dedication continue to advance the field of nutrition and metabolic research.

Silvia V. Conde and Fatima O. Martins
Guest Editors

nutrients

Article

Proteomic Profiling of Plasma- and Gut-Derived Extracellular Vesicles in Obesity

Pedro Baptista Pereira [1], Estefania Torrejón [1], Inês Ferreira [1], Ana Sofia Carvalho [2], Akiko Teshima [1], Inês Sousa-Lima [1], Hans Christian Beck [3], Bruno Costa-Silva [4], Rune Matthiesen [2], Maria Paula Macedo [1,*] and Rita Machado de Oliveira [1,*]

[1] Metabolic Diseases Research Group, iNOVA4Health, NOVA Medical School, Faculdade de Ciências Médicas, Universidade NOVA de Lisboa, 1169-056 Lisboa, Portugal; pedro.baptista.pereira@edu.nms.unl.pt (P.B.P.); estefania.torrejon@nms.unl.pt (E.T.); ines.ferreira@eithealth.eu (I.F.); akiko.teshima@nms.unl.pt (A.T.); ines.lima@nms.unl.pt (I.S.-L.)

[2] Computational and Experimental Biology Group, iNOVA4Health, NOVA Medical School, Faculdade de Ciências Médicas, Universidade NOVA de Lisboa, 1169-056 Lisboa, Portugal; ana.carvalho@nms.unl.pt (A.S.C.); rune.matthiesen@nms.unl.pt (R.M.)

[3] Centre for Clinical Proteomics, Department of Clinical Biochemistry, Odense University Hospital, DK-5000 Odense, Denmark; hans.christian.beck@rsyd.dk

[4] Champalimaud Physiology and Cancer Programme, Champalimaud Foundation, 1400-038 Lisboa, Portugal; bruno.costa-silva@research.fchampalimaud.org

* Correspondence: paula.macedo@nms.unl.pt (M.P.M.); rita.oliveira@nms.unl.pt (R.M.d.O.)

Abstract: Obesity entails metabolic alterations across multiple organs, highlighting the role of inter-organ communication in its pathogenesis. Extracellular vesicles (EVs) are communication agents in physiological and pathological conditions, and although they have been associated with obesity comorbidities, their protein cargo in this context remains largely unknown. To decipher the messages encapsulated in EVs, we isolated plasma-derived EVs from a diet-induced obese murine model. Obese plasma EVs exhibited a decline in protein diversity while control EVs revealed significant enrichment in protein-folding functions, highlighting the importance of proper folding in maintaining metabolic homeostasis. Previously, we revealed that gut-derived EVs' proteome holds particular significance in obesity. Here, we compared plasma and gut EVs and identified four proteins exclusively present in the control state of both EVs, revealing the potential for a non-invasive assessment of gut health by analyzing blood-derived EVs. Given the relevance of post-translational modifications (PTMs), we observed a shift in chromatin-related proteins from glycation to acetylation in obese gut EVs, suggesting a regulatory mechanism targeting DNA transcription during obesity. This study provides valuable insights into novel roles of EVs and protein PTMs in the intricate mechanisms underlying obesity, shedding light on potential biomarkers and pathways for future research.

Keywords: extracellular vesicles; obesity; prediabetes; metabolism; proteomics; post-translational modifications; protein acetylation; protein glycation

1. Introduction

Obesity is a growing global concern, with over 1 billion people affected, and unfortunately, the numbers continue to rise [1]. While genetics play a role, the increasing global prevalence of obesity is partially due to environmental factors, such as dietary habits, which are major contributors to the obesity epidemic [2–4]. The adoption of a Westernized lifestyle, which promotes greater consumption of calorie-rich and palatable foods while reducing physical activity, is a root cause of the obesity epidemic. For instance, overconsumption of foods with over 30% of total daily energy intake from fat, has contributed to the increase in obesity in the past years [3–5]. Additionally, the consumption of high-fat foods is strongly associated with a pro-inflammatory state, leading to deleterious consequences for organismal homeostasis [6]. Furthermore, excessive consumption of certain carbohydrates, particularly rapidly digestible carbohydrates, has been linked to suboptimal glycemic control in individuals with diabetes [7]. Therefore, obesity arises from an imbalance between energy intake and expenditure, resulting in excess energy being stored as fat in various organs, especially in adipose tissue [8]. This excess fat is associated with systemic complications, including hypertension, dyslipidemia, insulin resistance, and type 2 diabetes (T2D), which pose a significant challenge to healthcare [9–11].

Obesity impacts all organs; thus, its progression and associated comorbidities are modulated by inter-organ communication networks. Importantly, extracellular vesicles (EVs) play a crucial role in this complex inter-organ dialogue, intimately linked to the pathophysiology of obesity and diabetes [12,13]. EVs are bilayer vesicles that typically measure between 40 and 5000 nm in diameter, acting as carriers of diverse bioactive molecules, such as proteins, nucleic acids, and lipids [14–17]. Several studies have highlighted the impact of organ-derived EVs on metabolic homeostasis [16,18–20]. Importantly, growing evidence strongly suggests that EV-mediated crosstalk between adipose tissue, liver, and skeletal muscle is a key contributor to the development of insulin resistance [21–25]. Furthermore, EVs released by adipose tissue trigger abnormal activation of immune cells and endothelial dysfunction, which could explain vascular complications linked to obesity [11,26,27]. Crucially, when lean mice were administered EVs isolated from control mice but loaded with miRNAs associated with obesity, it resulted in increased glucose intolerance and hepatic steatosis [28]. Conversely, the administration of EVs from lean mice to obese mice ameliorated these conditions [29]. This highlights the significant role of miRNAs carried by EVs in triggering obesity-associated traits, even within a healthy context.

Although the miRNA content of EVs has received considerable attention, there is a knowledge gap concerning the protein cargo of these vesicles in the context of diet-induced obesity. Recently, we observed significant alterations in the protein content of gut-derived EVs isolated from a diet-induced obese mouse model [30]. The gut plays a significant role in maintaining metabolic homeostasis and is a key player in the development of obesity through various actions, including microbiota composition, hormone secretion and nutrient absorption, and its crucial connection to the central nervous system through the gut–brain axis [31,32]. Interestingly, bariatric surgery, the most effective intervention for weight loss, involves alterations in the anatomy of the small intestine and leads to T2D remission even before significant weight loss occurs [33,34]. Individuals with T2D who undergo bariatric surgery and achieve T2D resolution display distinct protein profiles in their circulating EVs, setting them apart from those in whom T2D persists post-surgery [35]. This underscores the importance of defining proteome profiles of gut EVs in the context of obesity and their potential role as predictors of resolution for obesity-related comorbidities following bariatric surgery.

Circulating EVs form a dynamic network that facilitates inter-organ communication by transporting a diverse array of organ-derived EVs, collectively forming a heterogeneous pool of EVs. These EVs act as messengers, transmitting vital molecular information across various cells and organs. It is worth noting that the number of small EVs is elevated in the circulation of patients with insulin resistance [36] and in individuals with T2D [37]. Notably, obese individuals exhibit a substantial (approximately 10-fold) increase in plasma EV levels compared to those maintaining a healthy weight [38–40]. On the contrary, interventions such as hypocaloric diets, exercise, and weight loss through bariatric surgery have been shown to reduce plasma EV levels [38,39,41].

Protein post-translational modifications (PTMs) play a significant role in obesity and diabetes by influencing insulin signaling and glucose metabolism, and thus, the development of associated complications [42–44]. Two significant post-translational modifications (PTMs) are acetylation and glycation. The first, acetylation, is closely linked to increased levels of acetyl-CoA and NAD+. The second, glycation, is connected to elevated blood glucose levels [45–49]. While lysine acetylation is thought to play a crucial role in maintaining energy homeostasis [50], limited reports suggest that acetylation impacts the sorting of specific proteins and/or RNA molecules into EVs [51]. Moreover, lysine acetylation is important in both immunological and metabolic pathways and helps in maintaining the equilibrium between energy storage and expenditure [52]. Glycation is a spontaneous chemical reaction between certain amino acids and reducing sugars, leading to the formation of advanced glycation end-products (AGEs) [53,54]. AGEs are considered to be the primary culprits behind various diabetic complications [55,56]. Importantly, one major precursor of AGE formation is methylglyoxal, a byproduct of glycolysis, strongly linking glycation to hyperglycemia, diabetes, and obesity [46,49]. Interestingly, glycated hemoglobin (HbA1c) is used as a marker for the diagnosis of diabetes [57,58]. Nevertheless, there is a clear need for further research delving into these PTMs in the context of metabolic diseases.

Here, we evaluate the protein content of EVs in obesity, providing crucial insights into proteomic changes and PTMs in plasma and gut EVs. This sheds light on the complex mechanisms of diet-induced obesity and prediabetes, highlighting potential biomarkers and pathways for future research.

2. Materials and Methods

2.1. Mouse Models

Six-week-old male C57Bl/6J mice were housed in a temperature-controlled room under a regular light/dark cycle of 12 h and with food and water ad libitum. For induction of prediabetes, mice were fed a high-fat diet (HFD) (OpenSource Diet, D12331) composed of 16.5% protein, 25.5% carbohydrate, and 58% fat with 13% addition of sucrose for 12 weeks. Control mice were fed a normal chow diet (NCD) (Special Diets Services, RM3) composed of 26.51% proteins, 62.14% carbohydrates, and 11.35% fat for the same period. After 12 weeks of diet, mice were subjected to an 8 h fasting period, followed by a 2 h feeding window, concluding with a 12 h fasting period. Mice were then anesthetized with isoflurane and euthanized. Blood was collected into a heparinized tube, which was then centrifuged at 500× *g* for 10 min. The supernatant was collected and centrifuged at 3000× *g* for 20 min to obtain plasma, which was stored at −80 °C for EV isolation. Three biological replicates were performed per experimental condition (diet-induced obesity and control groups). Each replica was generated by pooling plasma from 20 animals. Experimental protocols were approved by the ethics committee of the NOVA Medical School (nr.82/2019/CEFCM).

2.2. Intra-Peritoneal Glucose Tolerance Test (ipGTT)

In the 11th week of diet, an intra-peritoneal glucose tolerance test (ipGTT) was performed after an overnight fasting. Basal glycemia and weight were measured, and then an intra-peritoneal injection of glucose (Sigma Aldrich, St. Louis, MO, USA) at 2 g/kg body weight was administered. Blood glucose levels were measured by tail tipping at 15, 30, 60, 90 and 120 min after the injection using a OneTouch Ultra glucose meter (LifeScan Inc., Milpitas, CA, USA). The evaluation of the glycemic response was performed by calculating the total area under the whole glucose excursion, using the blood glucose concentration at timepoint 0 min as the baseline.

2.3. Hematoxylin–Eosin Staining

The liver was rinsed with phosphate buffered saline (PBS) and placed in 2% paraformaldehyde (PFA) with 20% sucrose overnight at 4 °C. After three PBS washes, the liver was transferred to 30% sucrose solution for 2 h at 4 °C, followed by an overnight incubation in optimal cutting temperature (OCT) compound with 30% sucrose solution (1:1 ratio). Next, the liver was embedded in molds using a mixture of OCT compound and 20% sucrose solution (3:1 ratio) and stored at −80 °C. Then, the liver was sliced into 6 µm thin sections in the cryostat and stored at −80 °C for following hematoxylin–eosin + Oil Red O staining.

2.4. Extracellular Vesicles Isolation

Plasma stored at −80 °C was centrifuged at 12,000× *g* for 20 min, supernatant was then subjected to ultracentrifugation (Beckman Ti70, rotor 70Ti, Brea, CA, USA) at 100,000× *g*, for 140 min. The EV-enriched pellet was collected. This pellet was resuspended in filtered PBS and layered on top of a sucrose solution. This sucrose solution was prepared with 30 g of protease-free sucrose (Sigma), 2.4 g of Tris-base (Sigma) in 100 mL of D_2O (Sigma); pH was adjusted to 7.4. Resuspended EVs over the sucrose cushion were centrifuged at 100,000× *g*, for 70 min. The fraction of sucrose cushion containing EVs was collected and transferred to a new tube with PBS. An overnight centrifugation was then performed at 100,000× *g*, and the EV pellet was collected and resuspended in filtered PBS.

2.5. Nanoparticle Tracking Analysis (Nanosight)

The concentration and size of EVs were analyzed in a NanoSight NS300 (NS3000) (Malvern Panalytical, Malvern, UK) following the manufacturer's guidelines. Briefly, EVs were diluted at a ratio of 1:1000 in filtered sterile PBS. Each sample was analyzed for 90 s, with measurements taken five times using the NanoSight automatic settings.

2.6. Protein Quantification

We prepared protein extracts using a lysis buffer composed of 20 mM Tris-HCl at pH 7.4, 5 mM EDTA at pH 8.0, 1% Triton-X 100, 2 mM Na_3VO_4, 100 mM NaF, and 10 mM $Na_4P_2O_7$, supplemented with protease inhibitors (cOmplete™, Mini, EDTA-free protein inhibitor cocktail tablets, Roche (Basel, Switzerland), Sigma). Plasma EVs were homogenized in lysis buffer and subjected to three rounds of sonication (Sonifier SFX 150, Branson, MO, USA), each lasting 10 s at 10 μm amplitude, with cooling on ice between each sonication. Lysates underwent centrifugation at 18,000× *g* for 10 min at 4 °C. The resulting supernatant was collected, and the total protein concentration was determined using the Pierce™ BCA Protein Assay kit (Thermo Fisher, Waltham, MA, USA).

2.7. Nano-LC-MS/MS Analysis

Each biological replicate was analyzed twice, yielding two technical replicates. Peptide samples were analyzed by nano-LC-MS/MS (Dionex RSLCnano 3000, Sunnyvale, CA, USA) coupled to an Exploris 480 Orbitrap mass spectrometer (Thermo Scientific, Hemel Hempstead, UK) virtually as previously described [59]. In brief, peptide samples (5 μL) were loaded onto a custom-made fused capillary pre-column (2 cm length, 360 μm OD, 75 μm ID) packed with ReproSil Pur C18 5.0 μm resin (Dr. Maish, Ammerbuch-Entringen, Germany) with a flow of 5 μL per minute for 6 minutes. Trapped peptides were separated on a custom-made fused capillary column (25 cm length, 360 μm outer diameter, 75 μm inner diameter) packed with ReproSil Pur C18 1.9-μm resin (Dr. Maish) with a flow of 250 nL per minute using a linear gradient from 89% A (0.1% formic acid) to 32% B (0.1% formic acid in 80% acetonitrile) over 56 min.

Mass spectra were acquired in positive ion mode applying an automatic data-dependent switch between one Orbitrap survey MS scan in the mass range of 350–1200 m/z followed by higher-energy collision dissociation (HCD) fragmentation and Orbitrap detection of fragment ions with a cycle time of 2 s between each master scan. MS and MSMS maximum injection time were set to "Auto", and HCD fragmentation in the ion routing multipole was performed at normalized collision energy of 30%, and ion selection threshold was set to 10,000 counts. Selected sequenced ions were dynamically excluded for 30 s. MS resolution was 120,000 and MS/MS resolution was 15,000.

2.8. Database Search

Mass accuracy was set to 5 ppm for peptides and 0.01 Da for ionized fragments. Trypsin cleavage allowing a maximum of four missed cleavages was used. Carbamidomethyl was set as a fixed modification. Methionine oxidation, lysine and N-terminal protein acetylation (Figure S1), lysine glycation (carboxymethyl, carboxyethyl, and pyrraline) (Figure S2), arginine glycation (argpyrimidine, glyoxal and methylglyoxal-derived hydroimidazolones, and tetrahydro pyrimidine), glutamine deamidation and asparagine deamidation were set as variable modifications. The MS/MS spectra were searched against all reviewed protein sequences available in a standard mouse proteome database from UniProt (UP000000589). For the search, all protein sequences were also concatenated in reverse order, with lysine and arginine residues maintained in their original positions. The data were searched and quantified with both MaxQuant [60] and VEMS [61].

2.9. Proteomic Functional Enrichment Analysis

The results from MaxQuant and VEMS searches were processed using Python and then subjected to functional enrichment analysis in R with the clusterProfiler [62] package. Spectral count [63] and intensity-based absolute quantification (iBAQ) [64] values were calculated for all identified proteins. Quantitative data were first preprocessed for normalization. This step was carried out using the 'normalize.quantiles' function from the 'preprocessCore' R package. Subsequently, the normalized data underwent a log2 transformation, incremented by one, to stabilize the variance and improve the analytical conditions for detecting differential expression. The differential expression analysis itself was conducted utilizing the 'limma' R package, setting the 'trend' set to FALSE [65]. To select proteins exclusively identified in one experimental group, the selection criteria were based on the spectral count of proteins in both groups, defined as the number of MS/MS spectra associated with peptides from a given protein. For instance, if a protein was associated with at least one spectrum in one group but none in the other group, it was categorized as exclusively identified in the former. Proteins with spectra identified in samples from both groups were considered shared between HFD-PDEV and NCD-PDEV. Genes corresponding to proteins identified with a false discovery rate (FDR) lower than 0.01 underwent over-representation analysis (ORA) [66], using the Gene Ontology [67,68] and KEGG [69–71] databases.

2.10. Statistical Analysis

Data were analyzed using GraphPad Prism Software, version 9.0.0 (GraphPad Software Inc., San Diego, CA, USA) and presented as mean values with the standard error of the mean (SEM). The significance of the differences between the groups was calculated by unpaired student's *t*-tests with Welch's correction. Differences were considered significant as p-value ≤ 0.05.

3. Results

3.1. Evaluation of Diet-Induced Obesity's Effects on Plasma Derived EVs

To investigate the impact of diet-induced obesity on protein content and PTMs within EVs, we established a murine model designed to replicate obesity hallmarks. A high fat and sugar diet (HFD) was initiated at 6 weeks of age and continued for a duration of 12 weeks to induce obesity in mice (Figure S3a). Mice on the HFD exhibited an increase in body weight and hepatic lipid accumulation compared to the control group (Figure S3b,c). To further assess the metabolic impact of obesity, at 11 weeks of diet, both HFD and NCD mice underwent an intra-peritoneal glucose tolerance test (ipGTT). Mice on the HFD displayed reduced glucose removal from the circulation compared to NCD mice, thus indicating impaired glucose tolerance as a consequence of obesity (Figure S3d,e).

Next, we analyzed the presence of proteins, as well as acetylation and glycation events within plasma EVs in the context of obesity. We isolated plasma EVs, which poses a substantial challenge, particularly when dealing with reduced volume samples from murine models. To address this limitation, we pooled samples from 20 animals for each biological replicate. Plasma EVs were characterized by nanoparticle tracking analysis (NTA) and total protein content was quantified by BCA. Both NCD and HFD plasma EVs presented mean sizes ranging between 100 and 150 nm (Figure S4), which falls well within the size range for small EVs, particularly exosomes [15]. We refrained from classifying these EVs as exosomes, since we did not analyze their intracellular origin. NTA and protein quantification revealed a higher particle number and total protein concentration in EVs isolated from HFD mice, resulting in a similar protein content per particle in both experimental groups (Figure 1a–c). Similarly, an increase in circulating EVs under obesity and diabetes was observed in human studies [38–40,72].

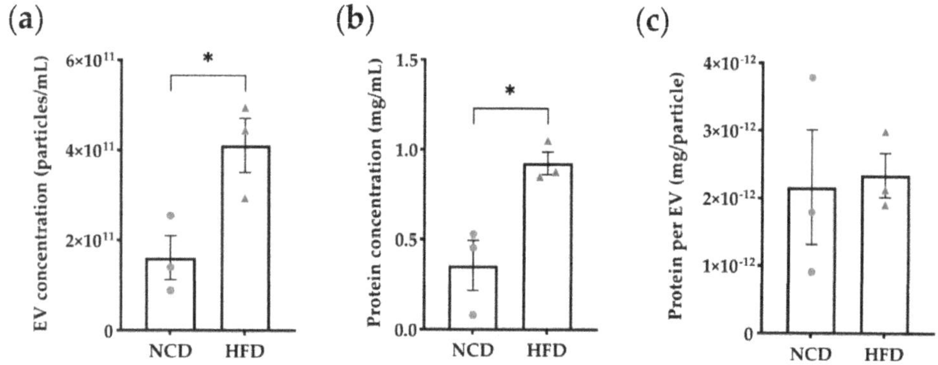

Figure 1. Diet-induced obesity increases the number of plasma extracellular vesicles. (**a**) Analysis of the number of particles per mL of sample from control (NCD) and diet-induced obese mice (HFD). (**b**) Protein quantification in plasma EVs in mg/mL, obtained by BCA. (**c**) Protein content per EV, represented in mg of protein per particle. All statistical analysis were performed using unpaired *t*-test with Welch's correction. All data are presented as mean $\pm$ standard error of the mean. * p-value < 0.05. Each n represents a sample of 20 animals, $n = 3$ for both groups, with blue dots representing NCD samples and orange triangles representing HFD samples.

3.2. Proteomic Analysis of Plasma Derived EVs

Next, we analyzed the protein content within plasma EVs isolated from HFD and NCD mice. Trypsinized samples were separated by liquid chromatography, ionized by electrospray ionization, and analyzed by tandem mass spectrometry (Figure 2a). A total of 340 proteins were identified with statistical significance (False Discovery Rate < 0.01). Among these proteins, five were exclusively identified in HFD EVs (Figure 2b,c), 112 proteins were exclusively identified in NCD EVs (Figure 2b, Table S1); and 223 were common to both groups (Figure 2b). Notably, the vast number of proteins exclusively identified in EVs isolated from NCD-fed animals accounts for a considerable portion of the overall identified proteins. Thus, we observed a decline in the number of protein species carried by plasma EVs in an obesity environment. The 223 proteins identified in both groups were subjected to intensity-based absolute quantification (iBAQ) and differential expression analysis. The resulting $\log_2$ (fold change) values and their respective $-\log_{10}$ (p-value) are depicted in Figure 2d. From those, 12 proteins were found to be regulated (p-value < 0.05). Within the regulated proteins (Figure 2e), two proteins were upregulated (immunoglobulins), and 10 proteins were downregulated (14-3-3 protein isoforms, proteasome subunits, annexin A7 among others).

Over-representation analysis (ORA) was conducted to gain a better understanding of the identified proteins. In this analysis, we evaluate whether proteins exclusively identified in one experimental group are more enriched in specific functional terms compared to all proteins in that same experimental group, whether exclusively identified or not. This approach allowed us to pinpoint functional terms that are particularly impacted by diet-induced obesity. ORA was conducted using the Gene Ontology database, encompassing three categories: biological processes, cellular components and molecular functions, as well as the KEGG database for both NCD- and HFD-exclusive proteins. However, we observed enrichment of functional terms meeting the significance threshold (adjusted p-value < 0.05) solely in the molecular functions ORA for proteins exclusively identified in NCD plasma

EVs (Figure 3). This enrichment pointed to an over-representation of functions related to protein folding, phosphatase activity, and hydrolase activity, suggesting the relevance of these functional terms in maintaining metabolic homeostasis. Interestingly, each of these enriched functions was associated with at least 10 out of 16 genes, indicating a high degree of interaction between these functional terms.

(a)

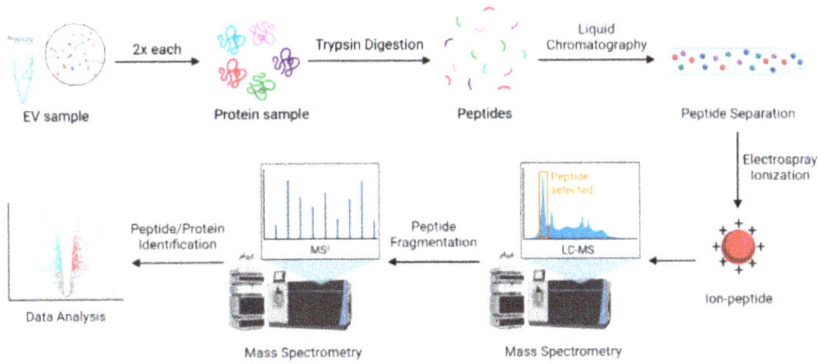

(b)

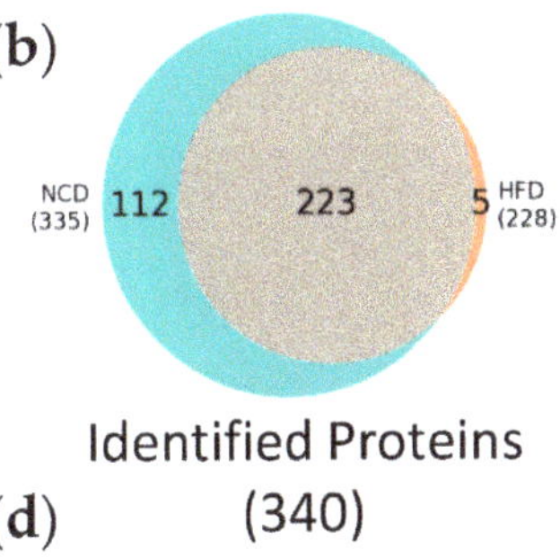

(c)

Protein	Gene	FDR
Apolipoprotein A-II	Apoa2	0
Argininosuccinate lyase	Asl	0.0097
Histamine H4 receptor	Hrh4	0.0093
Homogentisate 1,2-dioxygenase	Hgd	0.0094
Ig heavy chain V region VH558 A1/A4	Gm5629	0

(d)

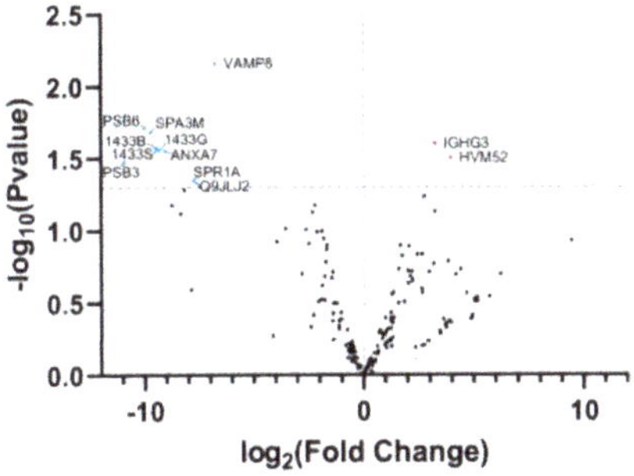

(e)

Protein	Gene	log2(FC)	P-value
14-3-3 protein beta/alpha	Ywhab	-9.4670	0.0266
14-3-3 protein gamma	Ywhag	-9.3241	0.0276
14-3-3 protein sigma	Sfn	-9.4054	0.0270
4-trimethylaminobutyraldehyde dehydrogenase	Aldh9a1	-7.6385	0.0461
Annexin A7	Anxa7	-9.0542	0.0279
Cornifin-A	Sprr1a	-7.7525	0.0450
Ig gamma-3 chain C region	-	3.2515	0.0250
Ig heavy chain V region VH558 A1/A4	Gm5629	3.9598	0.0316
Proteasome subunit beta type-3	Psmb3	-10.9268	0.0334
Proteasome subunit beta type-6	Psmb6	-10.0063	0.0191
Serine protease inhibitor A3M	Serpina3m	-9.7230	0.0205
Vesicle-associated membrane protein 8	Vamp8	-6.7579	0.0069

Figure 2. Comparative protein profile in plasma-derived extracellular vesicles from NCD and HFD-fed mice. (**a**) Schematic representation illustrating the experimental workflow for the analysis of plasma EVs. Each sample, obtained from mice fed either an NCD or an HFD, underwent two rounds of analysis using liquid chromatography-tandem mass spectrometry (LC-MS/MS). (**b**) Venn diagram representing the overlap of proteins identified in NCD and HFD plasma EVs with a false discovery rate (FDR) < 0.01. (**c**) List of proteins exclusively identified in HFD plasma EVs. (**d**) Volcano plot representing the fold change and *p*-value of regulated proteins shared between NCD and HFD plasma EVs. Dotted horizontal line indicates the threshold for a *p*-value < 0.05. A positive $\log_2$ (fold change) value indicates higher protein levels in diet-induced obese mice when compared to control mice. Down-regulated proteins are represented in light blue, while up-regulated proteins are indicated in light red. (**e**) List of significantly regulated proteins (*p*-value < 0.05) in plasma EVs from HFD mice compared to NCD mice.

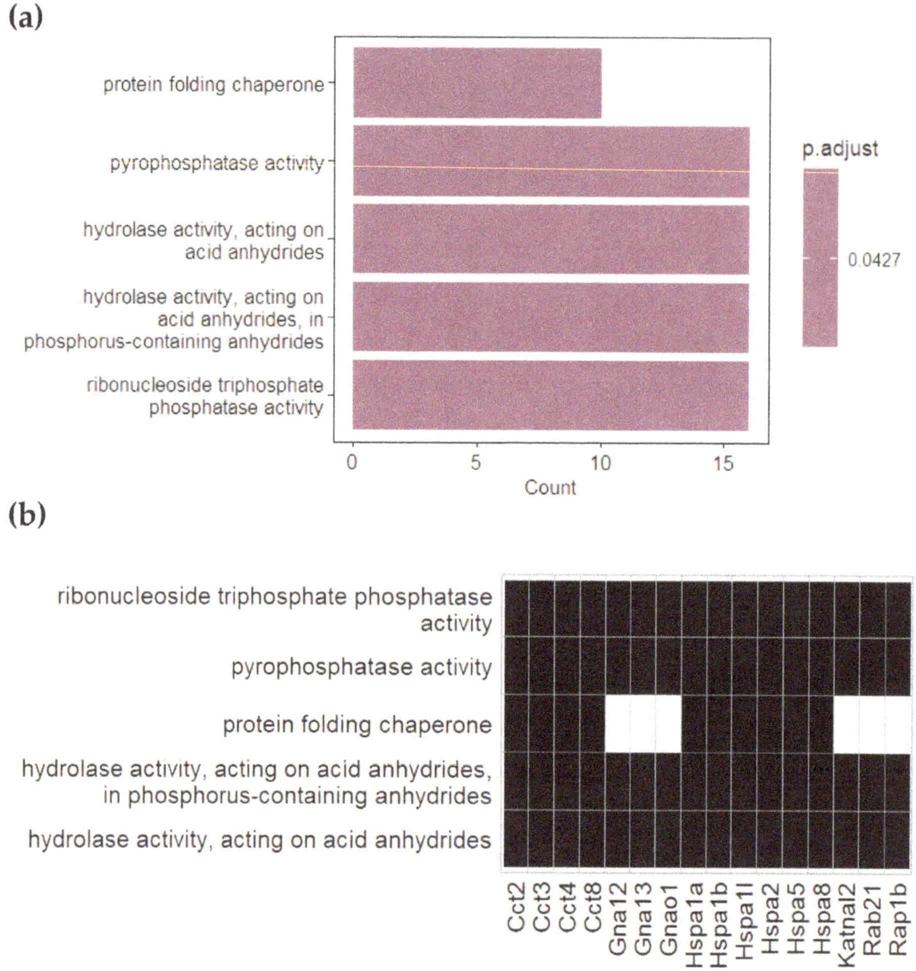

Figure 3. Gene Ontology over-representation analysis of plasma EVs proteins identified exclusively in NCD mice. (**a**) Bar plot and (**b**) table of enriched molecular functions and their related identified genes. Black rectangles indicate the association between the identified gene and the enriched term.

3.3. Plasma and Gut EVs Crosstalk

Previously, our group revealed changes in the protein content of gut EVs in the context of obesity-associated dysmetabolism [30]. These prior findings hinted at a key role of gut-derived EVs in the spread of the metabolic dysfunction. As such, understanding the intricate crosstalk between plasma and gut EVs becomes paramount to the identification of mechanisms involved in the development of obesity and importantly to the identification of possible biomarkers.

We analyzed the proteins that were shared between plasma and gut EVs (Figure 4a). While no proteins were exclusively detected in the HFD group, we identified four proteins that were exclusive to the control group of both gut and plasma EVs (Figure 4b) (complement factor H, mannose-binding Protein A, Ig heavy chain V region TEPC 1017, and Ig gamma-2A chain C region (secreted form)). The loss of these four proteins in both plasma and gut EVs with the onset of obesity suggests their crucial role in maintaining a healthy homeostatic state and their high potential as biomarkers. Future investigation should address the molecular mechanisms by which their absence may induce or result from obesity.

(a)

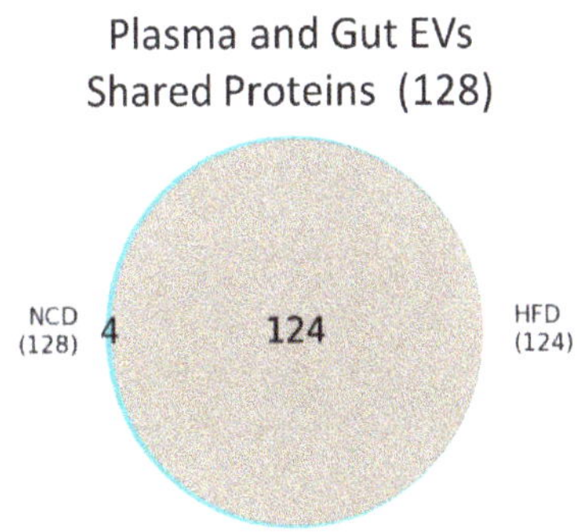

(b)

Protein	Gene	FDR Plasma	FDR Gut
Complement factor H	Cfh	0.0000	0.0053
Ig gamma-2A chain C region secreted form	-	0.0000	0.0033
Ig heavy chain V region TEPC 1017	-	0.0070	0.0035
Mannose-binding protein A	Mbl1	0.0000	0.0051

Figure 4. Obesity-associated changes in proteins shared between plasma and gut EVs. (**a**) Venn diagram displaying the intersection of proteins identified in HFD mice and NCD mice, from all proteins shared between PDEV and GDEV. All proteins presented an FDR < 0.01. (**b**) List of four proteins shared between PDEV and GDEV, present exclusively in NCD mice.

3.4. Post-Translational Modifications (PTMs) in Plasma and Gut EVs Proteins

PTMs act as precise molecular switches capable of modulating proteins' stability, localization, interactions, and activity. Hence, PTMs enhance the functional complexity of proteins beyond what can be determined solely from their amino acid sequence and folding. To evaluate shifts in PTMs' profiles, we limited our analysis to modified peptides from proteins shared between HFD and NCD EVs. In the case of plasma EVs, from the 223 proteins shared between both diet groups, nine proteins were acetylated with an FDR < 0.01 (Figure 5a), with four of these proteins being exclusively acetylated in the NCD group (Table S2). Additionally, we detected 11 glycated proteins with an FDR < 0.01 (Figure 5b), six of which were exclusively glycated in the NCD group (Table S3). Although the total number of proteins was larger for NCD plasma EVs, these findings suggest a decrease in both acetylation and glycation events in the context of obesity. However, validation of these results is necessary to draw further conclusions.

Regarding gut EVs, a total of 2446 proteins were identified under the significance threshold (FDR < 0.01). Among these proteins, 1804 were shared between HFD and NCD EVs. Next, following the same approach employed for plasma EVs, we focused on acetylated and glycated proteins within the shared set of 1804 proteins. Our analysis revealed 203 acetylated proteins (Figure 5c), with 41 proteins exclusively acetylated in NCD (Table S4) and 29 proteins exclusively acetylated in HFD (Table S5). Interestingly, these numbers indicate a decrease in acetylation events under obesity, mirroring the observations made for acetylated proteins in plasma EVs. For glycation, a total of 56 proteins were found to be glycated in gut EVs (Figure 5d), with 23 proteins exclusively glycated in NCD (Table S6) and 22 proteins exclusively glycated in HFD (Table S7).

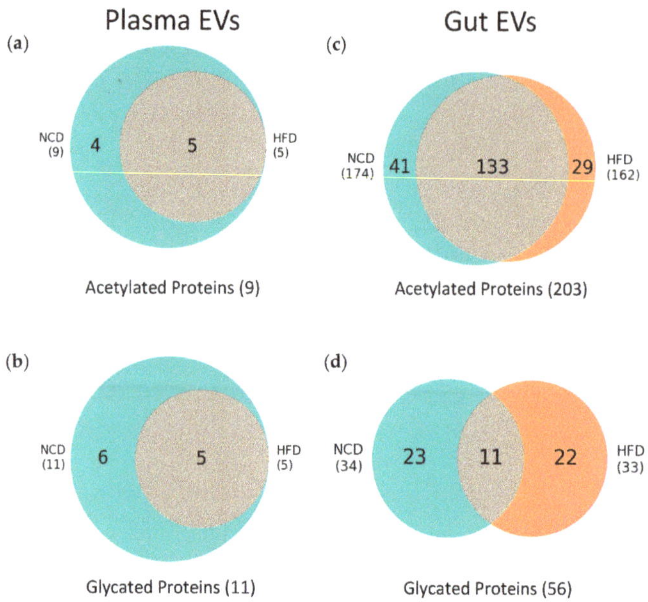

Figure 5. Analysis of protein acetylation and glycation in plasma and gut EVs. (**a**) Venn diagram representing the intersection of acetylated proteins identified in NCD and HFD plasma EVs. (**b**) Venn diagram representing the intersection of glycated proteins identified in NCD and HFD plasma EVs. (**c**) Venn diagram representing the intersection of acetylated proteins identified in NCD and HFD gut EVs. (**d**) Venn diagram representing the intersection of glycated proteins identified in NCD and HFD gut EVs. All proteins presented an FDR < 0.01. The identification of acetylated and glycated proteins was only conducted on proteins previously identified in both NCD and HFD samples (223 proteins for plasma EVs and 1804 proteins for gut EVs).

Acetylated and glycated proteins exclusively modified in either NCD or HFD gut derived EVs were subjected to ORA. For these analyses, the list of 1804 proteins identified in both NCD and HFD gut EVs was used as the background reference. Considering Gene Ontology for cellular compartments (Figure 6a), proteins exclusively acetylated in HFD gut EVs and proteins exclusively glycated in NCD gut EVs exhibited significant enrichment in protein-DNA complexes, nucleosome, chromatin, and polymerase complexes, while proteins exclusively glycated in HFD gut EVs were primarily associated with the cytoskeleton and actin filaments. Regarding Gene Ontology for molecular functions (Figure 6b), proteins exclusively acetylated in HFD gut EVs were enriched in structural constituents of the chromatin, while proteins exclusively glycated in NCD gut EVs exhibited a significant enrichment in structural constituents of the chromatin, DNA binding and protein heterodimerization. In addition, proteins exclusively glycated in HFD gut EVs showed significant enrichment in structural constituents related to the cytoskeleton, synapse, and post synapse.

Considering KEGG pathways (Figure 6c), proteins exclusively acetylated in HFD gut EVs and proteins exclusively glycated in NCD gut EVs exhibited significant enrichment in pathways related to viral carcinogenesis, systematic lupus erythematosus, alcoholism, and neutrophil extracellular trap formation. It is noteworthy that histones were associated with all four of these enriched KEGG pathways (Figure S5g,i). On the other hand, proteins exclusively glycated in HFD gut EVs were enriched in terms that were mainly related to apoptosis, tight junctions, motor proteins, and the phagosome. Importantly, in gut EVs, proteins exclusively glycated in HFD, the most enriched functional term across these three analyses were consistently linked to actin and tubulin (Figure S5b,e,h).

Altogether, these findings indicate that EVs transport proteins with several PTMs and this profile is altered by obesity. Our proteomic data can serve as the basis to develop future studies to understand how PTMs of proteins present in EVs can be modulated to contain obesity progression.

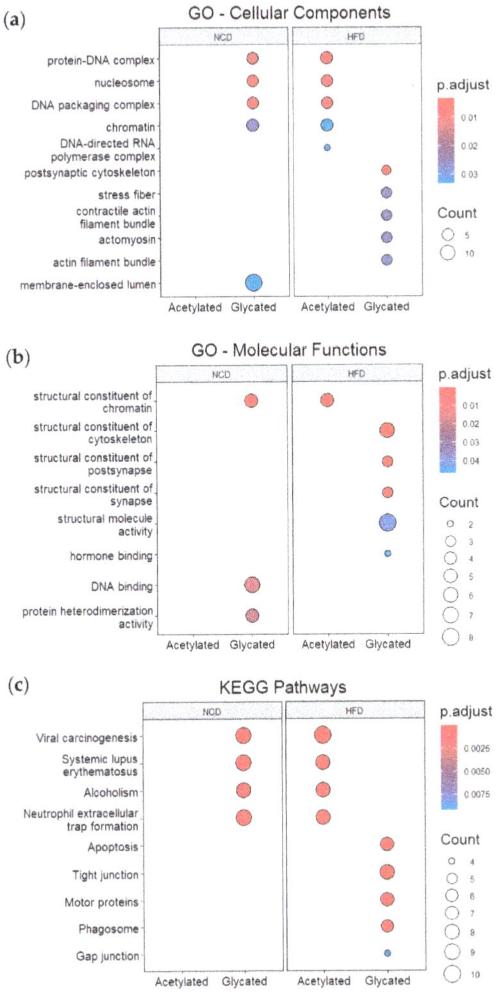

Figure 6. Over-representation analysis of gut EVs proteins featuring acetylation and glycation. (**a**) GO enrichment analysis according to cellular components. (**b**) GO enrichment analysis according to molecular functions. (**c**) KEGG enrichment analysis for KEGG pathways. Dot plots represent the five most significantly enriched terms in each group.

4. Discussion

Given the rapidly changing global dietary patterns, the prevalence of obesity and its associated health conditions continues to rise. Implementing interventions to combat obesity could serve as a potent strategy to mitigate multiple diseases that exert a significant socioeconomic impact. Hence, it is crucial to understand the molecular mechanisms underlying obesity for developing efficacious treatments. As obesity involves systemic interactions depending on inter-organ communication, EVs play a crucial role. Importantly, several studies have highlighted the relevance of EV cargo in maintaining metabolic homeostasis and enabling cell communication [16,18–20]. The protein content of EVs varies depending on the organ they originate from. As a result, EVs found in the bloodstream comprise a diverse mix originated from different organs, aiding in inter-organ communication. The possibility of isolating EVs from body fluids enhances their potential as biomarkers for diagnosis and/or prediction of treatment outcomes. Moreover, they serve as therapeutic tools for various diseases. In this study, we present the first proteomics analysis of plasma EVs isolated from obese animals and their overlap with gut EVs.

The obese male murine model we used not only exhibits increased body weight, but it also reflects obesity comorbidities, such as the hallmark features of prediabetes exemplified by glucose intolerance and liver steatosis indicated by hepatic lipid deposition (Figure S3). Although studies of gender impact on EV content are scarce and report inconsistent information, some have reported gender-related differences in protein, lipid, and RNA composition, and in physiological and pathological conditions. Therefore, the exclusive use of male mice limits the generalizability of our findings [73–77]. Consistent with findings from human studies [72], we observed an increase in the number of plasma EVs under obesity conditions (Figure 1). A total of 340 proteins were identified, with five being exclusive to HFD plasma EVs and 112 exclusive to NCD EVs (Figure 2). Although EV samples were isolated through a rigorous protocol that includes multiple rounds of ultracentrifugation and a sucrose cushion, complete elimination of contaminant proteins remains a challenge. We detected the presence of ApoA2 and immunoglobulins, both of which are present in high levels in plasma [78–80]. Nevertheless, our findings suggest a reduction in cargo diversity in diet-induced obesity, indicating that the messaging system becomes more refined and intense, with an increased number of circulating EVs, or that cells try to compensate the lack of protein diversity with heightened EVs secretion. Either way, this may represent an adaptive mechanism or a response to ongoing metabolic changes. The key finding here is that plasma EVs encapsulate proteins that are lost as the disease progresses, highlighting their crucial role in maintaining metabolic homeostasis. These proteins could provide insights into the mechanisms underlying obesity.

A further analysis using ORA unveiled a significant enrichment of proteins associated with protein-folding processes among the 112 proteins exclusively identified in plasma EVs isolated from control animals, particularly chaperonin-containing TCP1 and heat-shock proteins (Figure 3). Indeed, the functionality of a protein is intrinsically linked to its conformation, and anomalies in protein folding lead to a range of alterations associated with the etiology of various human diseases, including β-cell dysfunction [81–84]. Important examples of this association include the islet amyloid polypeptide, which forms toxic aggregates leading to β-cell dysfunction and death [82,83], and heat shock factor 1, a transcriptional activator of molecular chaperones involved in the development of T2D [84].

Regarding the 12 regulated proteins identified in plasma EVs common to both NCD and HFD (Figure 2), it is worth noting that the 14-3-3 proteins presented higher expression in NCD plasma EVs, whereas immunoglobulins presented higher expression in HFD plasma EVs (Figure 2d,e). The 14-3-3 proteins are molecular adaptors which regulate a broad spectrum of signaling pathways and, more importantly, are known to have a beneficial effect on β-cell function and survival, which may be comprised in prediabetes [85]. On the other hand, the increased expression of immunoglobulins in HFD plasma EVs could indicate the onset of subclinical chronic inflammation, characteristic of obesity [86].

One of the novel aspects of this study is the analysis of protein PTMs within EVs, aiming to identify signaling molecules that are acetylated and glycated. These molecules have the potential to directly influence signaling functions within the recipient cells, either enhancing or inhibiting them. PTMs act as a dynamic and finely tuned molecular switches that can significantly alter protein function, stability, and interactions [87,88]. Our investigation into acetylated and glycated proteins within plasma EVs revealed a trend—obesity was associated with decreased acetylation and glycation events (Figure 5a,b). This observation suggests that the regulation of these PTMs is altered by the onset of obesity. The similarity between proteins exclusively acetylated in gut EVs in the HFD group and proteins exclusively glycated in the NCD group suggests that these proteins, mainly histones, undergo regulatory shifts by PTMs under obesity conditions. Importantly, this regulation of PTMs is detectable within the intercellular communication network, indicating their relevance in the context of organ crosstalk in obesity. Given the well-established roles of PTMs in insulin signaling, glucose metabolism, and the development of associated complications [42–44], this finding opens up new avenues for understanding the molecular mechanisms underlying obesity. Importantly, it calls our attention to the need to not only to examine the presence of a given protein but also consider its PTM status. Further research into the functional consequences of these PTM alterations in the context of metabolic health and disease progression is warranted.

Expanding on our previous study of the protein cargo within gut EVs in obesity, we reanalyzed the data focusing on PTMs. We observed a decrease in acetylation events associated with obesity, indicating a potentially intricate regulatory mechanism for these PTMs that may involve interactions with other organs (Figure 5c). When analyzed by ORA, the 29 proteins exclusively acetylated in HFD gut EVs showed an enrichment in GO functional terms related to chromatin, nucleosome, and DNA transcription, with histones being the main enriched proteins (Figures 6a,b and S5). These proteins interact with DNA and provide structural support to chromatin, regulating its condensation according to their own PTM profile [89]. While these associations suggest a crucial role for acetylation, especially of histones, thus controlling transcription in regulating obesity, further research is needed to unravel the precise mechanisms and functional implications of these observations in the context of metabolic health and disease. More intriguingly, the enriched functional terms obtained from KEGG pathways include viral carcinogenesis, systemic lupus erythematosus, alcoholism, and neutrophil extracellular trap formation (Figure 6c). These four pathways have been associated with histone acetylation, suggesting their possible interconnection with the development of obesity [90–94].

After analyzing the 23 proteins exclusively glycated in NCD gut EVs using ORA, the results for GO and KEGG were very similar to those for proteins exclusively acetylated in HFD EVs (Figure 6), even with some proteins shared between the two conditions (Figure S5). This not only indicates that these enriched functions are impacted by obesity, but also strongly suggests that they are regulated by a shift in histones' PTM profile. More precisely, these changes could involve an interplay between two types of modifications: glycation and acetylation. This interplay suggests that histone acetylation and glycation may compete for the same sites, since both modifications primarily occur on lysine and arginine residues [95–97]. Exploring the specific glycation and acetylation sites of these proteins will be crucial to pinpoint the regulatory mechanism of inter-organ communication mediated by EVs under prediabetic conditions.

ORA enrichment of the 22 proteins exclusively glycated in HFD gut EVs revealed several GO terms related to cellular structural stability (Figure 6). The most enriched proteins were actin and tubulin (Figure S5). Glycation of both proteins leads to an increase in endothelial permeability, which compromises the gut barrier during T2D [98–100]. ORA enrichment also revealed KEGG pathways associated with cellular structure and integrity. The enriched pathways of amyotrophic lateral sclerosis, dilated cardiomyopathy and hypertrophic cardiomyopathy are potentiated by increases in blood central nervous system barrier permeability and sarcolemmal permeability, respectively [101–103]. It is possible that gut EVs are transporting glycated structural proteins and delivering them to other organs, thereby increasing the risk of developing such pathologies. Further research is needed to explore the mechanistic details and functional implications of these observations.

The crosstalk between plasma and gut EVs adds an extra layer of complexity to our understanding of obesity. In this study, we identified four proteins exclusively present in both plasma and gut EVs under healthy conditions (Figure 4a,b). Their absence in obesity raises questions about their role in maintaining metabolic homeostasis. These proteins possess high biomarker potential for a healthy physiological state, as their presence is lost in obesity. Notably, the term "immune response" emerged as significant when examining the function of individual proteins, suggesting that disruption in immune system regulation may play a crucial role in metabolic dysfunction [104]. Complement factor H, one of these proteins, has previously been linked to metabolic disorders, reinforcing its relevance in the context of prediabetes [105]. The detection of proteins shared between plasma and gut EVs highlights the possibility of a non-invasive method to evaluate gut health through the analysis of blood-derived EVs. However, most proteins identified in gut EVs were not found in plasma EVs. We attribute this primarily to anatomical and physiological factors. Gut EVs are released into the portal vein, which directly drains into the liver, where they may be up taken by liver cells, reducing their contribution for the overall population of EVs in systemic circulation. Moreover, the exact mechanisms by which these altered protein cargos and PTMs contribute to obesity progression remain elusive. Further research is needed to elucidate these mechanisms, potentially through functional studies.

5. Conclusions

This study illuminates the novel role of EVs in obesity and their potential as biomarkers for early metabolic dysfunction. The changes in protein composition and PTMs offer new perspectives into the molecular mechanisms behind obesity. These findings have the potential to pave the way for early intervention strategies to combat the rising prevalence of obesity, emphasizing the critical role of EVs in understanding and addressing these health challenge. Further research in this growing field promises to unravel more intricacies and insights, ultimately advancing our ability to prevent and treat obesity more effectively.

Supplementary Materials: The following supporting information can be downloaded at: https://www.mdpi.com/article/10.3390/nu16050736/s1, Figure S1: Examples of annotated raw spectrum of identified lysine acetylation site; Figure S2: Examples of annotated raw spectrum of identified lysine glycation site; Figure S3: Diet-induced obese mouse model characterization; Figure S4: Nanoparticle tracking analysis of plasma EVs; Figure S5: Enriched terms and their related identified genes, relative to Figure 6; Table S1: List of 112 protein exclusively identified in plasma EVs from NCD mice relative to Figures 2b and 3; Table S2: List of four proteins shared between NCD and HFD plasma EVs, but exclusively acetylated in NCD plasma EVs, relative to the Venn diagram in Figure 5a; Table S3: List of six proteins shared between NCD and HFD plasma EVs, but exclusively glycated in NCD plasma EVs, relative to the Venn diagram in Figure 5b; Table S4: List of 41 proteins shared between NCD and HFD gut EVs, but exclusively acetylated in NCD gut EVs, relative to the Venn diagram in Figure 5c; Table S5: List of 29 proteins shared between NCD and HFD gut EVs, but exclusively acetylated in HFD gut EVs, relative to the Venn diagram in Figures 5c and 6; Table S6: List of 23 proteins shared between NCD and HFD gut EVs, but exclusively glycated in NCD gut EVs, relative to the Venn diagram in Figures 5d and 6; Table S7: List of 22 proteins shared between NCD and HFD gut EVs, but exclusively glycated in HFD gut EVs, relative to the Venn diagram in Figures 5d and 6.

Author Contributions: Conceptualization, M.P.M. and R.M.d.O.; methodology, I.F., P.B.P. and R.M.d.O.; software, P.B.P. and R.M.; formal analysis, P.B.P. and R.M.; investigation, A.T., A.S.C., I.F., I.S.-L., H.C.B., P.B.P., R.M.d.O. and R.M.; writing—original draft preparation, E.T. and P.B.P.; writing—review and editing, R.M.d.O.; visualization, E.T., P.B.P. and R.M.d.O.; supervision, B.C.-S., M.P.M. and R.M.d.O.; funding acquisition, M.P.M. and R.M.d.O. All authors have read and agreed to the published version of the manuscript.

Funding: This research was funded by Fundação para a Ciência e a Tecnologia (PTDC/MEC-MET/29314/2017, and 2022.05764.PTDC), by the European Comission CORDIS Pas Gras Project (101080329) and by the European Union EVCA Twining Project (Horizon GA n° 101079264).

Institutional Review Board Statement: The animal study protocol was approved by the Ethics Committee of the NOVA Medical School (nr.82/2019/CEFCM).

Informed Consent Statement: Not applicable.

Data Availability Statement: Data are contained within the article and Supplementary Materials.

Acknowledgments: Schematic illustrations were created with BioRender.com.

Conflicts of Interest: The authors declare no conflicts of interest. The funders had no role in the design of the study; in the collection, analysis, or interpretation of data; in the writing of the manuscript; or in the decision to publish the results.

References

1. World Health Organization. *World Obesity Day 2022—Accelerating Action to Stop Obesity*; WHO: Geneva, Switzerland, 2022.
2. Bastías-Pérez, M.; Serra, D.; Herrero, L. Dietary Options for Rodents in the Study of Obesity. *Nutrients* **2020**, *12*, 3234. [CrossRef]
3. Hill, J.O.; Melanson, E.L.; Wyatt, H.T. Dietary Fat Intake and Regulation of Energy Balance: Implications for Obesity. *J. Nutr.* **2000**, *130*, 284S–288S. [CrossRef]
4. Jéquier, E. Pathways to Obesity. *Int. J. Obes.* **2002**, *26*, S12–S17. [CrossRef]
5. French, S.; Robinson, T. Fats and Food Intake. *Curr. Opin. Clin. Nutr. Metab. Care* **2003**, *6*, 629–634. [CrossRef] [PubMed]
6. Losacco, M.C.; de Almeida, C.F.T.; Hijo, A.H.T.; Bargi-Souza, P.; Gama, P.; Nunes, M.T.; Goulart-Silva, F. High-Fat Diet Affects Gut Nutrients Transporters in Hypo and Hyperthyroid Mice by PPAR-a Independent Mechanism. *Life Sci.* **2018**, *202*, 35–43. [CrossRef]
7. Hu, F.B. Globalization of Diabetes. *Diabetes Care* **2011**, *34*, 1249–1257. [CrossRef]
8. Shoelson, S.E.; Herrero, L.; Naaz, A. Obesity, Inflammation, and Insulin Resistance. *Gastroenterology* **2007**, *132*, 2169–2180. [CrossRef]
9. Fernández-Sánchez, A.; Madrigal-Santillán, E.; Bautista, M.; Esquivel-Soto, J.; Morales-González, Á.; Esquivel-Chirino, C.; Durante-Montiel, I.; Sánchez-Rivera, G.; Valadez-Vega, C.; Morales-González, J.A. Inflammation, Oxidative Stress, and Obesity. *Int. J. Mol. Sci.* **2011**, *12*, 3117–3132. [CrossRef]
10. Ouchi, N.; Parker, J.L.; Lugus, J.J.; Walsh, K. Adipokines in Inflammation and Metabolic Disease. *Nat. Rev. Immunol.* **2011**, *11*, 85–97. [CrossRef]
11. Pardo, F.; Villalobos-Labra, R.; Sobrevia, B.; Toledo, F.; Sobrevia, L. Extracellular Vesicles in Obesity and Diabetes Mellitus. *Mol. Aspects Med.* **2018**, *60*, 81–91. [CrossRef]
12. Garcia-Contreras, M.; Shah, S.H.; Tamayo, A.; Robbins, P.D.; Golberg, R.B.; Mendez, A.J.; Ricordi, C. Plasma-Derived Exosome Characterization Reveals a Distinct MicroRNA Signature in Long Duration Type 1 Diabetes. *Sci. Rep.* **2017**, *7*, 5998. [CrossRef] [PubMed]
13. Martínez, M.C.; Andriantsitohaina, R. Extracellular Vesicles in Metabolic Syndrome. *Circ. Res.* **2017**, *120*, 1674–1686. [CrossRef]
14. Colombo, M.; Raposo, G.; Théry, C. Biogenesis, Secretion, and Intercellular Interactions of Exosomes and Other Extracellular Vesicles. *Annu. Rev. Cell Dev. Biol.* **2014**, *30*, 255–289. [CrossRef]
15. Kalluri, R.; LeBleu, V.S. The Biology, Function, and Biomedical Applications of Exosomes. *Science* **2020**, *367*, eaau6977. [CrossRef]
16. Kalra, H.; Drummen, G.; Mathivanan, S. Focus on Extracellular Vesicles: Introducing the Next Small Big Thing. *Int. J. Mol. Sci.* **2016**, *17*, 170. [CrossRef] [PubMed]
17. van Niel, G.; D'Angelo, G.; Raposo, G. Shedding Light on the Cell Biology of Extracellular Vesicles. *Nat. Rev. Mol. Cell Biol.* **2018**, *19*, 213–228. [CrossRef]
18. Akbar, N.; Azzimato, V.; Choudhury, R.P.; Aouadi, M. Extracellular Vesicles in Metabolic Disease. *Diabetologia* **2019**, *62*, 2179–2187. [CrossRef]
19. Huang-Doran, I.; Zhang, C.-Y.; Vidal-Puig, A. Extracellular Vesicles: Novel Mediators of Cell Communication in Metabolic Disease. *Trends Endocrinol. Metab.* **2017**, *28*, 3–18. [CrossRef]
20. Li, C.-J.; Fang, Q.-H.; Liu, M.-L.; Lin, J.-N. Current Understanding of the Role of Adipose-Derived Extracellular Vesicles in Metabolic Homeostasis and Diseases: Communication from the Distance between Cells/Tissues. *Theranostics* **2020**, *10*, 7422–7435. [CrossRef]
21. Aswad, H.; Forterre, A.; Wiklander, O.P.B.; Vial, G.; Danty-Berger, E.; Jalabert, A.; Lamazière, A.; Meugnier, E.; Pesenti, S.; Ott, C.; et al. Exosomes Participate in the Alteration of Muscle Homeostasis during Lipid-Induced Insulin Resistance in Mice. *Diabetologia* **2014**, *57*, 2155–2164. [CrossRef]
22. Dang, S.-Y.; Leng, Y.; Wang, Z.-X.; Xiao, X.; Zhang, X.; Wen, T.; Gong, H.-Z.; Hong, A.; Ma, Y. Exosomal Transfer of Obesity Adipose Tissue for Decreased MiR-141-3p Mediate Insulin Resistance of Hepatocytes. *Int. J. Biol. Sci.* **2019**, *15*, 351–368. [CrossRef]
23. Jalabert, A.; Vial, G.; Guay, C.; Wiklander, O.P.B.; Nordin, J.Z.; Aswad, H.; Forterre, A.; Meugnier, E.; Pesenti, S.; Regazzi, R.; et al. Exosome-like Vesicles Released from Lipid-Induced Insulin-Resistant Muscles Modulate Gene Expression and Proliferation of Beta Recipient Cells in Mice. *Diabetologia* **2016**, *59*, 1049–1058. [CrossRef]
24. Wu, J.; Dong, T.; Chen, T.; Sun, J.; Luo, J.; He, J.; Wei, L.; Zeng, B.; Zhang, H.; Li, W.; et al. Hepatic Exosome-Derived MiR-130a-3p Attenuates Glucose Intolerance via Suppressing PHLPP2 Gene in Adipocyte. *Metabolism* **2020**, *103*, 154006. [CrossRef]
25. Yu, Y.; Du, H.; Wei, S.; Feng, L.; Li, J.; Yao, F.; Zhang, M.; Hatch, G.M.; Chen, L. Adipocyte-Derived Exosomal MiR-27a Induces Insulin Resistance in Skeletal Muscle Through Repression of PPARγ. *Theranostics* **2018**, *8*, 2171–2188. [CrossRef]
26. Deng, Z.; Poliakov, A.; Hardy, R.W.; Clements, R.; Liu, C.; Liu, Y.; Wang, J.; Xiang, X.; Zhang, S.; Zhuang, X.; et al. Adipose Tissue Exosome-Like Vesicles Mediate Activation of Macrophage-Induced Insulin Resistance. *Diabetes* **2009**, *58*, 2498–2505. [CrossRef]
27. Kranendonk, M.E.G.; Visseren, F.L.J.; van Balkom, B.W.M.; Nolte-'t Hoen, E.N.M.; van Herwaarden, J.A.; de Jager, W.; Schipper, H.S.; Brenkman, A.B.; Verhaar, M.C.; Wauben, M.H.M.; et al. Human Adipocyte Extracellular Vesicles in Reciprocal Signaling between Adipocytes and Macrophages. *Obesity* **2014**, *22*, 1296–1308. [CrossRef]

28. Castaño, C.; Kalko, S.; Novials, A.; Párrizas, M. Obesity-Associated Exosomal MiRNAs Modulate Glucose and Lipid Metabolism in Mice. *Proc. Natl. Acad. Sci. USA* **2018**, *115*, 12158–12163. [CrossRef] [PubMed]
29. Ying, W.; Riopel, M.; Bandyopadhyay, G.; Dong, Y.; Birmingham, A.; Seo, J.B.; Ofrecio, J.M.; Wollam, J.; Hernandez-Carretero, A.; Fu, W.; et al. Adipose Tissue Macrophage-Derived Exosomal MiRNAs Can Modulate In Vivo and In Vitro Insulin Sensitivity. *Cell* **2017**, *171*, 372–384.e12. [CrossRef] [PubMed]
30. Ferreira, I.; Machado de Oliveira, R.; Carvalho, A.S.; Teshima, A.; Beck, H.C.; Matthiesen, R.; Costa-Silva, B.; Macedo, M.P. Messages from the Small Intestine Carried by Extracellular Vesicles in Prediabetes: A Proteomic Portrait. *J. Proteome Res.* **2022**, *21*, 910–920. [CrossRef]
31. Breen, D.M.; Rasmussen, B.A.; Côté, C.D.; Jackson, V.M.; Lam, T.K.T. Nutrient-Sensing Mechanisms in the Gut as Therapeutic Targets for Diabetes. *Diabetes* **2013**, *62*, 3005–3013. [CrossRef]
32. Gurung, M.; Li, Z.; You, H.; Rodrigues, R.; Jump, D.B.; Morgun, A.; Shulzhenko, N. Role of Gut Microbiota in Type 2 Diabetes Pathophysiology. *EBioMedicine* **2020**, *51*, 102590. [CrossRef]
33. Evers, S.S.; Sandoval, D.A.; Seeley, R.J. The Physiology and Molecular Underpinnings of the Effects of Bariatric Surgery on Obesity and Diabetes. *Annu. Rev. Physiol.* **2017**, *79*, 313–334. [CrossRef]
34. Pareek, M.; Schauer, P.R.; Kaplan, L.M.; Leiter, L.A.; Rubino, F.; Bhatt, D.L. Metabolic Surgery. *J. Am. Coll. Cardiol.* **2018**, *71*, 670–687. [CrossRef]
35. Hillyer, R.; Sullivan, T.; Christ, K.; Hodarkar, A.; Hodge, M.B. 2117-P: Biomarkers Isolated from Extracellular Vesicles Prior to Bariatric Surgery May Be Associated with Postoperative Resolution of Type 2 Diabetes. *Diabetes* **2019**, *68*, 2117-P. [CrossRef]
36. Duarte, N.; Coelho, I.C.; Patarrão, R.S.; Almeida, J.I.; Penha-Gonçalves, C.; Macedo, M.P. How Inflammation Impinges on NAFLD: A Role for Kupffer Cells. *BioMed Res. Int.* **2015**, *2015*, 984578. [CrossRef] [PubMed]
37. Freeman, D.W.; Noren Hooten, N.; Eitan, E.; Green, J.; Mode, N.A.; Bodogai, M.; Zhang, Y.; Lehrmann, E.; Zonderman, A.B.; Biragyn, A.; et al. Altered Extracellular Vesicle Concentration, Cargo, and Function in Diabetes. *Diabetes* **2018**, *67*, 2377–2388. [CrossRef]
38. Eguchi, A.; Lazic, M.; Armando, A.M.; Phillips, S.A.; Katebian, R.; Maraka, S.; Quehenberger, O.; Sears, D.D.; Feldstein, A.E. Circulating Adipocyte-Derived Extracellular Vesicles Are Novel Markers of Metabolic Stress. *J. Mol. Med.* **2016**, *94*, 1241–1253. [CrossRef] [PubMed]
39. Murakami, T.; Horigome, H.; Tanaka, K.; Nakata, Y.; Ohkawara, K.; Katayama, Y.; Matsui, A. Impact of Weight Reduction on Production of Platelet-Derived Microparticles and Fibrinolytic Parameters in Obesity. *Thromb. Res.* **2007**, *119*, 45–53. [CrossRef]
40. Stepanian, A.; Bourguignat, L.; Hennou, S.; Coupaye, M.; Hajage, D.; Salomon, L.; Alessi, M.-C.; Msika, S.; de Prost, D. Microparticle Increase in Severe Obesity: Not Related to Metabolic Syndrome and Unchanged after Massive Weight Loss. *Obesity* **2013**, *21*, 2236–2243. [CrossRef]
41. Campello, E.; Zabeo, E.; Radu, C.M.; Spiezia, L.; Foletto, M.; Prevedello, L.; Gavasso, S.; Bulato, C.; Vettor, R.; Simioni, P. Dynamics of Circulating Microparticles in Obesity after Weight Loss. *Intern. Emerg. Med.* **2016**, *11*, 695–702. [CrossRef]
42. Stocks, B.; Zierath, J.R. Post-Translational Modifications: The Signals at the Intersection of Exercise, Glucose Uptake, and Insulin Sensitivity. *Endocr. Rev.* **2022**, *43*, 654–677. [CrossRef]
43. Wu, X.; Xu, M.; Geng, M.; Chen, S.; Little, P.J.; Xu, S.; Weng, J. Targeting Protein Modifications in Metabolic Diseases: Molecular Mechanisms and Targeted Therapies. *Signal Transduct. Target. Ther.* **2023**, *8*, 220. [CrossRef] [PubMed]
44. Yang, C.; Wei, M.; Zhao, Y.; Yang, Z.; Song, M.; Mi, J.; Yang, X.; Tian, G. Regulation of Insulin Secretion by the Post-Translational Modifications. *Front. Cell Dev. Biol.* **2023**, *11*, 1217189. [CrossRef]
45. Dieterich, I.A.; Lawton, A.J.; Peng, Y.; Yu, Q.; Rhoads, T.W.; Overmyer, K.A.; Cui, Y.; Armstrong, E.A.; Howell, P.R.; Burhans, M.S.; et al. Acetyl-CoA Flux Regulates the Proteome and Acetyl-Proteome to Maintain Intracellular Metabolic Crosstalk. *Nat. Commun.* **2019**, *10*, 3929. [CrossRef]
46. Maessen, D.E.M.; Stehouwer, C.D.A.; Schalkwijk, C.G. The Role of Methylglyoxal and the Glyoxalase System in Diabetes and Other Age-Related Diseases. *Clin. Sci.* **2015**, *128*, 839–861. [CrossRef]
47. Menzies, K.J.; Zhang, H.; Katsyuba, E.; Auwerx, J. Protein Acetylation in Metabolism—Metabolites and Cofactors. *Nat. Rev. Endocrinol.* **2016**, *12*, 43–60. [CrossRef] [PubMed]
48. Smith, K.T.; Workman, J.L. Introducing the Acetylome. *Nat. Biotechnol.* **2009**, *27*, 917–919. [CrossRef] [PubMed]
49. Sun, F.; Suttapitugsakul, S.; Xiao, H.; Wu, R. Comprehensive Analysis of Protein Glycation Reveals Its Potential Impacts on Protein Degradation and Gene Expression in Human Cells. *J. Am. Soc. Mass Spectrom.* **2019**, *30*, 2480–2490. [CrossRef]
50. Pessoa Rodrigues, C.; Chatterjee, A.; Wiese, M.; Stehle, T.; Szymanski, W.; Shvedunova, M.; Akhtar, A. Histone H4 Lysine 16 Acetylation Controls Central Carbon Metabolism and Diet-Induced Obesity in Mice. *Nat. Commun.* **2021**, *12*, 6212. [CrossRef]
51. Li, Z.; Zhuang, M.; Zhang, L.; Zheng, X.; Yang, P.; Li, Z. Acetylation Modification Regulates GRP78 Secretion in Colon Cancer Cells. *Sci. Rep.* **2016**, *6*, 30406. [CrossRef]
52. Iyer, A.; Fairlie, D.P.; Brown, L. Lysine Acetylation in Obesity, Diabetes and Metabolic Disease. *Immunol. Cell Biol.* **2012**, *90*, 39–46. [CrossRef]
53. Rabbani, N.; Thornalley, P.J. Glycation Research in Amino Acids: A Place to Call Home. *Amino Acids* **2012**, *42*, 1087–1096. [CrossRef]
54. Twarda-Clapa, A.; Olczak, A.; Białkowska, A.M.; Koziołkiewicz, M. Advanced Glycation End-Products (AGEs): Formation, Chemistry, Classification, Receptors, and Diseases Related to AGEs. *Cells* **2022**, *11*, 1312. [CrossRef]

55. Mengstie, M.A.; Chekol Abebe, E.; Behaile Teklemariam, A.; Tilahun Mulu, A.; Agidew, M.M.; Teshome Azezew, M.; Zewde, E.A.; Agegnehu Teshome, A. Endogenous Advanced Glycation End Products in the Pathogenesis of Chronic Diabetic Complications. *Front. Mol. Biosci.* **2022**, *9*, 1002710. [CrossRef]
56. Negre-Salvayre, A.; Salvayre, R.; Augé, N.; Pamplona, R.; Portero-Otín, M. Hyperglycemia and Glycation in Diabetic Complications. *Antioxid. Redox Signal.* **2009**, *11*, 3071–3109. [CrossRef]
57. International Diabetes Federation. *IDF Diabetes Atlas*, 10th ed.; International Diabetes Federation: Brussels, Belgium, 2021.
58. ElSayed, N.A.; Aleppo, G.; Aroda, V.R.; Bannuru, R.R.; Brown, F.M.; Bruemmer, D.; Collins, B.S.; Gaglia, J.L.; Hilliard, M.E.; Isaacs, D.; et al. 2. Classification and Diagnosis of Diabetes: Standards of Care in Diabetes—2023. *Diabetes Care* **2023**, *46*, S19–S40. [CrossRef]
59. Carvalho, A.S.; Baeta, H.; Henriques, A.F.A.; Ejtehadifar, M.; Tranfield, E.M.; Sousa, A.L.; Farinho, A.; Silva, B.C.; Cabeçadas, J.; Gameiro, P.; et al. Proteomic Landscape of Extracellular Vesicles for Diffuse Large B-Cell Lymphoma Subtyping. *Int. J. Mol. Sci.* **2021**, *22*, 1004. [CrossRef] [PubMed]
60. Cox, J.; Mann, M. MaxQuant Enables High Peptide Identification Rates, Individualized p.p.b.-Range Mass Accuracies and Proteome-Wide Protein Quantification. *Nat. Biotechnol.* **2008**, *26*, 1367–1372. [CrossRef]
61. Carvalho, A.S.; Ribeiro, H.; Voabil, P.; Penque, D.; Jensen, O.N.; Molina, H.; Matthiesen, R. Global Mass Spectrometry and Transcriptomics Array Based Drug Profiling Provides Novel Insight into Glucosamine Induced Endoplasmic Reticulum Stress. *Mol. Cell. Proteom.* **2014**, *13*, 3294–3307. [CrossRef]
62. Wu, T.; Hu, E.; Xu, S.; Chen, M.; Guo, P.; Dai, Z.; Feng, T.; Zhou, L.; Tang, W.; Zhan, L.; et al. ClusterProfiler 4.0: A Universal Enrichment Tool for Interpreting Omics Data. *Innovation* **2021**, *2*, 100141. [CrossRef] [PubMed]
63. Lundgren, D.H.; Hwang, S.-I.; Wu, L.; Han, D.K. Role of Spectral Counting in Quantitative Proteomics. *Expert Rev. Proteom.* **2010**, *7*, 39–53. [CrossRef]
64. Schwanhüusser, B.; Busse, D.; Li, N.; Dittmar, G.; Schuchhardt, J.; Wolf, J.; Chen, W.; Selbach, M. Global Quantification of Mammalian Gene Expression Control. *Nature* **2011**, *473*, 337–342. [CrossRef]
65. Ritchie, M.E.; Phipson, B.; Wu, D.; Hu, Y.; Law, C.W.; Shi, W.; Smyth, G.K. Limma Powers Differential Expression Analyses for RNA-Sequencing and Microarray Studies. *Nucleic Acids Res.* **2015**, *43*, e47. [CrossRef]
66. Boyle, E.I.; Weng, S.; Gollub, J.; Jin, H.; Botstein, D.; Cherry, J.M.; Sherlock, G. GO::TermFinder—Open Source Software for Accessing Gene Ontology Information and Finding Significantly Enriched Gene Ontology Terms Associated with a List of Genes. *Bioinformatics* **2004**, *20*, 3710–3715. [CrossRef]
67. Ashburner, M.; Ball, C.A.; Blake, J.A.; Botstein, D.; Butler, H.; Cherry, J.M.; Davis, A.P.; Dolinski, K.; Dwight, S.S.; Eppig, J.T.; et al. Gene Ontology: Tool for the Unification of Biology. *Nat. Genet.* **2000**, *25*, 25–29. [CrossRef]
68. Aleksander, S.A.; Balhoff, J.; Carbon, S.; Cherry, J.M.; Drabkin, H.J.; Ebert, D.; Feuermann, M.; Gaudet, P.; Harris, N.L.; Hill, D.P.; et al. The Gene Ontology Knowledgebase in 2023. *Genetics* **2023**, *224*, iyad031. [CrossRef]
69. Kanehisa, M.; Goto, S. KEGG: Kyoto Encyclopedia of Genes and Genomes. *Nucleic Acids Res.* **2000**, *28*, 27–30. [CrossRef]
70. Kanehisa, M. Toward Understanding the Origin and Evolution of Cellular Organisms. *Protein Sci.* **2019**, *28*, 1947–1951. [CrossRef] [PubMed]
71. Kanehisa, M.; Furumichi, M.; Sato, Y.; Kawashima, M.; Ishiguro-Watanabe, M. KEGG for Taxonomy-Based Analysis of Pathways and Genomes. *Nucleic Acids Res.* **2023**, *51*, D587–D592. [CrossRef]
72. Noren Hooten, N.; Evans, M.K. Extracellular Vesicles as Signaling Mediators in Type 2 Diabetes Mellitus. *Am. J. Physiol.-Cell Physiol.* **2020**, *318*, C1189–C1199. [CrossRef]
73. Sandau, U.S.; McFarland, T.J.; Smith, S.J.; Galasko, D.R.; Quinn, J.F.; Saugstad, J.A. Differential Effects of APOE Genotype on MicroRNA Cargo of Cerebrospinal Fluid Extracellular Vesicles in Females with Alzheimer's Disease Compared to Males. *Front. Cell Dev. Biol.* **2022**, *10*, 864022. [CrossRef]
74. Chung, D.D.; Mahnke, A.H.; Pinson, M.R.; Salem, N.A.; Lai, M.S.; Collins, N.P.; Hillhouse, A.E.; Miranda, R.C. Sex Differences in the Transcriptome of Extracellular Vesicles Secreted by Fetal Neural Stem Cells and Effects of Chronic Alcohol Exposure. *Biol. Sex Differ.* **2023**, *14*, 19. [CrossRef]
75. Perpiñá-Clérigues, C.; Mellado, S.; Galiana-Roselló, C.; Fernández-Regueras, M.; Marcos, M.; García-García, F.; Pascual, M. Novel Insight into the Lipid Network of Plasma Extracellular Vesicles Reveal Sex-Based Differences in the Lipidomic Profile of Alcohol Use Disorder Patients. *Biol. Sex Differ.* **2024**, *15*, 10. [CrossRef]
76. Perpiñá-Clérigues, C.; Mellado, S.; Català-Senent, J.F.; Ibáñez, F.; Costa, P.; Marcos, M.; Guerri, C.; García-García, F.; Pascual, M. Lipidomic Landscape of Circulating Extracellular Vesicles Isolated from Adolescents Exposed to Ethanol Intoxication: A Sex Difference Study. *Biol. Sex Differ.* **2023**, *14*, 22. [CrossRef]
77. Noren Hooten, N.; Byappanahalli, A.M.; Vannoy, M.; Omoniyi, V.; Evans, M.K. Influences of Age, Race, and Sex on Extracellular Vesicle Characteristics. *Theranostics* **2022**, *12*, 4459–4476. [CrossRef]
78. Ter-Ovanesyan, D.; Gilboa, T.; Budnik, B.; Nikitina, A.; Whiteman, S.; Lazarovits, R.; Trieu, W.; Kalish, D.; Church, G.M.; Walt, D.R. Improved Isolation of Extracellular Vesicles by Removal of Both Free Proteins and Lipoproteins. *Elife* **2023**, *12*, e86394. [CrossRef]
79. Sódar, B.W.; Kittel, Á.; Pálóczi, K.; Vukman, K.V.; Osteikoetxea, X.; Szabó-Taylor, K.; Németh, A.; Sperlágh, B.; Baranyai, T.; Giricz, Z.; et al. Low-Density Lipoprotein Mimics Blood Plasma-Derived Exosomes and Microvesicles during Isolation and Detection. *Sci. Rep.* **2016**, *6*, 24316. [CrossRef]

80. Smolarz, M.; Pietrowska, M.; Matysiak, N.; Mielańczyk, Ł.; Widłak, P. Proteome Profiling of Exosomes Purified from a Small Amount of Human Serum: The Problem of Co-Purified Serum Components. *Proteomes* **2019**, *7*, 18. [CrossRef]
81. Thomas, S.E. Unravelling the Story of Protein Misfolding in Diabetes Mellitus. *World J. Diabetes* **2011**, *2*, 114. [CrossRef]
82. Mukherjee, A.; Morales-Scheihing, D.; Butler, P.C.; Soto, C. Type 2 Diabetes as a Protein Misfolding Disease. *Trends Mol. Med.* **2015**, *21*, 439–449. [CrossRef]
83. Mukherjee, A.; Morales-Scheihing, D.; Salvadores, N.; Moreno-Gonzalez, I.; Gonzalez, C.; Taylor-Presse, K.; Mendez, N.; Shahnawaz, M.; Gaber, A.O.; Sabek, O.M.; et al. Induction of IAPP Amyloid Deposition and Associated Diabetic Abnormalities by a Prion-like Mechanism. *J. Exp. Med.* **2017**, *214*, 2591–2610. [CrossRef]
84. Klyosova, E.; Azarova, I.; Polonikov, A. A Polymorphism in the Gene Encoding Heat Shock Factor 1 (HSF1) Increases the Risk of Type 2 Diabetes: A Pilot Study Supports a Role for Impaired Protein Folding in Disease Pathogenesis. *Life* **2022**, *12*, 1936. [CrossRef]
85. Lim, G.E.; Piske, M.; Johnson, J.D. 14-3-3 Proteins Are Essential Signalling Hubs for Beta Cell Survival. *Diabetologia* **2013**, *56*, 825–837. [CrossRef]
86. Lontchi-Yimagou, E.; Sobngwi, E.; Matsha, T.E.; Kengne, A.P. Diabetes Mellitus and Inflammation. *Curr. Diabetes Rep.* **2013**, *13*, 435–444. [CrossRef]
87. Lin, H.; Du, J.; Jiang, H. Post-Translational Modifications to Regulate Protein Function. In *Wiley Encyclopedia of Chemical Biology*; Wiley: Hoboken, NJ, USA, 2008; pp. 1–31.
88. Bischoff, R.; Schlüter, H. Amino Acids: Chemistry, Functionality and Selected Non-Enzymatic Post-Translational Modifications. *J. Proteom.* **2012**, *75*, 2275–2296. [CrossRef]
89. Campos, E.I.; Reinberg, D. Histones: Annotating Chromatin. *Annu. Rev. Genet.* **2009**, *43*, 559–599. [CrossRef]
90. Ren, J.; Panther, E.; Liao, X.; Grammer, A.; Lipsky, P.; Reilly, C. The Impact of Protein Acetylation/Deacetylation on Systemic Lupus Erythematosus. *Int. J. Mol. Sci.* **2018**, *19*, 4007. [CrossRef]
91. Pandey, S.C.; Bohnsack, J.P. Alcohol Makes Its Epigenetic Marks. *Cell Metab.* **2020**, *31*, 213–214. [CrossRef]
92. Mews, P.; Egervari, G.; Nativio, R.; Sidoli, S.; Donahue, G.; Lombroso, S.I.; Alexander, D.C.; Riesche, S.L.; Heller, E.A.; Nestler, E.J.; et al. Alcohol Metabolism Contributes to Brain Histone Acetylation. *Nature* **2019**, *574*, 717–721. [CrossRef]
93. Hamam, H.; Khan, M.; Palaniyar, N. Histone Acetylation Promotes Neutrophil Extracellular Trap Formation. *Biomolecules* **2019**, *9*, 32. [CrossRef]
94. Mirzaei, H.; Ghorbani, S.; Khanizadeh, S.; Namdari, H.; Faghihloo, E.; Akbari, A. Histone Deacetylases in Virus-associated Cancers. *Rev. Med. Virol.* **2020**, *30*, e2085. [CrossRef]
95. Csizmok, V.; Forman-Kay, J.D. Complex Regulatory Mechanisms Mediated by the Interplay of Multiple Post-Translational Modifications. *Curr. Opin. Struct. Biol.* **2018**, *48*, 58–67. [CrossRef]
96. Zheng, Q.; Omans, N.D.; Leicher, R.; Osunsade, A.; Agustinus, A.S.; Finkin-Groner, E.; D'Ambrosio, H.; Liu, B.; Chandarlapaty, S.; Liu, S.; et al. Reversible Histone Glycation Is Associated with Disease-Related Changes in Chromatin Architecture. *Nat. Commun.* **2019**, *10*, 1289. [CrossRef]
97. Smith, B.C.; Denu, J.M. Chemical Mechanisms of Histone Lysine and Arginine Modifications. *Biochim. Biophys. Acta BBA—Gene Regul. Mech.* **2009**, *1789*, 45–57. [CrossRef]
98. Song, Y.; Brady, S.T. Post-Translational Modifications of Tubulin: Pathways to Functional Diversity of Microtubules. *Trends Cell Biol.* **2015**, *25*, 125–136. [CrossRef]
99. Terman, J.R.; Kashina, A. Post-Translational Modification and Regulation of Actin. *Curr. Opin. Cell Biol.* **2013**, *25*, 30–38. [CrossRef]
100. Horton, F.; Wright, J.; Smith, L.; Hinton, P.J.; Robertson, M.D. Increased Intestinal Permeability to Oral Chromium (^{51}Cr)—EDTA in Human Type 2 Diabetes. *Diabet. Med.* **2014**, *31*, 559–563. [CrossRef]
101. Monsour, M.; Garbuzova-Davis, S.; Borlongan, C.V. Patching Up the Permeability: The Role of Stem Cells in Lessening Neurovascular Damage in Amyotrophic Lateral Sclerosis. *Stem Cells Transl. Med.* **2022**, *11*, 1196–1209. [CrossRef]
102. Kawada, T.; Nakazawa, M.; Nakauchi, S.; Yamazaki, K.; Shimamoto, R.; Urabe, M.; Nakata, J.; Hemmi, C.; Masui, F.; Nakajima, T.; et al. Rescue of Hereditary Form of Dilated Cardiomyopathy by RAAV-Mediated Somatic Gene Therapy: Amelioration of Morphological Findings, Sarcolemmal Permeability, Cardiac Performances, and the Prognosis of TO-2 Hamsters. *Proc. Natl. Acad. Sci. USA* **2002**, *99*, 901–906. [CrossRef]
103. Kaye, D.; Pimental, D.; Prasad, S.; Mäki, T.; Berger, H.J.; McNeil, P.L.; Smith, T.W.; Kelly, R.A. Role of Transiently Altered Sarcolemmal Membrane Permeability and Basic Fibroblast Growth Factor Release in the Hypertrophic Response of Adult Rat Ventricular Myocytes to Increased Mechanical Activity in Vitro. *J. Clin. Investig.* **1996**, *97*, 281–291. [CrossRef]
104. Zmora, N.; Bashiardes, S.; Levy, M.; Elinav, E. The Role of the Immune System in Metabolic Health and Disease. *Cell Metab.* **2017**, *25*, 506–521. [CrossRef]
105. Ghosh, P.; Sahoo, R.; Vaidya, A.; Chorev, M.; Halperin, J.A. Role of Complement and Complement Regulatory Proteins in the Complications of Diabetes. *Endocr. Rev.* **2015**, *36*, 272–288. [CrossRef]

nutrients MDPI

Article

Daily Consumption of Golden Berry (*Physalis peruviana*) Has Been Shown to Halt the Progression of Insulin Resistance and Obesity in Obese Rats with Metabolic Syndrome

Alberto Ángel-Martín [1], Fabrice Vaillant [2,3] and Natalia Moreno-Castellanos [4,*]

1 Observatorio Epidemiológico de Nutrición y Enfermedades Crónicas, Nutrition School, Health Faculty, Universidad Industrial de Santander, Cra 32 # 29-31, Bucaramanga 680002, Colombia; angelmar@uis.edu.co
2 Colombian Corporation for Agricultural Research-Agrosavia, La Selva Research Center, Kilometer 7, Vía a Las Palmas, Vereda Llanogrande, Rionegro 054048, Colombia; fabrice.vaillant@cirad.fr
3 French Center for Agricultural Research for International Development (CIRAD), UMR Qualisud, 34398 Montpellier, France
4 Centro de Investigación en Ciencia y Tecnología de Alimentos, Department of Basic Sciences, Medicine School, Health Faculty, Universidad Industrial de Santander, Cra 27 calle 9, Bucaramanga 680002, Colombia
* Correspondence: nrmorcas@uis.edu.co

Abstract: In a study addressing the high risk of chronic diseases in people with diabetes and obesity linked to metabolic syndrome, the impact of a Golden Berry diet was investigated using a diabetic animal model. Obese rats with diabetic characteristics were fed a diet containing five percent Golden Berry for 16 days. This study focused on various parameters including organ weights, expression of metabolic genes, and urinary biomarkers. Post-Golden Berry intake, there was a notable decrease in the body, liver, pancreas, visceral, and subcutaneous adipose tissue weights in these obese, hyperglycemic rats. In contrast, an increase in brown adipose tissue (BAT) cell mass was observed. This diet also resulted in reduced blood glucose levels and normalized plasma biochemical profiles, including cholesterol, triglycerides, LDL, and HDL levels. Additionally, it modulated specific urinary biomarkers, particularly pipe-colic acid, a primary marker for type 2 diabetes. Bioinformatics analysis linked these dietary effects to improved insulin signaling and adipogenesis. Regular consumption of Golden Berry effectively prevented insulin resistance and obesity in rats, underscoring its significant health benefits and the protective role of an antioxidant-rich diet against metabolic syndrome. These findings offer promising insights for future therapeutic strategies to manage and prevent obesity and related chronic diseases.

Keywords: Golden Berry (*Physalis peruviana*); metabolic syndrome; insulin resistance; obesity; nutritional intervention

1. Introduction

Individuals with metabolic syndrome face an elevated risk of health problems like high blood glucose levels, abnormal lipid levels, high blood pressure, obesity, and insulin resistance [1]. Obesity typically stems from an excess intake of calories beyond the body's daily requirements [2]. The World Obesity Federation defines obesity as a chronic condition that harms health, raises disease risk, and increases worldwide healthcare costs [3].

Insulin resistance occurs when cells fail to respond to insulin, causing high blood sugar levels. To counteract this, the pancreas produces extra insulin. If insulin resistance persists, it may exhaust the pancreas, increasing the risk of developing type 2 diabetes [4].

Insulin resistance is tied to factors like obesity, sedentary behavior, poor dietary choices, genetics, and environmental factors. It is also connected to health issues like cardiovascular disease and metabolic syndrome. Managing and preventing insulin resistance usually requires lifestyle adjustments, such as adopting a healthy diet and regular exercise, and sometimes, medical intervention [5].

Due to the increasing prevalence of obesity and its associated problem of insulin resistance, there is a growing interest in researching natural foods with bioactive components that may help improve these metabolic issues. The primary aim of this research is to prevent or even reverse these conditions [5].

The Golden Berry (*Physalis peruviana*), native to South America, is an exotic fruit now grown in tropical and subtropical regions. It is a small fruit, about 4 to 5 g, with a zesty outer covering. As it matures, its acidity increases while the pH decreases [6]. This fruit is a nutritional powerhouse, containing vitamins A, C, B complex, polyunsaturated fatty acids, phytosterols, and minerals [7]. This fruit is notably rich in antioxidants such as carotenoids, flavonoids, and phenolic acids, known for their role in preventing chronic diseases, including metabolic disorders. Furthermore, both in vitro and in vivo studies indicate that *Physalis peruviana* may possess anti-inflammatory, antidiabetic, and antihyperlipidemic properties [8].

Studies confirm the positive impact of Golden Berry consumption on metabolic disorders. For example, when rats on a high-fat diet were given Golden Berry, they showed reductions in body weight, total cholesterol, triglycerides, and blood glucose levels. Additionally, there was an improvement in insulin sensitivity [9]. In a similar vein, diabetic rats treated with a Golden Berry extract displayed lower blood glucose levels, improved glucose tolerance, and decreased oxidative stress markers in comparison to a control group. In sum, Golden Berry consumption has demonstrated potential in managing metabolic disorders [10].

Therefore, this study's core objective is to investigate the effects of consuming Golden Berry on anatomical changes, plasma metabolites, genes, and urinary biomarkers in a rat model simulating obesity and insulin resistance.

2. Materials and Methods

2.1. Plant Material

Golden Berry "Dorada" is a variety selected by AGROSAVIA (Ref. ICA UCH-16-02) which are grown by the company Caribbean Exotic S.A.S. at the farm "La Bendición" (6.23779 latitude/75.32.45 longitude) in San Vicente Ferrer, Antioquia, Colombia [8,11].

2.2. Diets and Animals

This study involved 64 Wistar rats (*Rattus norvegicus*), both male and female, of the Wistar WI IOPS AF/Han type. These rats were housed in the Experimental Unit of the Industrial University of Santander. We confirmed that there were no exclusions in the study, ensuring that the sample represented the entire group of participating animals. Additionally, the assignment of the rats to the experimental control and treatment groups was conducted through a process of randomization, thereby minimizing potential biases and ensuring the reliability of the study's findings. The sample size was meticulously calculated using a specific equation to ensure statistical validity Equation (1).

$$n = 2(\delta Z_1 - \beta + Z_1 - \alpha/2)^2\sigma^2 \quad (1)$$

($\delta = 0.7$), ($Z_1 = 0.842$), ($\beta = 0.20$), ($\alpha = 0.05$), ($\sigma = 2$) (measured through the standard deviation), so that their product is close to 1.401; $n = 8$. The rats were maintained under controlled conditions, with ad libitum access to food and water. The research consisted of four experimental groups: (i) Standard Diet (SD), $n = 8$; conventional diet composed of 20% protein, 16% fat, 64% carbohydrates; (ii) Standard Diet with Golden Berry (SD-GB), $n = 8$; the conventional diet was supplemented with 8% (w/w) fresh *Physalis peruviana* fruit, finely chopped into small pieces, and added to the diet; (iii) High-Fat Diet (HFD), $n = 8$; composed of 19% protein, 43% fat, 38% carbohydrates; and (iv) High-Fat Diet with Golden Berry (HFD-GB), $n = 8$; the HFD was supplemented with 8% (w/w) fresh *Physalis peruviana* fruit, finely chopped into small pieces, and added to the diet, as shown in Table 1. Abdomen circumference, body weight, and animal length were measured each week [12]. Body mass index (BMI) was calculated using the equation BMI = Weight (g)/Length

(cm^2) [13]; urine and blood samples were collected at the beginning (pre-) and end of the experiment (post-day 16). Specifically, urine samples were collected and centrifuged at 7000× *g* rpm for 15 min and filtered under pressure with a microporous syringe filter of 0.22 µm (Minisart- RC25, Sartorius., Gotinga, Alemania). After animal sacrifices, liver, pancreas, visceral adipose tissue (VAT), brown adipose tissue (BAT), and subcutaneous adipose tissue (SAT) were collected and weighed. Blood samples were centrifuged, and glycemia and lipid profile, including cholesterol, triglycerides, high-density lipoprotein (HDL), and low-density lipoprotein (LDL), were measured following the manufacturer's protocols (Wiener Lab, Rosario, Argentina) and read at 280 nm with the spectrophotometer Synergy H1 (BioTek, Santa Clara, CA, USA) [14].

Table 1. The composition of rat diets.

Diet Type	Protein (%)	Fat (%)	Carbohydrates (%)	Kcal/g
SD	20	16	64	4.8
HFD	19	43	38	5.37
GB	Supplemented with 8% (*w*/*w*) fresh Golden Berry fruit, finely chopped into small pieces, and added to the diet.			

Standard Diet (SD), High-Fat Diet (HFD), Semi-purified diets were selected for their ability to provide consistent and precise nutrient intake. These diets were composed of refined ingredients, each sourced specifically for its nutritional content (proteins, fats, carbohydrates). Two commercial products were utilized for vivarium use. Stringent monitoring of raw material batches was conducted to ensure these Conventional Nutrition formulas were devoid of any biological, environmental, or physical contaminants. The consumption of Golden Berry (GB) was conducted using fresh fruit, which was carefully chopped into pieces to facilitate its intake by the animals.

2.3. Gene Expression

Gene extraction and amplification was performed on SAT, and peroxisome proliferator-activated receptor gamma (PPARγ), fatty acid synthase (FasN), insulin receptor (INSR), and lipoprotein lipase (LPL) gene expression was evaluated by real-time polymerase chain reaction (qPCR) [2]. Briefly, RNA was extracted with TRIzol (Ambion, Waltham, MA, USA), followed by quantification with NanoDrop (Thermo Fisher Scientific, Waltham, MA, USA); RNA samples were purified with DNAse I treatment (Promega, Madison, Wisconsin, USA), followed by retrotranscription with M-MLV reverse transcriptase (Invitrogen, Waltham, MA, USA). Real-time qPCR amplification was carried out using primers designed with Biosearch Technologies Software (https://www.genscript.com/tools/real-time-pcr-taqman-primer-design-tool?src=google&utm_source=google&utm_medium=cpc&utm_campaign=Oligo-qPCR_Probes_Design_Tool_NA&jiraid=12194&gad_source=1&gclid=Cj0KCQiAtaOtBhCwARIsAN_x-3K2QGTD4HnilgZbNEMyTviHGdBritqIPPt8v2te-pZ0hdBJw1sS6aYaAu2pEALw_wcB accessed on 6 April 2023) (Bio-Rad, Hercules, CA, USA): Forward (F) and Reverse (R) for INSR: F: GCGGGGTGAAGACGGTCAATG, R: TGACAGGTGAAGCCCTTCATG; FasN, F: GCTGCCGTGTCCTTCTTCTACTAC, R: GGTACTTGGCCTTGGGTGTTTATAC; PPARγ, F: GAACCCAGAGTCTGCTGCTGATCTCTG, R: TCAGCGGGAAGGACTTTATGTATG and, LPL, F: GGCTCTCTGCCTGAGTTGTAGAAAG, R: TCTTGGCTCTCTCTGACCTTGTTGA. All samples were amplified in thermal cycler- CFX96 (Bio-Rad, Hercules, CA, USA), and quantitative analyses were performed with Gen 5 Software (BioTek, Winooski, VT, USA) [14].

2.4. Analysis UPLC/ESI-Q-Orbitrap

A UPLC-ESI-Q-Orbitrap system was used to analyze diluted urine. Chromatographic separation was performed coupled to a pre-column (130 Å, 1.7 µm particle size, 2.1 mm × 5 mm, Waters). UHPLC (Thermo Fisher Scientific, Waltham, MA, USA) was coupled to an ESI-Q/Orbitrap mass detector (Thermo Fisher Scientific, Waltham, MA, USA). The mobile phases were solvent A (water + 0.1% formic acid) and solvent B (acetonitrile + 0.1% formic acid). The elution gradient started at 0% of B. The high-resolution mass detector ESI-Q-Orbitrap system was used under a complete scan from 100 to 1000 Da in continuous mode. The column was kept at 40 °C. For untargeted metabolomics, mass spectrometry analyses were performed with a constant collision energy of 6 eV. The raw data were analyzed by progenesis QI software (https://www.waters.com/nextgen/us/en/products/informatics-and-software/mass-spectrometry-software/progenesis-qi-software/progenesis-qi.html accessed on 28 July 2023). (Waters Inc., Milford, MA, USA). Subsequently, the data drift was corrected using the Metabodrift [15]. For UPLC/ESI-Q-Orbitrap analysis, multivariate data analysis was conducted with software SIMCA Version 15.0.2 (Sartorius Stedim Data Analytical, Goettingen, Germany) in order to identify metabolites between groups with and without Golden Berry consumption (baseline) [16].

2.5. *Discovery of Biological Association Networks*

Ingenuity Pathway Analysis (IPA) software (https://analysis.ingenuity.com/pa/installer/select accessed on 25 August 2023).(Ingenuity, Redwood City, CA, USA) was used for the discovery of association networks of the urine metabolites [12]. The metabolites KEGG IDs and log2 FC (fold change) were used to identify possible interactions with biological molecules. The p-values were calculated using Fisher's exact test to determine the probability of association between the metabolites in the dataset; the logarithm (p value) > 2 was taken as the threshold and a Z score > 2 was defined as the threshold of significant inhibition for canonical pathway, disease, and function analysis.

2.6. *Statistical Analysis*

Statistical analysis was performed using GraphPad Prism® version 8.0 (GraphPad Software, San Diego, CA, USA) and SPSS/Windows software version 15.0 (SPSS Inc., Chicago, IL, USA). All results were expressed as mean ± standard error (SEM). One-way analyses of variance (ANOVA) with Tukey's test were applied for the analysis of more than 2 groups. In all cases, a p-value < 0.05 indicated a statistically significant difference between tested groups.

3. Results

3.1. *Anatomical Measurements*

The results in Table 2 show anatomical parameters showing how rat body weight changed after treatments with or without Golden Berry compared to their initial measurements. The post-diet (HFD-post) shows a significant increase in weight compared to its initial state (HFD-pre) in both males (p = 0.0014) and females (p = 0.0004). Among the post-treatment values, HFD-post exhibits the most significant differences compared to others, notably in males (p = 0.000012) and females (p = 0.0148) concerning SD, SD-GB (males p = 0.000012, females p = 0.00089), and HFD-GB only in males (p = 0.0077), as females showed no significant differences (p = 0.90008). Furthermore, when comparing the effects of all post-treatments, it was observed that HFD-post had significant differences compared to SD-post, but only in males (p = 0.0001) and not in females (p = 0.1440). However, HFD-post showed significant differences when compared to SD-GB-post (males p = 0.0020, females p = 0.0083) and HFD-GB-post (males p = 0.0277, females p = 0.0236).

Table 2. Pre and post results of anatomical and biochemical parameters in treatments with and without Golden Berry applied to the in vivo model of male and female Wistar rats.

SEX	Anatomical Parameters	SD		SD-GB		HFD		HFD-GB	
		Pre	Post	Pre	Post	Pre	Post	Pre	Post
MALES	Weight	370.3 ± 17.89	397.2 ± 17.49	395 ± 20.88	414.4 ± 16.5	418.8 ± 1.448	527 ± 18.45 ^{b,d,e}	422.2 ± 9.42	457.1 ± 9.44 ^{f}
	BMI	0.218 ± 0.006	0.221 ± 0.014	0.206 ± 0.002	0.195 ± 0.006	0.218 ± 0.003	0.343 ± 0.035 ^{b,d,e}	0.258 ± 0.005	0.258 ± 0.003 ^{f}
FEMALES	Weight	286 ± 9.77	311.8 ± 8.46	271.1 ± 3.79	296.1 ± 1.10	279.3 ± 0.54	342.1 ± 8.53 ^{d,e}	281.4 ± 10.41	310.5 ± 2.53 ^{f,g}
	BMI	0.191 ± 0.012	0.191 ± 0.012	0.182 ± 0.002	0.181 ± 0.004	0.184 ± 0.002	0.217 ± 0.007 ^{b,d,e}	0.212 ± 0.009	0.219 ± 0.003 ^{b,d}

Table 2. *Cont.*

SEX	Biochemical Parameters	SD		SD-GB		HFD		HFD-GB	
		Pre	Post	Pre	Post	Pre	Post	Pre	Post
MALES	GLY	177 ± 959	209 ± 10.17	176.5 ± 11.96	208.3 ± 10.38	131 ± 9.64	253.3 ± 4.63 [b,d]	129.5 ± 7.03	99.25 ± 7.98 [b,d,f]
	CHO	31 ± 3.97	40.14 ± 5.87	31.17 ± 4.21	38.67 ± 5.68	29.33 ± 3.48	119 ± 9.41 [b,d,e]	27.5 ± 2.78	26 ± 2 [f]
	TG	53.23 ± 3.97	66 ± 1.41	58.67 ± 2.33	29.67 ± 1.45 [b]	39.61 ± 4.82	91.4 ± 9.51 [b,d,e]	58.67 ± 3.93	35 ± 1.85 [b,f]
	LDL	198.5 ± 38.78	284.8 ± 25.08 [a]	77.6 ± 0.76	236.4 ± 16.48 [c]	91.6 ± 10.29	168 ± 21.3 [b,e]	78.93 ± 1.07	254.5 ± 11.01 [g]
	HDL	240.1 ± 39.88	346.9 ± 21.12 [a]	200.3 ± 39.71	278.7 ± 15.01	132.8 ± 6.55	302.5 ± 12.19 [e]	122.9 ± 2.23	275.1 ± 7.52 [g]
FEMALES	GLY	190.2 ± 2.65	176 ± 10.97	171.6 ± 15.96	186.7 ± 16.22	144.9 ± 14.04	231 ± 10.73 [b,e]	165.3 ± 17.45	191 ± 13.61 [f]
	CHO	26.33 ± 1.20	27.67 ± 0.88	24.5 ± 1.26	25.67 ± 1.76	17.67 ± 1.45	53 ± 3.79 [b,d,e]	27 ± 1.16	32.75 ± 1.38 [f]
	TG	72.75 ± 1.49	72.2 ± 9.48	65.8 ± 5.01	35.33 ± 1.67 [b,d]	74.25 ± 2.39	125.3 ± 8.29 [b,d,e]	88.2 ± 10.61	57.33 ± 3.33 [f]
	LDL	262.9 ± 3.37	260.1 ± 8.60	117.1 ± 43.51	279.6 ± 5.03 [c]	101.1 ± 4.02	209.3 ± 10.28 [e]	33.33 ± 4.35	215.7 ± 4.90 [g]
	HDL	215.7 ± 35.49	340 ± 15.49 [a]	159.8 ± 32.39	304 ± 7.78 [c]	240.1 ± 56.74	312.4 ± 9.52 [d]	76.29 ± 1.47	284.9 ± 7.71 [g]

SD: Standard Diet; SD-GB: Standard Diet with Golden Berry; HFD: High-Fat Diet; HFD-GB: High-Fat Diet with Golden Berry; pre: before; and post: after treatment. BMI: body mass index; GLY: glycemia; CHO: cholesterol; TG: triglycerides; LDL: (low-density lipoprotein) cholesterol; HDL: (high-density lipoprotein) cholesterol. [a] vs. SD pre; [b] vs. SD post; [c] vs. SD-GB pre; [d] vs. SD-GB pre; [e] vs. HDF pre; [f] vs. HDF post; [g] vs. HDF-GB.

3.2. Biochemical Parameters

The results presented encompass pre- and post-evaluations of various biochemical parameters under different dietary conditions, showing statistically significant differences between genders and dietary treatments. These results are divided into five categories, as shown in Table 2.

Glycemia response to diet: The HFD-post led to significantly higher levels compared to the initial HFD state (males $p = 0.000001$, females $p = 0.000077$) and compared to the SD-post (males $p = 0.0074$, females $p = 0.0304$). Interestingly, for the addition of Golden Berry (HFD-GB-post), significant differences were observed only in males ($p < 0.000001$) and females ($p = 0.04966$).

Cholesterol levels and dietary influence: Cholesterol levels spiked significantly in the HFD group across genders (males $p < 0.000001$ and females $p < 0.000001$). Compared to other diets, SD (males $p < 0.000001$, females $p < 0.000001$), SD-GB (males $p < 0.000001$, females $p < 0.000001$), and HFD-GB (males $p < 0.000001$, females $p = 0.000003$). This elevation was markedly higher, suggesting a strong impact of high-fat content on cholesterol levels. The Golden Berry inclusion in diets (both SD-GB and HFD-GB) showed an apparent moderating effect.

Triglyceride dynamics across diets: Post-HFD triglyceride levels significantly surpassed initial levels for both genders (males $p < 0.000001$, females $p = 0.0027$). In comparison to other dietary treatments, SD (males $p = 0.00552$, females $p = 0.0010$), SD-GB (males $p < 0.000001$, females $p = 0.000002$), and HFD-GB (males $p < 0.000001$, females $p = 0.000173$). This increase was the most pronounced, illustrating the impactful role of high-fat content on triglyceride levels.

LDL-cholesterol variations: Each treatment exhibited significant differences in post-treatment concentrations compared to its baseline: SD-post (males $p = 0.0185$), SD-GB-post (males $p = 0.0042$), females $p = 0.0004$), HFD-post (males $p = 0.0350$, females $p = 0.0352$), and HFD-GB-post (males $p = 0.0018$, females $p = 0.000098$). Notably, males in the SD-post group showed a unique response when compared to HFD-post ($p = 0.0070$), indicating a differential gender-specific impact of these diets on LDL levels.

HDL-cholesterol trends followed a similar pattern, showing significant differences between treatments. SD-post exhibited significant differences in comparison to its initial state (males $p = 0.0166$, females $p = 0.0155$), and SD-GB-post had significant differences compared to its baseline, but only in females ($p = 0.0051$). Differences were also observed between HFD-post and its initial state exclusively in males ($p = 0.000083$), as well as in both males and females when comparing HFD-GB-post with its initial state (males $p = 0.000005$, females $p < 0.000001$). The response of HDL levels to the diets was gender-specific and varied across different dietary compositions, particularly highlighting the role of Golden Berry.

Interconnected implications: This study reveals a complex interaction between diet composition, gender, and metabolic health. High-fat diets consistently led to adverse effects across all parameters, which were somewhat mitigated by the inclusion of Golden Berry. This indicates the potential of dietary modifications in managing diet-induced metabolic changes. Gender-specific responses further emphasize the need for personalized dietary recommendations. The cross-linking of these parameters underlines a multi-faceted approach to understanding dietary impacts on health, suggesting a broader narrative of diet and metabolic interplay.

3.3. Impact of Dietary Variations on Organ and Adipose Tissue Weights in Wistar Rats: The Role of Golden Berry Supplementation

The results in Table 3 show the tissue weights of rats exposed to different diets. Liver: Significant differences were observed when comparing liver weights between rats in the HFD treatment and the SD group (males $p < 0.000001$, females $p < 0.000001$), SD-GB (males $p < 0.000001$, females $p < 0.000001$), and HFD-GB (males $p = 0.00004$, females $p < 0.000001$). Additionally, SD showed significant differences in comparison to SD-GB ($p = 0.00124$) and HFD-GB ($p = 0.000739$), but only in females, while SD-GB had significant differences in relation to HFD-GB in both (males $p = 0.00140$ and females $p < 0.000001$). The liver weight, particularly in the SD-GB and HFD-GB groups, indicated a slight yet significant reduction when compared to their respective control groups. This points to a potential beneficial impact of Golden Berry supplementation in modulating liver weight, a crucial factor in metabolic health.

Pancreas: HFD resulted in higher organ weights in both males and females compared to SD (males $p < 0.000001$, females $p = 0.000011$), SD-GB (males $p < 0.000001$, females $p = 0.000008$), and HFD-GB (males $p < 0.000001$, females $p = 0.01205$). Furthermore, SD-GB had differences compared to HFD-GB (males $p = 0.000004$, females $p = 0.003009$), and in females, SD and SD-GB showed significance in relation to HFD-GB ($p = 0.01075$ and $p = 0.003009$, respectively). Regarding pancreatic weight, the variations observed across the groups were minimal, suggesting a relative stability of this organ's mass under varying dietary conditions. This finding contributes to the understanding of the pancreas's resilience to dietary changes.

Adipose tissues: Both VAT and SAT in males and females exhibited similar trends. HFD led to higher VAT and SAT weights in males compared to SD (VAT: $p < 0.000001$, SAT: $p < 0.000001$), SD-GB (VAT: $p < 0.000001$, SAT: $p < 0.000001$), and HFD-GB (VAT: $p < 0.000001$, SAT: $p < 0.000001$). SD also showed significant differences compared to SD-GB (VAT: $p = 0.00066$, SAT: $p = 0.000051$) and HFD-GB (VAT: $p < 0.000001$, SAT: $p < 0.000001$). Notably, when evaluating SD-GB in comparison to HFD-GB, significant differences were observed (VAT: $p < 0.000001$, SAT: $p < 0.000001$). For females, the pattern was similar to males. HFD resulted in higher VAT and SAT weights compared to SD (VAT: $p < 0.000001$, SAT: $p < 0.000001$), SD-GB (VAT: $p < 0.000001$, SAT: $p < 0.000001$), and HFD-GB (VAT: $p = 0.002743$, SAT: $p < 0.000015$). However, HFD-GB showed a reduction in adipose tissue weights compared to SD (VAT: $p < 0.000001$, SAT: $p < 0.000001$) and SD-GB (VAT: $p < 0.000001$, SAT: $p < 0.000001$), with SD-GB only displaying significant differences on SAT in comparison to SD ($p = 0.001214$). A notable observation was made in the context of visceral adipose tissue (VAT) weights. In the HFD group, a substantial increase in VAT was documented. However, in the presence of Golden Berry supplementation, this increase was significantly mitigated, implying a protective effect of this dietary component against the fat accumulation typically associated with high-fat diets.

This study also shed light on the dynamics of brown adipose tissue (BAT). A decrease in BAT weight was noted in the HFD group, whereas an increase was observed in the Golden Berry-supplemented groups. This suggests a role for Golden Berry in promoting BAT mass, a tissue known for its metabolic benefits.

Similar trends were observed in subcutaneous adipose tissue (SAT), where the HFD group exhibited increased weights, but the introduction of Golden Berry appeared to counteract this effect, especially in male rats. This finding underscores the potential of Golden Berry in regulating adipose tissue distribution and growth.

Table 3. Mean and standard error of relative organ weights in different treatment groups with and without Golden Berry in an in vivo model of Wistar rats (g/kg body weight).

SEX	Organ	SD (g/kg)	SD-GB (g/kg)	HFD (g/kg)	HFD-GB (g/kg)
MALES	Lv	29.25 ± 4.80	25.14 ± 3.93	30.61 ± 16.77 [a,b]	28.59 ± 12.63 [a,b,c]
	Pc	1.72 ± 0.27	1.51 ± 0.70	1.63 ± 0.99	1.63 ± 0.61
	VAT	15.70 ± 0.0569	8.9 ± 0.1263	39.0 ± 0.9886 [a,b]	24.19 ± 0.3004 [a,b,c]
	BAT	4.11 ± 0.0546	4.76 ± 0.0320	0.93 ± 0.0356 [a,b]	2.19 ± 0.0121 [a,b,c]
	SAT	18.32 ± 0.0832	10.76 ± 0.1515	40.13 ± 0.2659 [a,b]	32.46 ± 0.6124 [a,b,c]
FEMALES	Lv	26.49 ± 8.97	25.24 ± 99.72	31.54 ± 17.05 [a,b]	29.04 ± 15.41 [a]
	Pc	1.64 ± 0.50	1.55 ± 16.27	2.37 ± 8.49 [a]	2.11 ± 1.69 [a]
	VAT	13.67 ± 0.0815	12.71 ± 0.0498	53.46 ± 0.917 [a,b]	45.15 ± 0.9088 [a,b,c]
	BAT	3.70 ± 0.0105	5.66 ± 0.0210	1.39 ± 0.0132 [a,b]	3.02 ± 0.0119 [a,b,c]
	SAT	17.33 ± 0.0717	13.30 ± 0.1447	49.80 ± 0.4652 [a]	44.89 ± 0.6132 [a]

Values are presented as mean ± standard error in grams per kilogram of body weight. SD: Standard Diet; SD-GB: Standard Diet with Golden Berry; HFD: High-Fat Diet; HFD-GB: High-Fat Diet with Golden Berry. Lv: liver; Pc: pancreas. VAT: visceral adipose tissue; BAT: brown adipose tissue; SAT: subcutaneous adipose tissue. [a] vs. SD; [b] vs. SD-GB; [c] vs. HFD.

3.4. *Comparative Analysis of Key Gene Expressions Influenced by Dietary Variations*

The results in Figure 1a display the gene expression of INSR and FasN. Notably, the HFD-post diet presented a significant reduction in INSR gene expression with respect to SD-post males ($p = 0.0316$) and females ($p = 0.0002$), SD-GB-post only in females ($p = 0.01342$), and HFD-GB-post males ($p = 0.0019$) and females ($p = 0.000005$). HFD-GB-post in relation to SD-GB in males ($p = 0.0213$); and in females in relation to SD ($p = 0.003642$) and SD-GB ($p = 0.000087$). As for standard diets, only females presented significant differences between SD-post and SD-GB-post ($p = 0.02213$).

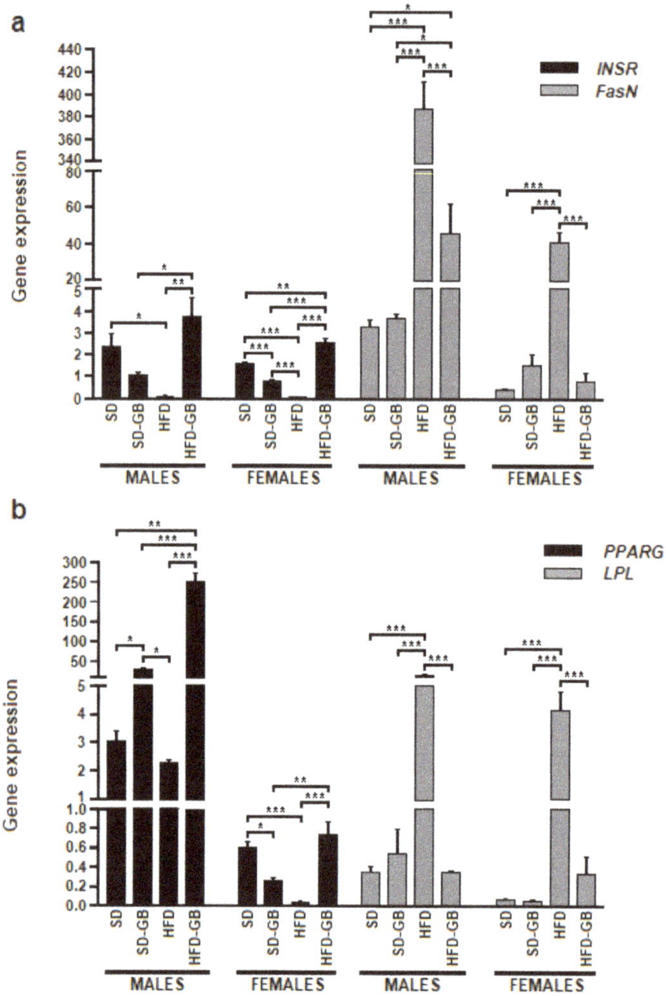

Figure 1. Gene expression (**a**) INSR and FasN and (**b**) PPARG and LPL in males and females in treatments with and without Golden Berry. Effects of Golden Berry supplementation on gene expression influenced by dietary variations in Wistar rats. (**a**). Gene expression of INSR and FasN: The figure shows the expression of INSR and FasN in different diets. A significant increase in INSR gene expression was observed in the HFD-GB diet compared to the HFD diet in both males ($p = 0.0316$) and females ($p = 0.0002$). Golden Berry increases INSR expression and decreases FasN expression in Wistar rats fed with HFD ($n = 8$; one-way ANOVA followed by Tukey's multiple comparison test; ** $p < 0.01$) over a 16-week intervention period compared to rats fed with SD. (**b**). Gene expression of PPARγ and LPL: The figure illustrates the increase in PPARγ expression and the decrease in LPL, showing significant differences compared to HFD in both males ($p = 0.04062$) and females ($p = 0.02009$) and with HFD-GB. Groups are represented as (i) Standard Diet (SD), $n = 8$; (ii) Standard Diet with Golden Berry (SD-GB), $n = 8$; the conventional diet is supplemented with 8% (w/w) Golden Berry. (iii) High-Fat Diet (HFD), $n = 8$; and (iv) High-Fat Diet with Golden Berry (HFD-GB), $n = 8$; the HFD diet is supplemented with 8% (w/w) Golden Berry. * $p < 0.05$, ** $p < 0.01$, *** $p < 0.001$.

The results in Figure 1b show the expression of PPARγ for all diets tested. The highest gene expression of PPARγ was observed in the HFD-GB diet, with a significant increase compared to SD in males ($p < 0.000001$), SD-GB in both males ($p < 0.000001$) and females ($p = 0.001193$), and HFD in both males ($p < 0.000001$) and females ($p = 0.000026$). The treatment with high PPARγ expression was SD, showing significant differences compared to SD-GB in males ($p = 0.04062$) and females ($p = 0.02009$) and with respect to HFD in females ($p = 0.00040$).

LPL exhibited maximum expression in the HFD diet, which, when compared to the other diets, demonstrated high significance with respect to SD in males ($p < 0.000001$) and females ($p = 0.000017$), SD-GB in males ($p < 0.000001$) and females ($p = 0.000028$), and HFD-GB in males ($p < 0.000001$) and females ($p = 0.00005$).

These results indicate that dietary variations, especially the incorporation of Golden Berry in high-fat diets, significantly influence the expression of genes related to metabolism and insulin signaling in Wistar rats. The data suggest a notable impact of diet on metabolic gene regulation, with implications for understanding dietary influences on health.

3.5. Metabolic Profiling and Biomarker Discovery in Rat Urine Post-Golden Berry Consumption: An OSC-PLS-DA Approach

In the analysis of rat urine samples before and after Golden Berry consumption, significant urinary biomarkers were successfully identified with low standard deviation values in quality control pools. An OSC-PLS-DA model was applied to differentiate between urine samples of rats with and without Golden Berry consumption. This model generated a list of the most important ions based on their VIP scores, as shown in Table 4. The model's performance was considered acceptable, with a CV-ANOVA value of 2×10^{-5}, an R2X value of 0.46, an R2Y value of 0.99, and an RQ2 value of 0.83. Furthermore, permutation tests for cross-validation did not reveal signs of overfitting.

Unidentified compounds: The primary discriminant ions were identified as m/z 432.34 [M + H] and m/z 430.333. The first discriminant metabolites, which differ only by hydrogenation, could not be identified. The second-most discriminant metabolite had an m/z value of 380.1914 [M + H], and its fragmentation pattern corresponded to the loss of NH3 (resulting in a fragment at m/z 363.1639 with a mass loss of 17.02 amu), hydration (yielding a fragment at m/z 345.1534 with a mass loss of 18.01 amu), and glucuronidation (producing a fragment at m/z 169.11219 with a mass loss of 176.03 amu). Subsequently, phase I and II transformations led to a predominant fragment at m/z 169.11219, which displayed a fragmentation pattern consistent with geranic acid ($C_{10}H_{16}O_2$), a compound previously observed among the flavor compounds in Golden Berry [17].

Table 4. Main urinary biomarkers of Golden Berry consumption in rats.

Obs *m/z* [M + H]	Rt (min)	FORMULA	VIP Score	Z	Tentative Identification	HMDB	Error (ppm)	MS2 Measured (%peak)	*p*-Value	Average Relative Intensity	
			TU vs. TS						B/A	Before	After GC
432.3470	12.55	$C_{27}H_{45}NO_3$	3.29	1	Unknown			98.0965 (100); 81.07 (10)	9.5×10^{-3}	<10	5,008,467
430.3314	12.46		2.95	1	Unknown			98.0965 (100); 81.07 (10)		<10	300,000
380.1914	11.73	$C_{10}H_{16}O_2$-glu-H_2O-NH_3	2.62	1	Geranic acid	0303972	2	345.1534 (100); 169.1219 (60) 151.1114 (30)	3.2×10^{-4}	<10	188
364.1963	13.24	$C_{10}H_{16}O$-glu-H_2O-NH_3	2.01	1	*p*-menth-4(8)-ene-1,2-diol	0035706	3	329.1594 (20); 153.1270 (100) 135.1165 (100)	1.3×10^{-2}	<10	190,928
184.0968	11.59	C_7H_8O-Glcycine	2.62	1	Benzyl-alcohol glycine	0003119		109.0640 (80) 81.07 (100)	31×10^{-2}	<10	77,000
158.1175	12.44	$C_8H_{15}NO_2$	2.02	1	Unknown	0004827	3		4.1×10^{-3}	20,000	5000
130.0651	14.29	C_9H_7N	1.80	1	Unknown	0004827	0	103.05 (100);	8.4×10^{-4}	94,000	50,000

Table 4. *Cont.*

Obs m/z [M + H]	Rt (min)	FORMULA	VIP Score	Z	Tentative Identification	HMDB	Error (ppm)	MS2 Measured (%peak)	p-Value	Average Relative Intensity	
			TU vs. TS						B/A	Before	After GC
167.1067	15.12	$C_{10}H_{14}O_2$	1.75	1	Perillic acid	0000070	3	149.008 (30) 121.028 (20) 93.07 (80) 79.05 (100)	3.4×10^{-2}	5000	15,000
374.2526	12.53	$C_{19}H_{35}NO_6$	1.72	1	Dodecanedioyl-L-carnitine	0013327	4	231.1585 (50); 130.05 (40);85.02 (35)	1×10^{-3}	44,264	29,738
164.0367	12.51	$C_5H_9NO_3S$	1.58	1	Unknown	0001890	5	122.0269 (100); 76.02 (70)	2×10^{-2}	60,000	30,000
245.1595	1.04	$C_8H_{18}N_4O_2$ + [C_2H_2O]	1.55	1	Dimethylarginine acetlylated	0001539	3	203.14 (30); 158.08 (100); 115.08 (70); 70.06 (50)	2×10^{-3}	5000	8000
146.0600	14.29	$C_9H_7NO_2$- [O]	1.36	1	Indolecarboxylic acid	0002285	1	118.65 (100) 91.054 (70) 65.039 (60) 128.049 (30)	2.4×10^{-2}	50,000	25,000

The main urinary biomarkers identified in rats after Golden Berry consumption. Observations include mass-to-charge ratio (m/z), retention time (Rt), chemical formula, Variable Importance in Projection (VIP) Score, charge state (Z), tentative identification, Human Metabolome Database (HMDB) reference, mass error in parts per million (ppm), MS2 measured peak intensities, p-value for significance, and average relative intensity before (B) and after (A) Golden Berry consumption. The table focuses on biomarkers related to metabolic health, highlighting Golden Berry's influence on metabolic pathways and providing insights into its potential health benefits.

3.6. Identification and Analysis of Key Urinary Metabolites in Rats following Golden Berry Consumption

In the comprehensive metabolomic analysis focusing on the impact of Golden Berry consumption in rats, several key urinary metabolites were identified, each revealing a facet of the metabolic alterations induced by this dietary intervention. B-Cyclocitral Glucuronide, a significant monoterpene hydrocarbon in Golden Berries, emerged as a notable metabolite with its elevated presence in urine samples. This finding aligns with the high amounts of β-Cyclocitral reported in Golden Berries, underscoring its metabolic relevance, as shown in Figure 2a.

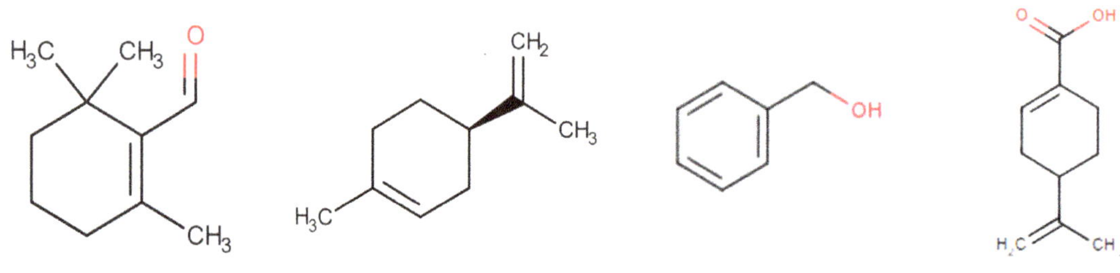

(a) (b) (c) (d)

Figure 2. Main urinary biomarkers in rats associated with Golden Berry fruit consumption. UPLC/ESI-Q-Orbitrap metabolomic analysis of Golden Berry consumption in rats. This analysis utilized a UPLC-ESI-Q-Orbitrap system for examining diluted urine samples to assess the impact of Golden Berry (Cape Gooseberry) consumption. Chromatographic separation was achieved using a pre-column (130 Å, 1.7 µm particle size, 2.1 mm × 5 mm, Waters, Rydalmere NSW, Australia) coupled with UHPLC (Thermo Fisher Scientific, Waltham, MA, USA) and an ESI-Q/Orbitrap mass detector (Thermo Fisher Scientific, Waltham, MA, USA). The high-resolution mass detector operated in a full scan mode from 100 to 1000 Da. Untargeted metabolomic analyses were conducted with a constant collision energy of 6 eV. Data processing utilized Progenesis QI software (Waters Inc., Milford, MA, USA). Multivariate data analysis was performed with SIMCA software Version 15.0.2 (Sartorius Stedim Data Analytical, Goettingen, Germany). The key urinary metabolites identified in this study, revealing various facets of the metabolic alterations induced by Golden Berries, include β-Cyclocitral Glucuronide (**a**), Limonene Glucuronide (**b**), Benzyl Alcohol (**c**), and Perillic Acid (**d**).

Pipecolic Acid ($C_6H_{11}NO_2$), which is particularly prominent in male rats on an obesogenic diet, showed a marked decrease in urinary excretion following Golden Berry consumption, highlighting the berry's potential role in modulating amino acid metabolism in high-fat diets. Limonene Glucuronide, tentatively identified in glucuronide form and shown in Figure 2b, and Benzyl-Alcohol Glycine, a glycosidically bound volatile in Golden Berries, were both notable for their distinct excretion patterns. These compounds underscore the metabolic processing of Golden Berry constituents, as shown in Figure 2c.

The tentative identification of D-alpha-Cyclohexylglycine, though with low confidence, adds to the diversity of metabolites influenced by Golden Berry intake. Quinoline and Perillic Acid, a significant metabolite of limonene, further illustrate the broad spectrum of metabolic changes, with Perillic Acid notably increasing in rats consuming Golden Berries. Dodecanoyl-L-Carnitine, identified with characteristic acyl-carnitine fragments, points to alterations in lipid metabolism, a finding supported by previous observations in human studies, as shown in Figure 2d.

Lastly, Acetylated Dimethylarginine, with a slight increase post-consumption, adds to the array of nitrogenous compounds affected by Golden Berry intake. Collectively, these findings provide a multifaceted view of the metabolic impact of Golden Berry consumption, revealing specific metabolites and their altered excretion patterns in rats. This intricate metabolic portrait, although requiring further validation, paves the way for a deeper understanding of the nutritional influence of Golden Berries. It is worth noting that these identifications are tentative and may require further validation and confirmation.

3.7. Elucidating Metabolic Networks and Biological Pathways Influenced by Golden Berry Consumption using Ingenuity Pathway Analysis

The results in Figure 3 show the integrative association between metabolomic study in a biological context, this analysis, and the application of advanced analytical tools like the Ingenuity Pathway Analysis (IPA) software, which was adeptly utilized to dissect the intricate metabolic association network stemming from the effects of Golden Berry consumption on cell death and survival and renal and urological disease networks (p-value = 2.02×10^{-6}).

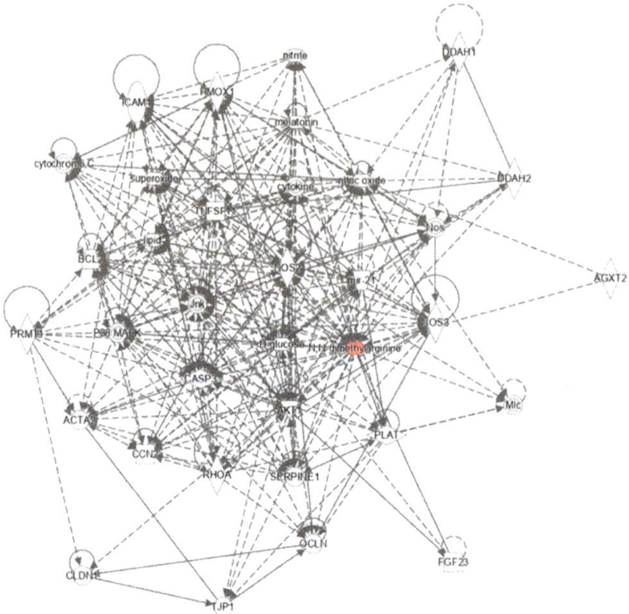

Figure 3. Significant biological network: impact of Golden Berry consumption. Figure 3 illustrates the significant biological network identified from the integrative association between metabolomic studies in a biological context. Utilizing the advanced analytical capabilities of Ingenuity Pathway Analysis (IPA) software, this analysis adeptly unraveled the complex metabolic association network arising from the effects of Golden Berry consumption, particularly focusing on cell death and survival, and the network of renal and urological diseases (p-value = 2.02×10^{-6}). In this network representation, metabolites are depicted as nodes, with the biological relationship between two nodes indicated by a line. The colored symbols represent metabolites that not only appeared in our data but also correlate with pathways in the Ingenuity® database. In contrast, transparent entries are molecules included in the Ingenuity knowledge base. Metabolites highlighted in red indicate increased levels, whereas those in green show decreased levels. Solid lines between molecules signify direct physical relationships, while dotted lines suggest indirect functional relationships.

4. Discussion

In this groundbreaking study, we have thoroughly explored the significant impact of Golden Berry on in vivo models of Wistar rats, focusing on how this fruit alters the response to standard and high-fat diets. Our findings reveal a notable resistance to rapid weight gain and improvements in biochemical profiles when Golden Berry is incorporated into an HFD, highlighting its potential as a natural agent for weight management and metabolic control. Additionally, we have observed significant changes in the weight of key organs such as the liver and pancreas, along with alterations in gene expression related to glucose metabolism and fat synthesis. These results not only provide new insights into the benefits of Golden Berry but also open promising avenues for future research on its role in the prevention and management of metabolic disorders. In the following sections, we will detail these findings and discuss their relevance in the broader context of metabolic health and nutrition.

4.1. Golden Berry's Role in Counteracting Obesity and Supporting Weight Management

An HFD is widely recognized as a contributor to obesity. However, when Golden Berry was introduced alongside the HFD, it became evident that there was a notable resistance to rapid weight gain. These results are particularly fascinating as they suggest that Golden

Berry might hold promise for weight management, potentially providing valuable support for individuals striving to maintain a healthy body weight [18,19].

4.2. Beneficial Biochemical Effects of Golden Berry in Diet: Improved Glucose and Lipid Metabolism

Adding Golden Berry to the diet provides several health benefits, such as improved blood glucose levels and reduced insulin resistance, resulting in lower glycemia levels [20,21]. HFDs led to a significant increase in cholesterol levels, while standard diets (SD and SD-GB) contributed to substantial reductions in cholesterol levels, indicating potential heart-protective properties [22]. Golden Berry consumption in obese animals has beneficial effects. Specifically, the findings related to triglyceride levels underscore the positive impact on lipid metabolism [23,24]. Adding Golden Berry to a SD effectively controls weight gain, even in individuals with a normal body weight, by preventing rapid weight gain and the onset of chronic obesity [25,26]. Conversely, incorporating Golden Berry into the diet delays weight gain and results in a lower BMI, indicating its potential in mitigating obesity-related issues [27].

4.3. Golden Berry's Impact on Reducing Organ Weights and Preventing Metabolic Diseases

Observations from the standard diet cohort reveal that organ weights are at the lower end of the spectrum, aligning with expectations for a diet lower in calories. The incorporation of Golden Berry into a standard diet indicates a trend towards stabilizing or slightly reducing average organ weights relative to the standard diet. This trend emphasizes Golden Berry's potential to subtly influence organ weight, validating its application within a standard dietary framework. This benchmark establishes a crucial reference for evaluating the nuanced impacts of Golden Berry supplementation. Adding Golden Berry into the HFD diet notably reduced liver weight. This contrasts with the HFD diet, which tends to increase liver weight [28]. Excessive weight gain and high blood glucose levels can harm the liver, potentially leading to liver fibrosis [29]. The HFD regimen is starkly contrasted by its inclination to increase organ weights, notably affecting adipose tissues like VAT and SAT. This correlation starkly delineates the high-fat diet's role in promoting body weight gain and fat accumulation. The weight gain induced by the HFD indicated a risk of liver damage. However, introducing Golden Berry into the diet significantly reduced liver weight [30]. This effect is attributed to the high levels of polyphenols, as well as vitamins A (α-carotene, β-carotene, and β-cryptoxanthin) and vitamin C found in Golden Berry [31].

Interestingly, the HFD-GB led to a decrease in pancreas size and weight, even when part of an HFD. Excessive weight gain in the pancreas, as observed in the HFD, may be linked to potential insulin resistance [32]. When it comes to adipose tissues, the data on anatomical parameters and biochemical levels strongly indicate that the HFD has a considerable impact on these measurements. This impact is most evident in the liver, pancreas, and the accumulation of VAT and SAT. However, the incorporation of Golden Berry into a standard diet indicates a trend towards stabilizing or slightly reducing average organ weights relative to the standard diet. This trend emphasizes Golden Berry's potential to subtly influence organ weight, validating its application within a standard dietary framework, and the introduction of Golden Berry into the daily diet helps normalize these measurements [23].

The biochemical parameters in Table 2 are linked to the organ weight results in Table 3. It was previously mentioned that excessive weight gain might be associated with the rapid expansion of WAT, including visceral and subcutaneous fat [33]. The inclusion of Golden Berry in the diet led to a decrease in both organ and tissue weights and an overall improvement in biochemical profiles. Adding Golden Berry into the diet offers a preventive strategy for managing and avoiding metabolic diseases.

The integration of Golden Berry into a high-fat diet intriguingly appears to mitigate the usual organ weight increases associated with a high-fat diet. This observation suggests

Golden Berry's efficacy in countering some of the detrimental effects of an HFD, particularly in terms of fat accumulation.

4.4. Influence of Golden Berry on the Reactivity of White and Brown Adipose Tissue

Adipose tissues, both VAT and SAT in males and females, showed similar trends. The HFD led to higher weights compared to the SD. However, the HFD-GB showed a reduction in adipose tissue weight compared to the SD. A notable observation was made in the context of VAT weights. A substantial increase in VAT was documented in the HFD group. However, in the presence of Golden Berry supplements, this increase was significantly mitigated, implying a protective effect of this dietary component against fat accumulation typically associated with HFDs. Adipose tissues exhibit significant weight fluctuations, highlighting their sensitivity to dietary changes and supplementation. This responsiveness is key to understanding the metabolic outcomes of different dietary patterns and interventions.

This study notably reveals the dynamics of BAT, with a marked decrease in BAT weight in the HFD group, contrasted by an increase in groups receiving Golden Berry supplementation. This observation points to Golden Berry's potential in augmenting BAT mass, a tissue renowned for its metabolic virtues. BAT is integral for thermogenesis and functions as an endocrine organ that influences fat and carbohydrate metabolism, which could enhance calorie burning.

This comprehensive analysis illuminates Golden Berry's potential in regulating organ weight and attenuating fat deposition, especially in the context of an HFD. These initial findings lay a foundation for further exploration into Golden Berry's characteristics and its prospective applications in nutrition and health. This research opens up a promising path for future inquiries into the therapeutic and nutritional potentials of Golden Berry.

4.5. Golden Berry's Influence on Gene Expression in Metabolic Regulation and Health

Golden Berry consumption impacts gene expression related to metabolic health, focusing on genes involved in glucose metabolism, fat synthesis, and adipocyte differentiation.

4.5.1. INSR Gene Expression Modulation by Golden Berry

Under typical conditions, the INSR gene, integral to glucose metabolism and insulin signaling, demonstrates diminished expression in response to HFDs. This decrease in expression potentially impairs insulin signaling pathways, escalating the risk of developing metabolic disorders, notably type 2 diabetes and metabolic syndrome [34]. Crucially, the consumption of Golden Berry correlates with a noticeable upregulation of INSR expression. This upregulation signifies enhanced insulin signaling, which could serve as a protective mechanism against the onset and progression of these metabolic conditions [34]. The association of Golden Berry intake with increased INSR activity highlights its potential role in mitigating adverse metabolic effects induced by HFD patterns. This finding places Golden Berry as a significant dietary component that may offer therapeutic benefits in the context of metabolic health regulation.

4.5.2. FasN Expression and the Impact of Golden Berry

FasN, an enzyme pivotal for synthesizing long-chain fatty acids and storing surplus glucose as fat within adipose tissues, typically exhibits heightened expression in response to an HFD. This overexpression is closely linked to an increased risk of obesity and associated metabolic disorders [35]. Intriguingly, the consumption of Golden Berry appears to play a modulatory role in FasN expression. The intake of Golden Berry correlates with a reduction in FasN expression, suggesting a meaningful impact on metabolic pathways. This reduction points to a potential decrease in the synthesis and accumulation of fat. Such an effect of Golden Berry consumption presents an important avenue in understanding its potential in mitigating obesity and related metabolic complications, particularly those exacerbated by HFDs.

4.5.3. PPARγ Expression and Its Role in Metabolic Regulation: The Impact of Golden Berry Supplementation

PPARγ, a nuclear receptor, plays an instrumental role in adipocyte differentiation, lipid storage, and glucose metabolism. Its significant therapeutic potential, particularly in the management of type 2 diabetes and metabolic syndrome, is noteworthy. PPARγ's involvement in these metabolic processes positions it as a key target for therapeutic strategies. In the context of dietary influence, specifically under the regime of a HFD supplemented with Golden Berry (HFD-GB), there is a notable increase in the expression of PPARγ. This heightened expression is seen as beneficial, contributing to healthier fat storage patterns and improved insulin sensitivity. Such an increase, prompted by Golden Berry intake, underscores the potential of dietary components in modulating key metabolic pathways [36,37].

The augmented expression of PPARγ in the HFD-GB group highlights the significant modulatory impact of Golden Berry on crucial metabolic functions. This finding is particularly promising in the context of metabolic conditions characterized by disrupted fat and glucose handling. It showcases the potential of targeted dietary interventions, such as Golden Berry supplementation, in the effective management and potential alleviation of metabolic disorders.

4.5.4. Golden Berry's Influence on LPL Gene Expression and Metabolic Health

Lipoprotein Lipase (LPL), a gene intimately connected to lipid metabolism, demonstrates significantly elevated expression levels in animals subjected to an HFD. This heightened expression is symptomatic of metabolic imbalances, typically characterized by escalated blood glucose levels and a disrupted lipid profile. Such a profile is a common precursor to various metabolic disorders. Intriguingly, the introduction of Golden Berry into the diet manifests beneficial effects on both fat and glucose metabolism. This is evidenced by a marked reduction in LPL expression within adipose tissues. This modulation suggests a potential improvement in insulin signaling pathways and the adipogenesis process, both of which are critical components in maintaining metabolic equilibrium [38]. The consumption of Golden Berry, therefore, emerges as a promising dietary intervention. Its capacity to attenuate LPL expression, and consequently its role in reshaping metabolic profiles, positions it as a valuable nutritional tool. These findings, as indicated in this study, illuminate the potential of Golden Berry in enhancing metabolic health, particularly by mitigating the adverse effects associated with HFDs.

The modulation of key metabolic genes through Golden Berry consumption, including INSR, PPARγ, FasN, and LPL, marks a significant impact on metabolic health. These findings underscore Golden Berry's potential as a strategic nutritional intervention to counterbalance the adverse effects associated with an HFD. This insight positions Golden Berry as a promising candidate for dietary approaches aimed at improving metabolic profiles and combating diet-induced metabolic disorders.

4.6. *Metabolomic Insights: Golden Berry's Effect on Metabolic Biomarkers and Potential in Diabetes Management*

The results in Table 4, which show a metabolomic analysis of rat urine before and after Golden Berry consumption, reveal several significant metabolites.

The discovery of Geranic Acid in Golden Berry, denoted by a specific fragment at m/z 169.11219, stands out as a notable aspect of the fruit's metabolomic profile. As a primary compound within Golden Berry, the unique ion pattern of Geranic Acid underscores its significance in the berry's metabolic processes. This finding enhances our understanding of the metabolomic intricacies of Golden Berry. Additionally, the identification of *p*-menth-4(8)-ene-1,2-diol in Golden Berry's volatile fraction, characterized by fragments at m/z 329.158 and m/z 153.127, further emphasizes the complexity and richness of its metabolomic composition. These specific compounds contribute to the distinctive characteristics of Golden Berry, revealing a multifaceted metabolic profile that warrants further exploration [39]. This qualitative assessment not only highlights the diverse nature of

Golden Berry's metabolites but also suggests a broader implication of these compounds in the fruit's potential health benefits. The presence of such unique components like Geranic Acid and *p*-menth-4(8)-ene-1,2-diol illustrates the intricate biochemistry of Golden Berry, contributing to its growing recognition in nutritional and health-related research.

A key discovery in the nutritional analysis of Golden Berry is the high concentration of glycine, an amino acid identified by a specific fragment at 109.0680. Glycine stands out for its abundant presence in Golden Berry, distinguishing this fruit as a significant source of this crucial nutrient. Glycine plays an essential role in glycine conjugation, a vital metabolic process. This process is instrumental in the excretion of aromatic acids, compounds that are crucial in various bodily functions. Glycine's involvement in this mechanism underlines its importance in maintaining metabolic health and facilitating detoxification processes in the body [40]. This insight into Golden Berry's glycine content adds to the understanding of its nutritional value, emphasizing the fruit's potential benefits in supporting metabolic functions and overall health. This finding underscores the importance of considering the comprehensive nutritional profile of foods in dietary planning and health management.

Perillic Acid ($C_{10}H_{14}O_2$), identified by an ion at m/z 167.1067, emerges as a significant metabolite associated with the consumption of Golden Berry. As a primary circulating metabolite derived from limonene, which is abundantly present in the aroma of Golden Berry, the notable increase in its excretion in rat urine underscores the fruit's substantial impact on limonene metabolism. This observation, as illustrated in Figure 2, not only accentuates the metabolic processing of limonene in the body but also highlights the unique biochemical footprint of Golden Berry consumption. The presence and increased excretion of Perillic Acid signify the metabolic transformation of limonene, pointing towards an enhanced biotransformation process influenced by the ingestion of Golden Berry. This finding enriches our understanding of the metabolic pathways activated by the fruit and underscores its potential role in modulating specific metabolite profiles within the body.

4.6.1. Evaluation of Metabolic Impact through Acylcarnitines Post-Golden Berry Consumption

The metabolic influence of Golden Berry consumption extends to lipid metabolism, as evidenced by the increase in acylcarnitines both in rat and human urine. This is highlighted by the identification of Dodecanoyl-L-carnitine through an ion at m/z 374.35 [M + H], which exhibited characteristic acyl-carnitine fragments. The observed elevation in acylcarnitines following Golden Berry intake suggests a significant modulation of lipid metabolism [8,41]. This particular increase in acylcarnitines, such as Dodecanoyl-L-carnitine, not only indicates an enhanced lipid metabolism but also hints at a broader metabolic reconfiguration. Acylcarnitines, known for their role in transporting fatty acids into mitochondria for beta-oxidation, are crucial markers in understanding the metabolic shifts induced by dietary changes. The fact that these changes are consistent across both rat and human models further underscores the potential of Golden Berry as a significant dietary component influencing lipid metabolism. In light of these findings, Golden Berry emerges as a dietary factor with the potential to influence lipid metabolism pathways significantly. This observation opens up new perspectives on the metabolic impacts of Golden Berry consumption, suggesting a role beyond basic nutrition and pointing towards its potential utility in managing metabolic health.

4.6.2. Acetylated Dimethylarginine and Golden Berry's Metabolic Impact

The provisional identification of Acetylated Dimethylarginine, marked by an ion at 245.15 m/z [M + H], and the observed subtle elevation in its excretion after consuming Golden Berry introduce an additional facet to our comprehension of the berry's metabolic effects. This nuanced observation suggests that Golden Berry intake might influence the body's arginine metabolism, potentially impacting nitric oxide synthesis pathways. Acetylated Dimethylarginine, a known inhibitor of nitric oxide synthase, plays a critical role in regulating vascular function and endothelial health. The slight increase in its excretion could imply a modulatory effect of Golden Berry on this important metabolic pathway, shedding light on its broader implications for cardiovascular health and metabolic processes. This finding, while preliminary, underscores the complexity of Golden Berry's influence on metabolic health and warrants further investigation to fully elucidate its impact on arginine metabolism and related physiological functions.

4.6.3. Golden Berry's Influence on Metabolic Pathways Linked to Diabetes

The observed reduction in 2-Indolecarboxylic Acid, as indicated by the ion at m/z 146.06, posits an intriguing aspect of Golden Berry's impact on metabolic processes. This particular change points towards Golden Berry's potential role in modulating metabolic pathways that are closely associated with diabetes. The presence of 2-Indolecarboxylic Acid is a notable marker in metabolic studies, especially in the context of glucose regulation and insulin sensitivity. The specific reduction of this compound, as evidenced by the ion at m/z 146.06, could be indicative of biochemical shifts brought about by Golden Berry consumption. This reduction aligns with the potential regulatory effects of Golden Berry on pathways implicated in diabetes pathogenesis. It suggests a modulation of metabolic processes that could have implications for managing blood sugar levels and improving insulin response.

The data point to Golden Berry's capacity to influence key metabolic pathways, reinforcing its potential as a dietary component beneficial for metabolic health, particularly in the context of diabetes prevention and management. The qualitative evidence of Golden Berry's influence on 2-Indolecarboxylic Acid levels provides a foundation for further exploratory studies. These findings could pave the way for more in-depth research into how Golden Berry consumption might contribute to the modulation of diabetes-related metabolic pathways.

Overall, these findings collectively provide significant insights into the metabolic changes induced by Golden Berry consumption. The array of identified metabolites, ranging from amino acids to volatile compounds, underscores the fruit's potential in influencing metabolic processes, particularly those relevant to diabetes management. This integrated metabolomic profile offers a valuable perspective on the biochemical pathways affected by Golden Berry, paving the way for further exploration of its therapeutic potential.

4.7. Golden Berry's Multifaceted Impact on Metabolic Health and the Potential for Personalized Nutritional Therapy

The integration of metabolomic data into a biological context, as illustrated in Figure 3, employs advanced analytical methods. This analysis uncovers the metabolic associations resulting from Golden Berry consumption and reveals profound insights into the connections among specific discriminatory metabolites.

Golden Berry consumption activates pathways related to endothelial function and cardiovascular health. This encompasses molecules associated with oxidative stress response, inflammation, apoptosis, and endothelial function, suggesting potential cardiovascular health benefits [42].

The antioxidants in Golden Berry, along with compounds like NAC, may ameliorate inflammation and oxidative stress, thereby enhancing insulin sensitivity. This effect could safeguard insulin-producing cells and promote effective glucose regulation, offering therapeutic potential for conditions like type 2 diabetes mellitus [43,44].

Golden Berry may bolster the body's defense against oxidative stress by elevating NAC levels, contributing to the neutralization of reactive oxygen species [45].

While increased ADMA levels might initially seem detrimental due to their inhibitory effect on nitric oxide production, in contexts like type 2 diabetes where oxidative stress is high, this increase might be protective. Golden Berry's antioxidant properties could help counterbalance these effects, potentially protecting against cardiovascular complications [46].

Regulators such as AMPK and mTOR signaling pathways may orchestrate a comprehensive metabolic response to Golden Berry consumption. This response entails enhanced antioxidant defenses, inflammation modulation, and complex effects on cell survival and apoptosis [42,47].

Golden Berry's consumption results in complex metabolic effects with implications for cardiovascular health, insulin sensitivity, and antioxidant defense. This positions it as a significant subject for nutrition and potential therapeutic applications. Moreover, this integrated analysis presents opportunities for personalized nutritional strategies tailored to an individual's genetic makeup [48].

The dietary recommendation suggests that a 60 kg individual should consume 85 g of Golden Berry daily as part of a balanced nutrition regimen. In contrast, findings from animal studies indicate that to achieve the same metabolic outcomes observed in the pathological rat model of obesity, the same individual would need to ingest a substantial 4.8 kg of fresh Golden Berry. Such a significant discrepancy points to the raw fruit's comparatively lower concentration of the active ingredients found in its pharmaceutical counterpart. This difference further accentuates the prospective advantages of formulating pharmaceutical-grade concentrates or supplements derived from Golden Berry. Future studies building on this foundation can further demystify the therapeutic potentials and health benefits of this important fruit.

4.8. Expanding the Horizon: New Insights into Golden Berry's Role in Metabolic Regulation

While existing research has laid the groundwork for understanding Golden Berry's impact on metabolism, our study makes considerable advances in expanding this knowledge. We distinguish our research in several key areas, offering novel insights into the multifaceted role of Golden Berry in metabolic regulation. Our study delves into the intricate mechanisms through which Golden Berry exerts its metabolic benefits. By identifying specific enzymes and pathways influenced by Golden Berry consumption, we provide a more detailed and comprehensive understanding of its role in metabolic processes. This in-depth mechanistic exploration helps to clarify the precise biochemical interactions involved, shedding light on the therapeutic potential of Golden Berry

Venturing beyond established impacts, we investigated Golden Berry's effects on additional metabolic disorders or parameters not previously studied. Our findings reveal that Golden Berry extends its positive influence to newly studied aspects of metabolic health. This expansion of its known benefits underscores the fruit's versatility and broadens the scope of its potential applications in treating and managing diverse metabolic conditions.

Distinguishing our research from prior studies, we conducted a comparative analysis between Golden Berry and other treatments or interventions used in metabolic disorder management. This comparative approach provides invaluable insights into how Golden Berry stacks up against other options, highlighting its efficacy and potential advantages in a therapeutic context.

An essential component of our study is the examination of the long-term effects and safety profile of Golden Berry consumption, particularly in the context of metabolic disorders. This aspect of our research addresses a gap in existing studies, offering a vital perspective on the sustainability and safety of Golden Berry as a long-term dietary inclusion for individuals with metabolic concerns.

5. Conclusions

Golden Berry's emergence as a potent agent in reversing obesity and insulin resistance pathologies in animal models marks a significant breakthrough in metabolic health research. This comprehensive study underscores its profound influence on critical metabolic parameters, including the regulation of glucose, cholesterol, triglycerides, and both LDL and HDL levels.

Beyond its nutritional value, Golden Berry demonstrates a remarkable ability to modulate gene expression and influence urinary excretion patterns of essential metabolites.

This study offers pioneering insights into the complex mechanisms governing insulin signaling, lipid metabolism enzymes, lipid mobilization, and the adipogenesis process. The wide-ranging impact of Golden Berry on these metabolic pathways highlights its potential as a key player in addressing chronic health conditions, especially metabolic syndrome.

Golden Berry supplementation's effect on reducing organ weights and its influence on the reactivity of both white and brown adipose tissue in Wistar rats further suggests its role in obesity management and weight control.

Our findings confirm the significant impact of Golden Berry on crucial metabolic regulatory genes like INSR, FasN, PPARγ, and LPL. This indicates its ability to modulate key biological pathways, enhancing its therapeutic potential.

By identifying significant urinary metabolites and elucidating metabolic networks affected by Golden Berry consumption, our study contributes valuable insights into its potential role in diabetes management. The evaluation of metabolic impact through acylcarnitines and other biomarkers post-Golden Berry consumption further underscores its effect on pathways linked to diabetes.

This research not only significantly enriches our current knowledge but also opens new horizons for personalized nutrition and therapeutic strategies, representing a trailblazing advancement in nutrigenomics and metabolomics and integrating anatomical measurements, biochemical parameters, and the effects of dietary variations. By unraveling the multifaceted effects of superfoods like Golden Berry, we move towards a future where nutrition transcends conventional guidelines, evolving into a personalized science that caters to individual metabolic needs and responses. The comprehensive impact of Golden Berry on metabolic health underscores its potential role in personalized nutritional therapy, offering effective, targeted approaches to health and wellness.

Author Contributions: Conceptualization, A.Á.-M., F.V. and N.M.-C.; methodology, A.Á.-M., F.V. and N.M.-C.; writing—original draft preparation, A.Á.-M., F.V. and N.M.-C.; writing—review and editing, A.Á.-M., F.V. and N.M.-C. All authors have read and agreed to the published version of the manuscript.

Funding: This research was funded by "ANALDEX", a research support program for off-balance sheet (OBS) projects, under Project 8755. This research was supported by the Universidad Industrial de Santander under registration code 3723. Finally, we also acknowledge the contributions of the international mobility program for professors of the "Vicerrectoría de Investigación of the Universidad Industrial de Santander", Colombia.

Institutional Review Board Statement: The animal study protocol was approved by the Ethics Committee from the University Industrial de Santander (code: 4110 on 20 September 2019).

Informed Consent Statement: Not applicable.

Data Availability Statement: Data are contained within the article.

Acknowledgments: We would like to express our deepest gratitude to Jennifer Melissa Giorgi Ortiz for her unwavering support. The author thanks Raquel Elvira-Ocazionez of the Universidad Industrial de Santander, Colombia, for infrastructure support.

Conflicts of Interest: The authors declare no conflict of interest.

References

1. Fahed, G.; Aoun, L.; Bou Zerdan, M.; Allam, S.; Bou Zerdan, M.; Bouferraa, Y.; Assi, H.I. Metabolic syndrome: Updates in pathophysiology and management in 2021. *Int. J. Mol. Sci.* **2022**, *23*, 786. [CrossRef]
2. Hall, K.D.; Farooqi, I.S.; Friedman, J.M.; Klein, S.; Loos, R.J.; Mangelsdorf, D.J.; Rahilly, S.; Ravussin, E.; Redman, L.M.; Ryan, D.H.; et al. The energy balance model of obesity: Beyond calories in, calories out. *Am. J. Clin. Nutr.* **2022**, *115*, 1243–1254. [CrossRef]
3. Fazal, F.; Saleem, T.; Ur Rehman, M.E.; Haider, T.; Khalid, A.R.; Tanveer, U.; Mustafa, H.; Tanveer, J.; Noor, A. The rising cost of healthcare and its contribution to the worsening disease burden in developing countries. *Ann. Med. Surg.* **2022**, *82*, 104683. [CrossRef]

4. Mengstie, M.A.; Chekol Abebe, E.; Behaile Teklemariam, A.; Tilahun Mulu, A.; Agidew, M.M.; Teshome Azezew, M.; Zewde, E.A.; Agegnehu, T.A. Endogenous advanced glycation end products in the pathogenesis of chronic diabetic complications. *Front. Mol. Biosci.* **2022**, *9*, 1002710. [CrossRef] [PubMed]
5. Li, Y.; Liu, Y.; Liu, S.; Gao, M.; Wang, W.; Chen, K.; Huang, L.; Liu, Y. Diabetic vascular diseases: Molecular mechanisms and therapeutic strategies. *Signal Transduct. Target. Ther.* **2023**, *8*, 152. [CrossRef]
6. Pillon, N.J.; Loos, R.J.F.; Marshall, S.M.; Zierath, J.R. Metabolic consequences of obesity and type 2 diabetes: Balancing genes and environment for personalized care. *Cell* **2021**, *184*, 1530–1544. [CrossRef]
7. Ramadan, M.F.; Mörsel, J.-T. Oil Goldenberry (*Physalis peruviana* L.). *J. Agric. Food Chem.* **2003**, *51*, 969–974. [CrossRef] [PubMed]
8. Vaillant, F.; Corrales-Agudelo, V.; Moreno-Castellanos, N.; Ángel-Martín, A.; Henao-Rojas, J.C.; Muñoz-Durango, K.; Poucheret, P. Plasma Metabolome Profiling by High-Performance Chemical Isotope-Labelling LC-MS after Acute and Medium-Term Intervention with Golden Berry Fruit (*Physalis peruviana* L.), Confirming Its Impact on Insulin-Associated Signaling Pathways. *Nutrients* **2021**, *13*, 3125. [CrossRef]
9. Novelli, E.L.B.; Diniz, Y.S.; Galhardi, C.M.; Ebaid, G.M.X.; Rodrigues, H.G.; Mani, F.; Fernandes, A.A.H.; Cicogna, A.C.; Novelli, J.L.V.B. Anthropometrical parameters and markers of obesity in rats. *Lab. Anim.* **2007**, *41*, 111–119. [CrossRef]
10. Imbernon, M.; Whyte, L.; Diaz-Arteaga, A.; Russell, W.R.; Moreno, N.R.; Vazquez, M.J.; Gonzalez, C.R.; Ruiz, A.; Lopez, M.; Malagon, M.M.; et al. Regulation of GPR55 in rat white adipose tissue and serum LPI by nutritional status, gestation, gender and pituitary factors. *Mol. Cell. Endocrinol.* **2014**, *383*, 159–169. [CrossRef] [PubMed]
11. ICONTEC. Fresh Fruits Cape Gooseberry. Specifications. Colombian Standard. NTC 4580. Bogota, Colombia. p. 15. 1999. Available online: https://kontii.files.wordpress.com/2012/10/ntc-4580.pdf (accessed on 16 January 2024).
12. Bonilla-Carvajal, K.; Stashenko, E.E.; Moreno-Castellanos, N. *Lippia alba* chemotype carvone essential oil (Verbenaceae) regulates lipid mobilization and adipogenesis in adipocytes. *Curr. Issues Mol. Biol.* **2022**, *44*, 5741–5755. [CrossRef]
13. Thonusin, C.; IglayReger, H.B.; Soni, T.; Rothberg, A.E.; Burant, C.F.; Evans, C.R. Evaluation of intensity drift correction strategies using MetaboDrift, a normalization tool for multi-batch metabolomics data. *J. Chromatogr. A* **2017**, *1523*, 265–274. [CrossRef]
14. Triba, M.N.; Le Moyec, L.; Amathieu, R.; Goossens, C.; Bouchemal, N.; Nahon, P.; Rutledge, D.N.; Savarin, P. PLS/OPLS models in metabolomics: The impact of permutation of dataset rows on the K-fold cross-validation quality parameters. *Mol. BioSystems* **2015**, *11*, 13–19. [CrossRef] [PubMed]
15. Suleiman, J.B.; Mohamed, M.; Bakar, A.B.A. A systematic review on different models of obesity induction in animals: Advantages and limitations. *J. Adv. Vet. Anim. Res.* **2019**, *7*, 103–114. [CrossRef] [PubMed]
16. Hassanien MF, R. *Physalis peruviana*: A Rich Source of Bioactive Phytochemicals for Functional Foods and Pharmaceuticals. *Food Rev. Int.* **2011**, *27*, 259–273. [CrossRef]
17. Mayorga, H.; Knapp, H.; Winterhalter, P.; Duque, C. Glycosidically Bound Flavor Compounds of Cape Gooseberry (*Physalis peruviana* L.). *J. Agric. Food Chem.* **2001**, *49*, 1904–1908. [CrossRef]
18. Panchal, S.K.; Poudyal, H.; Iyer, A.; Nazer, R.; Alam, A.; Diwan, V.; Kauter, K.; Sernia, C.; Campbell, F.; Ward, L.; et al. High-carbohydrate high-fat diet–induced metabolic syndrome and cardiovascular remodeling in rats. *J. Cardiovasc. Pharmacol.* **2011**, *57*, 51–64. [CrossRef] [PubMed]
19. Fokunang, C.N.; Mushagalusa, F.K.; Tembe, E.; Ngoupayo, J.; Ngameni, B.; Njinkio, L.N.; Kadima, J.N.; Kechia, F.A.; Tiedeu, B.; Mbacham, W.F.; et al. Phytochemical and zootechnical studies of *Physalis peruviana* L. leaves exposured to streptozotocin-induced diabetic rats. *J. Pharmacogn. Phytother.* **2017**, *9*, 123–130. [CrossRef]
20. Valenzuela, A.; Ronco, A.M. Fitoesteroles y fitoestanoles: Aliados naturales para la protección de la salud cardiovascular. *Rev. Chil. Nutr.* **2004**, *31* (Suppl. S1), 161–169. [CrossRef]
21. Reyes-Beltrán, M.E.D.; Guanilo-Reyes, C.K.; Ibáñez-Cárdenas, M.W.; García-Collao, C.E.; Idrogo-Alfaro, J.J.; Huamán-Saavedra, J.J. Efecto del consumo de *Physalis peruviana* L. (aguaymanto) sobre el perfil lipídico de pacientes con hipercolesterolemia. *Acta Médica Peru.* **2015**, *32*, 195–201. Available online: http://www.scielo.org.pe/scielo.php?script=sci_arttext&pid=S1728-59172015000400002&lng=es&tlng=es (accessed on 24 April 2023). [CrossRef]
22. Balcázar, H.G.; de Heer, H.; Rosenthal, L.; Aguirre, M.; Flores, L.; Puentes, F.A.; Cardenas, V.M.; Duarte, M.O.; Ortiz, M.; Schulz, L.O. A promotores de salud intervention to reduce cardiovascular disease risk in a high-risk Hispanic border population, 2005–2008. *Prev. Chronic Dis.* **2010**, *7*, A28.
23. Erman, F.; Kirecci, O.; Ozsahin, A.; Erman, O.; Kaya, T.; Yilmaz, O. Effects of *Physalis peruviana* and *Lupinus albus* on malondialdehyde, glutathione, cholesterol, vitamins and fatty acid levels in kidney and liver tissues of diabetic rats. *Prog. Nutr.* **2018**, *20* (Suppl. S1), 218–230. [CrossRef]
24. Bazalar Pereda, M.S.; Nazareno, M.A.; Viturro, C.I. Nutritional and antioxidant properties of *Physalis peruviana* L. fruits from the Argentinean Northern Andean region. *Plant Foods Hum. Nutr.* **2019**, *74*, 68–75. [CrossRef]
25. Bogdanos, D.P.; Gao, B.; Gershwin, M.E. Liver Immunology. *Compr. Physiol.* **2013**, *3*, 567. [CrossRef]
26. Ramadan, M.F. *Physalis peruviana* pomace suppresses highcholesterol diet-induced hypercholesterolemia in rats. *Grasas Aceites* **2012**, *63*, 411–422. [CrossRef]
27. Fuente, F.P.; Nocetti, D.; Sacristán, C.; Ruiz, P.; Guerrero, J.; Jorquera, G.; Uribe, E.; Bucarey, J.L.; Espinosa, A.; Puente, L. *Physalis peruviana* L. Pulp Prevents Liver Inflammation and Insulin Resistance in Skeletal Muscles of Diet-Induced Obese Mice. *Nutrients* **2020**, *12*, 700. [CrossRef]

28. Garcia-Compean, D.; Jaquez-Quintana, J.O.; Gonzalez-Gonzalez, J.A.; Maldonado-Garza, H. Liver cirrhosis and diabetes: Risk factors, pathophysiology, clinical implications and management. *World J. Gastroenterol.* **2009**, *15*, 280–288. [CrossRef]
29. Moussa, S.A.A.; Ibrahim, F.A.A.; Elbaset, M.A.; Aziz, S.W.; Morsy, F.A.; Abdellatif, N.; Attia, A.; El Toumy, S.A.; Salib, J.Y.; Bashandy, S.A.E. Goldenberry (*Physalis peruviana*) alleviates hepatic oxidative stress and metabolic syndrome in obese rats. *J. Appl. Pharm. Sci.* **2022**, *12*, 138–150. [CrossRef]
30. Obregón-La Rosa, A.J.; Contreras-López, E.; Flores Juárez, E.; Gonzales Barrón, Ú.; Muñoz, A.M.; Ramos-Escudero, F. Nutritional and antioxidant profile of the Physalis fruit grown in three Andean regions of Peru. *Rocz. Panstw. Zakl. Hig.* **2023**, *74*, 49–57. [CrossRef]
31. Eibl, G.; Cruz-Monserrate, Z.; Korc, M.; Petrov, M.S.; Goodarzi, M.O.; Fisher, W.E.; Habtezion, A.; Lugea, A.; Pandol, S.J.; Hart, P.A.; et al. Diabetes Mellitus and Obesity as Risk Factors for Pancreatic Cancer. *J. Acad. Nutr. Diet.* **2018**, *118*, 555–567. [CrossRef] [PubMed]
32. Longo, M.; Zatterale, F.; Naderi, J.; Parrillo, L.; Formisano, P.; Raciti, G.A.; Beguinot, F.; Miele, C. Adipose Tissue Dysfunction as Determinant of Obesity-Associated Metabolic Complications. *Int. J. Mol. Sci.* **2019**, *20*, 2358. [CrossRef]
33. Bol, V.V.; Reusens, B.M.; Remacle, C.A. Postnatal Catch-up Growth After Fetal Protein Restriction Programs Proliferation of Rat Preadipocytes. *Obesity* **2008**, *16*, 2760–2763. [CrossRef] [PubMed]
34. Makhijani, P.; Basso, P.J.; Chan, Y.T.; Chen, N.; Baechle, J.; Khan, S.; Furman, D.; Tsai, S.; Winer, D.A. Regulation of the immune system by the insulin receptor in health and disease. *Front. Endocrinol.* **2023**, *14*, 1128622. [CrossRef]
35. Čolak, E.; Pap, D. The role of oxidative stress in the development of obesity and obesity-related metabolic disorders. *J. Med. Biochem.* **2021**, *40*, 1–9. [CrossRef]
36. Monsalve, F.A.; Pyarasani, R.D.; Delgado-Lopez, F.; Moore-Carrasco, R. Peroxisome proliferator-activated receptor targets for the treatment of metabolic diseases. *Mediat. Inflamm.* **2013**, *2013*, 549627. [CrossRef] [PubMed]
37. Lv, Y.; Zhao, P.; Pang, K.; Ma, Y.; Huang, H.; Zhou, T.; Yang, X. Antidiabetic Effect of a Flavonoid-Rich Extract from *Sophora alopecuroides* L. in HFD- and STZ- Induced Diabetic Mice through PKC/GLUT4 Pathway and Regulating PPARα and PPARγ expression. *J. Ethnopharmacol.* **2020**, *2020*, 113654. [CrossRef] [PubMed]
38. Kitajima, S.; Morimoto, M.; Liu, E.; Koike, Y.; Higaki, Y.; Taura, K.; Mamba, K.; Itamoto, K.; Watanabe, T.; Tsutsumi, K.; et al. Overexpression of lipoprotein lipase improves insulin resistance induced by a high-fat diet in transgenic rabbits. *Diabetologia* **2004**, *47*, 1202–1209. [CrossRef]
39. Kupska, M.; Wasilewski, T.; Jędrkiewicz, R.; Gromadzka, J. Determination of Terpene Profiles in Potential Superfruits. *Int. J. Food Prop.* **2016**, *19*, 2726–2738. [CrossRef]
40. Yilmaztekin, M. Characterization of Potent Aroma Compounds of Cape Gooseberry (*Physalis peruviana* L.) Fruits Grown in Antalya Through the Determination of Odor Activity Values. *Int. J. Food Prop.* **2013**, *17*, 469–480. [CrossRef]
41. Vaillant, F.; Llano, S.; Ángel, A.; Moreno, N. Main urinary biomarkers of golden berries (*Physalis peruviana*) following acute and short-term nutritional intervention in healthy human volunteers. *Food Res. Int.* **2023**, *173*, 113443. [CrossRef]
42. Sierra, J.A.; Escobar, J.S.; Corrales-Agudelo, V.; Guzman, O.J.; Mejia, E.P.; Rojas, J.C.; Quintero, A.; Vaillant, F.; Durango, K. Consumption of golden berries (*Physalis peruviana* L.) might reduce biomarkers of oxidative stress and alter gut permeability in men without changing inflammation status or the gut microbiota. *Food Res. Int.* **2022**, *162*, 111949. [CrossRef]
43. Jeremias, A.; Soodini, G.; Gelfand, E.; Xu, Y.; Stanton, R.C.; Horton, E.S.; Cohen, D.J. Effects of N-acetyl-cysteine on endothelial function and inflammation in patients with type 2 diabetes mellitus. *Heart Int.* **2009**, *4*, e7. [CrossRef] [PubMed]
44. Unuofin, J.O.; Lebelo, S.L. Antioxidant Effects and Mechanisms of Medicinal Plants and Their Bioactive Compounds for the Prevention and Treatment of Type 2 Diabetes: An Updated Review. *Oxid. Med. Cell. Longev.* **2020**, *2020*, 1356893. [CrossRef]
45. Unnithan, A.S.; Jiang, Y.; Rumble, J.L.; Pulugulla, S.H.; Posimo, J.M.; Gleixner, A.M.; Leak, R.K. N-acetyl cysteine prevents synergistic, severe toxicity from two hits of oxidative stress. *Neurosci. Lett.* **2014**, *560*, 71–76. [CrossRef]
46. Sibal, L.; Agarwal, S.C.; Home, P.D.; Boger, R.H. The Role of Asymmetric Dimethylarginine (ADMA) in Endothelial Dysfunction and Cardiovascular Disease. *Curr. Cardiol. Rev.* **2010**, *6*, 82–90. [CrossRef]
47. Tan, V.P.; Miyamoto, S. Nutrient-sensing mTORC1: Integration of metabolic and autophagic signals. *J. Mol. Cell. Cardiol.* **2016**, *95*, 31–41. [CrossRef] [PubMed]
48. Ulaszewska, M.; Garcia-Aloy, M.; Vázquez-Manjarrez, N.; Soria-Florido, M.T.; Llorach, R.; Mattivi, F.; Manach, C. Food intake biomarkers for berries and grapes. *Genes Nutr.* **2020**, *15*, 17. [CrossRef] [PubMed]

nutrients MDPI

Article

Overnutrition during Pregnancy and Lactation Induces Gender-Dependent Dysmetabolism in the Offspring Accompanied by Heightened Stress and Anxiety

Gonçalo M. Melo †, Adriana M. Capucho †, Joana F. Sacramento , José Ponce-de-Leão, Marcos V. Fernandes, Inês F. Almeida, Fátima O. Martins and Silvia V. Conde *

iNOVA4Health, NOVA Medical School, Faculdade de Ciências Médicas, Universidade NOVA de Lisboa, Rua Camara Pestana, 6, Edificio 2, 1150-082 Lisboa, Portugal; goncalo.melo@nms.unl.pt (G.M.M.); adriana.capucho@nms.unl.pt (A.M.C.); joana.sacramento@nms.unl.pt (J.F.S.); jose.poncedeleao@nms.unl.pt (J.P.-d.-L.); viniciusmar3@gmail.com (M.V.F.); ines.almeida@nms.unl.pt (I.F.A.); fatima.martins@nms.unl.pt (F.O.M.)

* Correspondence: silvia.conde@nms.unl.pt

† These authors contributed equally to this work.

Abstract: Maternal obesity and gestational diabetes predispose the next generation to metabolic disturbances. Moreover, the lactation phase also stands as a critical phase for metabolic programming. Nevertheless, the precise mechanisms originating these changes remain unclear. Here, we investigate the consequences of a maternal lipid-rich diet during gestation and lactation and its impact on metabolism and behavior in the offspring. Two experimental groups of Wistar female rats were used: a control group (NC) that was fed a standard diet during the gestation and lactation periods and an overnutrition group that was fed a high-fat diet (HF, 60% lipid-rich) during the same phases. The offspring were analyzed at postnatal days 21 and 28 and at 2 months old (PD21, PD28, and PD60) for their metabolic profiles (weight, fasting glycemia insulin sensitivity, and glucose tolerance) and euthanized for brain collection to evaluate metabolism and inflammation in the hypothalamus, hippocampus, and prefrontal cortex using Western blot markers of synaptic dynamics. At 2 months old, behavioral tests for anxiety, stress, cognition, and food habits were conducted. We observed that the female offspring born from HF mothers exhibited increased weight gain and decreased glucose tolerance that attenuated with age. In the offspring males, weight gain increased at P21 and worsened with age, while glucose tolerance remained unchanged. The offspring of the HF mothers exhibited elevated levels of anxiety and stress during behavioral tests, displaying decreased predisposition for curiosity compared to the NC group. In addition, the offspring from mothers with HF showed increased food consumption and a lower tendency towards food-related aggression. We conclude that exposure to an HF diet during pregnancy and lactation induces dysmetabolism in the offspring and is accompanied by heightened stress and anxiety. There was sexual dimorphism in the metabolic traits but not behavioral phenotypes.

Keywords: overnutrition; offspring; dysmetabolism; metabolic diseases; stress; anxiety

1. Introduction

Metabolic syndrome (MS) is a serious public health problem affecting about 25% of the general population worldwide [1]. The definition of this syndrome has recently evolved to include a group of at least three of the five cardio-metabolic abnormalities, including high blood pressure, central obesity, insulin resistance (IR), elevated blood triglycerides, and atherogenic dyslipidemia, which, together, lead to an increased risk of cardio-metabolic pathologies [1]. The presence of MS-related disorders also has an impact on the central nervous system, causing neurological and neurodegenerative diseases [2]. The intricate relationship between dysmetabolism and neurodegeneration has been described to be relevant not only to aging-associated comorbidities but also in mild cognitive impairment in younger ages and during the initial steps of an organism's development. In fact, it is known that the nutritional status of the mother can prematurely influence the onset of diseases in the offspring, and there is significant research focused on understanding its impact on the metabolic status and nervous system of the offspring [3]. Studies in animals have shown that maternal obesity can increase the likelihood of metabolic and neurodevelopmental diseases in the offspring, such as cognitive impairment, autism spectrum disorders, and attention deficit hyperactivity disorder [2–4]. Moreover, maternal obesity and dysmetabolism are linked to a heightened risk of psychiatric disorders in the offspring, encompassing conditions such as anxiety, depression, schizophrenia, psychosis, eating disorders, and food addiction [5]. Several mechanisms have been postulated to underlie these neurodevelopment and psychiatric conditions, including oxidative stress and inflammation-induced malprogramming; the deregulation of insulin, glucose, and leptin signaling; the dysregulation of synaptic transmission; or even the deregulation of the gut–brain axis [6,7]. Nevertheless, the precise ones are not completely understood.

Recent evidence indicates that many components of a mother's diet are pivotal in shaping aspects of the offspring's health, including the microbiome and the neonatal immune system [8]. Research suggests that the maternal diet can influence the development of the offspring's brain, endocrine system, and long-term behavior [9]. Therefore, maintaining a balanced and nutritious diet during pregnancy is crucial to meet the increased demands on the mother's body, as adequate nutrient intake is essential to support the offspring's growth and ensure a healthy birth weight.

During the perinatal period, maternal obesity can elevate the risk of gestational diabetes and hypertension, potentially affecting placental function and fetal energy metabolism [10]. Obesity during pregnancy is also associated with complex alterations in the neuroendocrine, metabolic, and inflammatory processes, which can potentially influence fetal hormonal exposure and nutrient supply [11–13].

Throughout pregnancy, the mother's body undergoes significant metabolic changes, which, if left untreated, can contribute to certain health disparities. In early pregnancy, there is an increase in maternal insulin secretion to facilitate the transportation of glucose and amino acids to the developing fetus [12,13]. However, this increased insulin secretion, combined with high levels of circulating lipids and the consequent increase in visceral adiposity, often results in the development of insulin resistance in maternal cells [10–13]. When lactation begins, a metabolic shift occurs, redirecting resources from storage milk synthesis, with this process being intricately regulated by lactogenic hormones such as insulin, prolactin, and glucocorticoids, as well as by cytokines, growth factors, and substrate availability [14]. The alterations promoted by lactation causes several positive changes in the metabolic processes. This leads to better glucose regulation, reducing insulin production, improving insulin sensitivity, and decreasing β-cell proliferation. In addition, lipid metabolism becomes less active in certain tissues, and stored lipids are mobilized to facilitate the transport of lipids to the mammary gland for milk production [11,13,14]. As a result of these metabolic changes during lactation, there is a reduction in postpartum adiposity, potentially decreasing the risk of obesity. Based on this evidence, Stuebe and Rich-Edwards proposed the "reset hypothesis", which suggests that lactation plays a crucial role in resetting the metabolic processes that take place during labor [15].

However, when maternal overnutrition continues, the offspring's metabolism as well as neurological development may be impacted. Determining the mechanisms behind the expected impact of hypercaloric diet-fed mothers on the offspring's metabolism and neurological development is the main goal of this study. Additionally, since it is well known that sex influences both metabolic [16] and neurodegenerative disorders [17], development, and progression, we will also focus our attention on the differences of the impact of maternal overnutrition on male vs. female offspring metabolism and stress and anxiety behaviors.

2. Materials and Methods

2.1. Animals

Experiments were performed on 12-week-old female and 24-week-old male Wistar rats (200–400 g), obtained from Charles River Laboratories (Barcelona, Spain) and maintained at the NOVA Medical School animal facility. Animals were kept under temperature and humidity control (21 ± 1 °C; 55 ± 10% humidity) and a regular light (08.00–20.00 h) and dark (20.00–08.00 h) cycle, with food and water provided ad libitum. After randomization, animals were divided into a normal chow diet group (NC) or a high-fat (HF) diet group. The NC group was fed a standard diet (7.4% fat + 75% carbohydrates (4% sugar) + 17% protein, SDS diets RM1, Probiológica, (Lisboa, Portugal), and the HF diet group was fed a 60% lipid-rich diet (61.6% fat + 20.3% carbohydrate + 19.1% protein, TestDiet, St. Louis, MO, USA). After confirming that the females were pregnant, they stayed in cages, with 1 animal per cage. During the breastfeeding period, the pup's body weight was weekly motorized. On the weaning day—21 days postnatal (PD21)—both male and female offspring were separated from the mothers and kept in cages, with 4 animals per cage. They were randomly divided, and part of them were sacrificed at PD21 and 28 days postnatal (PD28), and the other group was kept alive until 45 days postnatal, where metabolic and behavioral parameters were assessed. Metabolic evaluations were performed in the mothers by using the insulin tolerance test (ITT) and oral glucose tolerance test (OGTT) (Figure 1).

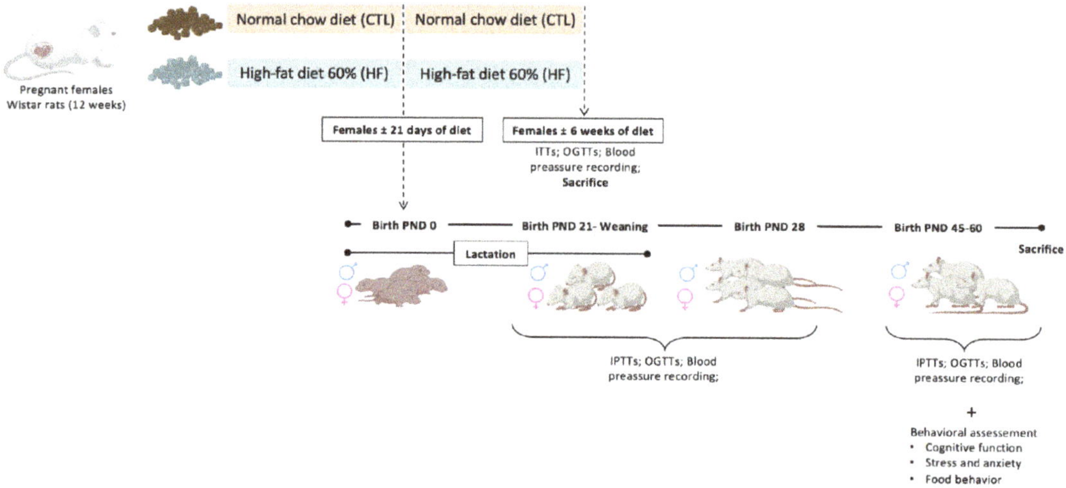

Figure 1. Schematic illustration of the protocol of the study. CTL—control; HF—High-fat diet; PND—post-natal day. ♀—female; ♂—male.

Pups were also tested for insulin sensitivity through the ITT and for glucose tolerance through the OGTT. Behavior was assessed by open field, Y maze, elevated plus maze (EPM), light dark box, sucrose competition, food competition, and novel object recognition (NOR) tests. At the time of sacrifice, PD21, PD28, and PD60, the animals were anesthetized and sacrificed with pentobarbital (60 mg/kg i.p.), and the brains were dissected and frozen at −80 °C for further protein analysis.

All experiments and animal care were performed in accordance with the European Union Directive for Protection of Vertebrates Used for Experimental and Other Scientific Ends (2010/63/EU) and with the ARRIVE guidelines. Experimental protocols were approved in 2021 by the Ethics Committee of NOVA Medical School/Faculdade de Ciências Médicas (no. 94/2021/CEFCM) and by the Portuguese Authority for Animal Health (DGAV, Ref no. 0421/000/000/2021).

2.2. Metabolic Evaluation

2.2.1. Intravenous Insulin Tolerance Test (ITT)

Insulin sensitivity was assessed in conscious mothers using the intravenous ITT, which provides an estimate of overall insulin sensitivity [18]. In brief, following an overnight fast, an insulin bolus (0.1 U/Kg, Humulin Regular, Lilly, Alcobendas, Spain) was administered in the tail vein, and the subsequent 15 min decline in blood sugar levels was measured. Blood samples were collected by tail tipping, and glycemia was assessed using a glucometer (Precision Xtra Meter, Abbott Diabetes Care, Amadora, Portugal) and test strips (Abbott Diabetes Care, Amadora, Portugal). The constant rate of decline in plasma glucose (K_{ITT}) was calculated as previously outlined [19].

2.2.2. Intraperitoneal Insulin Tolerance Test (ipITT)

In the offspring, insulin sensitivity was assessed using the intraperitoneal ITT. Pups were fasted for approximately 6 h with free access to water. Basal blood glucose was measured, and 0.1 U/Kg of insulin (Humulin Regular, Lilly, Alcobendas, Spain) was administered intraperitoneally. Glucose was measured from blood collected from the tip of the tail vein at 5, 10, 15, 30, 45, 60, and 120 min after the injection with a glucometer (Precision Xtra Meter, Abbott Diabetes Care, Amadora, Portugal) and test strips (Abbott Diabetes Care, Amadora, Portugal). Glucose excursion curves from the ITT were used to calculate the area under the curve and therefore to evaluate insulin sensitivity.

2.2.3. Oral Glucose Tolerance Test (OGTT)

Glucose tolerance was assessed through the OGTT. Both mothers and pups that were fasted overnight for approximately 12–15 h were administered with a glucose solution (2 g/kg in a 10 uL/g body weight volume) by gavage after the measurement of basal glycemia. Glucose levels were measured at 0 and 15, 30, 60, and 120 min after the oral gavage by tail tipping using a glucometer (Precision Xtra Meter, Abbott Diabetes Care, Amadora, Portugal) and test strips (Abbott Diabetes Care, Amadora, Portugal). The glucose excursion curves were used to calculate the area under the curve.

2.3. Behavioral Assessment

2.3.1. Open Field (OF)

The OF test is used to assess gross motor activity, anxiety, and willingness to explore [20]. In this test, animals were placed for 5 min in a square arena (70 cm × 70 cm × 75 cm). Their behaviors were recorded at 30 frames per second and analyzed using Bonsai software (version 7.0). For that, an inner and central zone in the maze was defined (40 cm × 40 cm), and the following were measured: the total distance covered, distance covered in the inner zone, total immobility time, immobility time in the inner zone, average velocity, and average velocity in the inner zone. It was considered that the animal was in the inner zone when more than half of its body was in the 40 cm area. Immobility was considered when the animal was stationary. Data presented are results from a single trial for each rat.

2.3.2. Y Maze

To assess short-term spatial memory, animals were challenged to a Y-shaped maze with 120° between each arm with a 10 cm width and 30 cm height. First, animals were submitted to a training session in which a rat was placed in the start point (A arm) of the Y maze with a closed arm and allowed to freely explore it for 5 min. The experimental session occurred 1 h after training where the "novel" arm was open. The animal freely explored the maze for 5 min. The exploratory capacity of the animal was evaluated as well as the time spent both in the novel arm and in the arm that was always open. The first arm choice when both arms were open was evaluated, as well as the number of triads (i.e., ABC, CAB, or BCA) and entries. The spontaneous alternative behavior score (%) for each rat was calculated as the ratio of the number of alternations to the possible number (total number of arm entries minus two) multiplied by 100. Videos were recorded as explained in Section 2.4.1.

2.3.3. Elevated Plus Maze (EPM)

EPM test is used to characterize anxiety-related behavior. The apparatus is composed of a small central platform with four arms angled 90° from each other radiating outwards, and it is raised above the ground to a height of 112 cm; it is 50 cm in length and 12 cm in width. Alternating arms are enclosed by high opaque walls of 41.5 cm in height with open tops. Animals were placed in the center of the maze and allowed to freely explore for 5 min for a single trial and recorded using an overhead camera. The time spent on open arms and the number of entries were measured. Data presented are results from a single trial for each rat. Videos were recorded and analyzed as described in Section 2.3.1.

2.3.4. Light Dark Box (LDB)

The LDB test is used to evaluate anxiety responses in rodents. The apparatus is composed of two compartments. The large light compartment (2/3 of the box) is brightly lit and open, while the small dark compartment (1/3 of the box) is covered and dark. These two compartments are connected by a door of 10 cm. Rats were placed in the corner of the brightly illuminated chamber and were allowed to freely explore the maze for 5 min. Parameters such as the time spent in the light/dark compartment and the number of transitions were evaluated. Data presented are results from a single trial for each rat. Videos were recorded and analyzed as described in Section 2.3.1.

2.3.5. Novel Object Recognition (NOR)

NOR test is commonly used in rodents to evaluate cognition, particularly recognition memory. This test was performed in the same maze as the OF test, with two different objects. The objects were different in shape and appearance. This test is composed of a train where the animals are allowed to explore the maze with two identical objects placed at an equal distance. One hour later, rats are allowed to explore the arena in the presence of the familiar object and a novel object to test long-term recognition memory. The time spent exploring each object and the ratio between the novel/novel+ familiar (interaction time) were analyzed. Data presented are results from a single trial for each rat. Videos were recorded and analyzed as described in Section 2.3.1.

2.3.6. Block Test

Block test is used to assess olfactory function. This test evaluates sensitivity to social smells, which is an ability in rats [21]. Housed animals were exposed to five wood blocks (7 cm × 2 cm × 2 cm) (Ultragene, Santa Comba Dão, Portugal) placed inside each cage for 1 week. Each rat underwent two training sessions. In the trial test, one block was replaced by a block that was originally in a cage with a different set of animals. Rat was videotaped for 1 min. Videos were recorded and analyzed as described in Section 2.3.1. Parameters such as the time that animals take to recognize the novel block and the time spent to sniff the novel block were evaluated. The animal was only considered to be sniffing the block when its nose was touching/very close to the block. Data presented are results from a single trial for each rat.

2.3.7. Food Competition

Food competition is a test that not only measures the antagonistic and dominance behaviors of offspring but also assesses voracity and food preferences. The animals were fasted for 24 h. Following the fasting period, the animals were randomly paired with a same-sex partner and placed in the maze (44 cm wide, 49 cm long, and 20 cm high box). Inside this maze, on one of its sides, there was an integrated dispenser with a retractable door, where the food was stored. The pair was placed inside the maze, and for 2 min, the door was kept closed. After 2 min, the dispenser's door was opened to allow the rats access to the food. During the following 8 min, observations were made regarding the amount of food consumed, the occurrence of antagonistic behaviors, their duration, as well as the frequency of grooming behaviors exhibited by the rats. Data presented are results from a single trial for each pair.

2.3.8. Water/Sucrose Competition

Similarly, to the food competition, the water/sucrose competition measures antagonistic and dominance behaviors in offspring, as well as their liquid consumption preferences.

The animals were subjected to a 24 h liquid deprivation. During this period, only food was available ad libitum. After the fasting period, the animals were randomly paired as described in Section 2.3.7. Inside this maze, on one of its faces, there was a dispenser set at a 45° angle capable of holding a water bottle. This dispenser featured a removable door, controlling the rats' access to the liquid. The dispenser closed for 2 min. After that, for 8 min, the bottles became available, and the rats had access to it. Observations were made regarding the liquid intake, the occurrence of antagonistic behaviors, their duration, as well as the frequency of grooming behaviors exhibited by the rats.

The antagonistic behaviors included stealing pellets from another individual's mouth or paws (food competition), pushing the other's head while drinking from the water bottle (water/sucrose competition), and aggression, such as biting, moving the other individual away, or just fighting.

The data presented are results from a single trial for each pair. After the water trial, the same pair was exposed to the sucrose trial with the same characteristics, but the sucrose level was 1 g/L.

2.4. Ex Vivo Analysis

After 21, 28, and 60 days, animals were sacrificed, and certain brain areas—frontal cortex, hypothalamus, and hippocampus—were dissected and harvested for protein analysis.

2.4.1. Tissue Lysate Preparation and Western Blot Analysis

For Western blot analysis, brain samples (20 mg) were homogenized in RIPA buffer (50 mM Tris-HCl pH 7.4, 150 mM NaCl, 2 mM EDTA, 0.1% SDS, 0.25% sodium deoxycholate) and sonicated 3 times in cycle of 30 s. Protein extracts were centrifuged for 10 min at 13,000 rpm at 4 °C, and the supernatant was collected.

Proteins (8 μg) were separated by electrophoresis in 10% sodium dodecylsulfate-polyacrylamide gel electrophoresis (SDS-PAGE), followed by transference to a nitrocellulose membrane (BioRad, Dusseldorf, Germany). After transference, the membrane was incubated with blocking solution composed of milk or bovine serum albumin (BSA) at room temperature for 1 h. Primary antibody incubations were carried out overnight at 4 °C using the following concentrations: phosphorylated insulin receptor (anti-IR-p, rabbit 1:500, abcam, Cambridge, UK); total insulin receptor (anti-IR T, mouse 1:1000, Santa Cruz Biotechnology, Heidelberg, Germany); phosphorylated mitogen-activated protein kinase (AMPK) (anti-pAMPK 1:500 rabbit, Cell Signaling, Beverly, MA, USA) and total AMPK (anti-AMPK rabbit 1:1000, Cell Signaling, Beverly, MA, USA); synaptosome-associated protein (SNAP25) (anti-SNAP25 1:500, Santa Cruz Biotechnology, Heidelberg, Germany); postsynaptic density protein (PSD95) (anti-PSD95 1:1000, Invitrogen, Waltham, MA, USA); vesicular glutamate transporter 1 (vGLuT) (anti-vGLuT 1:1000, MilliporeSigma, Burlington, MA, USA); glial fibrillary acidic protein (GFAP) (anti-GFAP 1:2000, Palex, Oeiras, Portugal); and tumor necrosis factor alpha (TNF-α) (anti-TNF-α 1:200, Sicgen, Cantanhede, Portugal). Secondary antibody was incubated at room temperature for 1.5 h. Detection procedures were carried out according to ECL system (GE Healthcare Life Sciences; Little Chalfont, UK), and the signal was detected using a ChemiDocTM Imaging System (Bio-Rad, Hercules, CA, USA). Membranes were re-probed for the β-actin (anti-β-actin 1:2000, Sicgen, Cantanhede, Portugal).

2.5. Data Analysis

Data were analyzed using GraphPad Prism Software, version 8.0.2 (GraphPad Software Inc., San Diego, CA, USA), and presented as mean values with the standard error of the mean (SEM). The significance of the differences between the groups was calculated using Student's *t* test and one- and two-way ANOVA with Dunnett's and Bonferroni and Sidak's multiple comparison tests. Differences were considered significant at $p \leq 0.05$. Experimental groups constituted 8–10 animals.

3. Results

3.1. Effect of Overnutrition during Pregnancy and Lactation on Metabolic Parameters in the Mothers

Mothers that were overnourished for 6 weeks under an HF diet were evaluated for body weight and glucose metabolism, assessed using basal glycemia and OGTT; insulin sensitivity, evaluated using the ipITT; and body composition, assessed using the liver and AT depots' weight measurements.

As expected, feeding an HF diet to the mothers for 6 weeks promoted an increase in basal glycemia (basal glycemia: NC mothers = 69.14 ± 1.792; HF mothers = 78.88 ± 3.573, mg/dL, $p < 0.05$) and a drastic decrease in insulin sensitivity (k_{ITT}: NC mothers = 4.864 ± 0.3957; HF mothers = 1.849 ± 0.3188, % glucose/min, $p < 0.0001$) without significantly affecting glucose tolerance (Table 1). Moreover, the HF diet did not alter the body weights or liver and AT depots' weights in the mothers, which can be attributed to the pregnancy and offspring delivery during that period (Table 1).

Table 1. Impact of high-fat diet on metabolic parameters in mothers.

	NC Mothers	HF Mothers
Body Weight (g)	243.2 ± 5.122	236.8 ± 2.577
Basal Glycaemia (mg/dL)	69.14 ± 1.792	78.88 ± 3.573 *
k_{ITT} (% glucose/min)	4.864 ± 0.3957	1.849 ± 0.3188 ****
AUC Glycaemia (mg/dL·min)	18,138 ± 417.4	19,612 ± 594.9
Liver (g/kg)	0.0280 ± 0.001	0.0289 ± 0.001
Visceral Adipose Tissue (g/kg)	0.0057 ± 0.002	0.0066 ± 0.002
Perinephric Adipose Tissue (g/kg)	0.0153 ± 0.002	0.0171 ± 0.004
Genital Adipose Tissue (g/kg)	0.0232 ± 0.002	0.0254 ± 0.002
Brown Adipose Tissue (g/kg)	0.0007 ± 0.000	0.0016 ± 0.000

Data represent mean ± SEM of 5 NC mothers and 5 HF mothers. * $p < 0.05$, **** $p < 0.0001$. Student's *t* test was used to compare values in animals submitted to normal chow (NC) and animals submitted to high-fat diet (HF). K_{ITT}—constant of the insulin tolerance test; AUC—area under the curve of the glucose excursion curves obtained from the oral glucose tolerance test.

3.2. Effect of Overnutrition during Pregnancy and Lactation on Metabolic Function in the Offspring

3.2.1. Insulin Action and Glucose Homeostasis

Female and male offspring born from overnourished mothers during pregnancy and lactation periods were evaluated for body weight, basal glycemia, insulin sensitivity, and glucose tolerance (Figure 2).

HF feeding to mothers promoted a significant increase in the body weights of both the female and male descendants during all the postnatal periods analyzed (Figure 2A). The body weights of the female offspring from the HF mothers increased by 65.1%, 33.3%, and 9.6% at PD 21, 28, and 60, respectively (Figure 2A, left panel). The male offspring born from the overnourished mothers showed significant body weight increases of 63.9%, 45.6%, and 20.5% at PD 21, 28, and 60, respectively (Figure 2A, right panel).

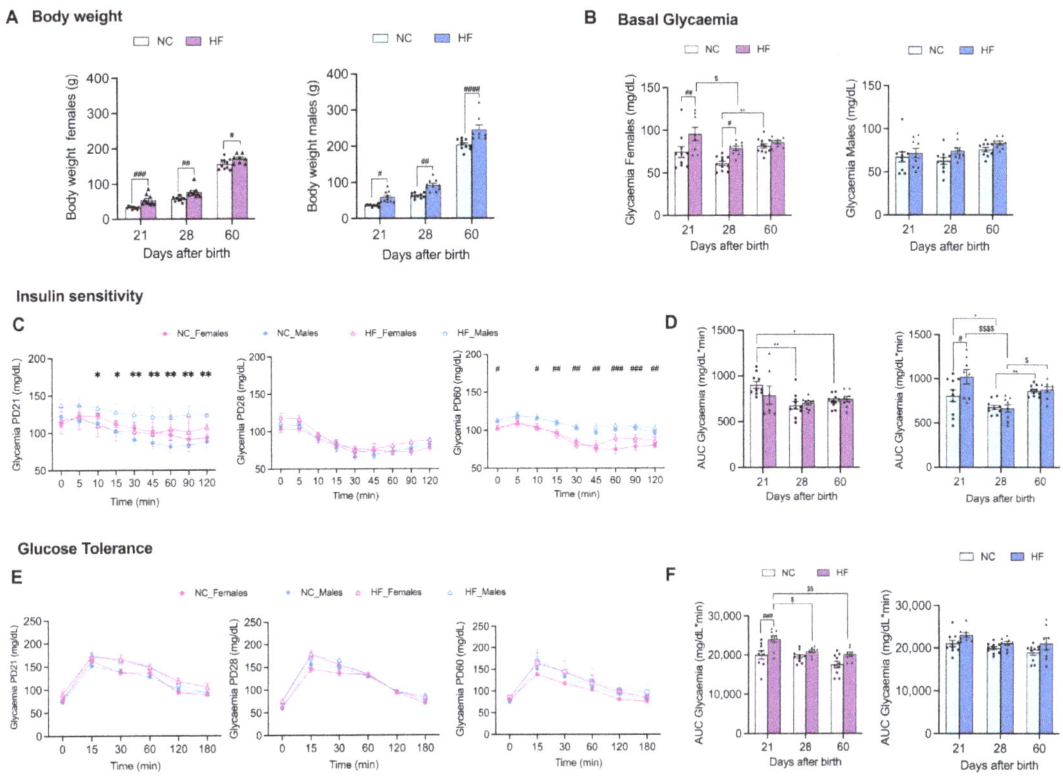

Figure 2. Effect of overnutrition during pregnancy and lactation on metabolic phenotype of the offspring, evaluated by body weight gain (**A**), basal glycemia (**B**), insulin sensitivity (**C**,**D**), and glucose tolerance (**E**,**F**) at postnatal days 21 (PD21), 28 (PD28), and 60 (PD60). (**A**) From the left to the right, body weight in g during offspring development for females and males. (**B**) From the left to the right, basal glycemia in mg/dL during early development for females and males. (**C**,**D**) Insulin sensitivity represented by the glycemia profiles during ipITT (**C**) and by the area under the curve of those glycemia curves during early descendants' development for females and males, from the left to the right, respectively (**D**). (**E**,**F**) Glucose tolerance represented by glycemia profiles during OGTT (**E**) and by the area under the curve of those glycemia curves during early descendants' development for females and males, from the left to the right, respectively (**F**). PD21—postnatal day 21; PD28—postnatal day 28; PD60—postnatal day 60. Data represent mean ± SEM of 10 NC females, 10 NC males, 9 HF females, and 9 HF males per time-point. Black circles correspond to offspring born form mothers submitted to NC diet and black triangles to offspring born form mothers submitted to HF diet. Two-way ANOVA with Bonferroni and Sidak's multiple comparison tests; * $p < 0.05$ and ** $p < 0.01$ comparing NC animals with different ages; # $p < 0.05$, ## $p < 0.01$, ### $p < 0.001$ and #### $p < 0.0001$ comparing NC vs. HF animals with the same age; and \$ $p < 0.05$, \$\$ $p < 0.01$, and \$\$\$\$ $p < 0.0001$ comparing HF animals with different ages.

Maternal overnutrition during pregnancy and lactation significantly increased the basal glycemia of the female offspring at PD21 and 28 in comparison to the female offspring from the NC mothers (basal glycemia: NC PD21 = 74.5 ± 6.5, HF PD21 = 95.6 ± 7.8, $p < 0.01$; NC PD28 = 61.0 ± 2.4, HF PD28 = 78.8 ± 2.6, $p < 0.05$), and the basal glycemia did not change at PD60 (Figure 2B, left panel). Interestingly, we found that during litter development, the NC female offspring showed an increase in basal glycemia that was indeed significant at PD60 (basal glycemia: NC PD21 = 81.3 ± 2.6, $p < 0.01$) when compared to the PD28 values, and in contrast, the HF female offspring had decreased basal glycemia from PD21 to PD28 ($p < 0.05$), and it was maintained at PD60 (basal glycemia: HF PD60 = 85.6 ± 2.0, $p < 0.01$). Basal glycemia was not altered by HF maternal feeding in male offspring in the period of development that was analyzed (Figure 2B, right panel).

Figure 2C,D represent the curves of glycemia during the ipITT performed on the offspring (Figure 2C) and the correspondent AUC of those curves (Figure 2D). From the curves, it can be seen that at PD21, the male offspring from the HF mothers presented a significant increase in glycemia levels throughout all of the ipITTs (Figure 2C, left panel), demonstrating a decrease in insulin sensitivity. At PD28, neither HF maternal feeding nor sex impacted the glycemia profile during the ipITT (Figure 2C, central panel). At PD60, the glycemia profiles during the ipITT showed that the male offspring had a significant increase in glycemia throughout the test, independently to the mothers' diets during pregnancy and lactation (Figure 2C, right panel). Through the analysis of the glycemia AUC from the ipITT during offspring development in the initial 60 days, insulin sensitivity significantly increased in the female descendants of the NC mothers (AUC glycemia during ipITT: NC PD21 = 900.4 ± 36.2; NC PD28 = 677.2 ± 41.6, $p < 0.01$; NC PD60 = 724.8 ± 22.5, $p < 0.05$) (Figure 2D, left panel). In the male offspring, the glycemia AUC during the ipITT decreased from PD21 to PD28 and increased from PD28 to PD60 in both the NC and HF descendants (AUC glycemia during ipITT: NC PD21 = 807.8 ± 66.4; NC PD28 = 677.6 ± 23.9, $p < 0.01$; NC PD60 = 862.6 ± 19.5, $p < 0.05$; HF PD21 = 1022.2 ± 81.6; HF PD28 = 663.9 ± 41.2, $p < 0.01$; HF PD60 = 878.6 ± 34.9, $p < 0.05$) (Figure 2D, right panel), showing a significant increase in insulin sensitivity from PD21 to PD28 and a significant decrease in insulin sensitivity from PD28 to PD60 independent of maternal diet.

Finally, the impacts of maternal diet and offspring sex on the descendants' glucose tolerance was evaluated using the OGTT (Figure 2E,F). The glycemia profiles showed that neither the mothers' diets nor offspring sex impacted glucose tolerance at each PD that was evaluated (Figure 2E). However, the female offspring from the HF mothers showed a significant decrease in glucose tolerance in comparison to the female offspring of the NC mothers at PD21 (AUC glycemia during OGTT: NC PD21 = 19,955.2 ± 1033.4; HF PD21 = 23,999.4 ± 857.5; $p < 0.05$), an effect that is attenuated during offspring development (AUC glycemia during OGTT: HF PD21 = 23,999.4 ± 857.5; HF PD28 = 20,866.1 ± 345.6; HF PD60 = 20,076.8 ± 598.9; $p < 0.05$) (Figure 2F, left panel). Male offspring glucose tolerance was not altered by the diet fed to the mothers during development (Figure 2F, right panel).

3.2.2. Liver and Adipose Tissue Depots' Weights

In Table 2 are the liver and adipose tissue depots' weights of descendants of mothers submitted to a NC or HF diet during the gestation and lactation periods. Maternal overnutrition during pregnancy and lactation impacted both the male and female offspring AT depots' weights mainly at PD60.

In the female offspring, overnutrition in mothers significantly increased visceral and perinephric AT at PD60 (visceral AT weight: NC females PD60 = 0.406 ± 0.038; HF females PD60 = 0.527 ± 0.055, $p < 0.001$; perinephric AT weight: NC females PD60 = 0.196 ± 0.017; HF females PD60 = 0.374 ± 0.068, $p < 0.01$, mg).

Regarding the male offspring, maternal overnutrition promoted an increase in liver weight at PD21 (NC males PD21 = 2.286 ± 0.086; HF males PD21 = 2.717 ± 0.126, $p < 0.01$) and at PD60 (NC males PD60 = 6.377 ± 0.326; HF males PD60 = 7.218 ± 0.465, $p < 0.05$). Also, an HF diet in mothers during pregnancy and lactation promoted an increase in the weight of visceral AT (NC males PD60 = 1.216 ± 0.182; HF males PD60 = 1.864 ± 0.152, $p < 0.01$) and genital AT (NC males PD60 = 2.324 ± 0.375; HF males PD60 = 3.642 ± 0.503, $p < 0.01$) in male descendants at PD60.

The brown AT weight was not altered significantly in neither the female nor male descendants of the HF mothers, at least until PD60.

Table 2. Impact of overnutrition in mothers during pregnancy and lactation on the weights (mg) of the liver and of the different adipose tissue (AT) depots in male and female descendants at postnatal days 21, 28, and 60.

	Liver	Visceral AT	Perinephric AT	Genital AT	Brown AT
NC Females PD21	0.963 ± 0.066	0.262 ± 0.026	0.105 ± 0.017	0.128 ± 0.015	0.087 ± 0.010
HF Females PD21	1.646 ± 0.139	0.278 ± 0.039	0.222 ± 0.026	0.318 ± 0.098	0.130 ± 0.021
NC Females PD28	0.978 ± 0.056	0.257 ± 0.021	0.139 ± 0.012	0.114 ± 0.012	0.103 ± 0.009
HF Females PD28	1.965 ± 0.115	0.344 ± 0.042	0.303 ± 0.053	0.286 ± 0.050	0.143 ± 0.015
NC Females PD60	1.892 ± 0.118	0.406 ± 0.038	0.196 ± 0.017	0.214 ± 0.033	0.135 ± 0.009
HF Females PD60	2.670 ± 0.158	0.527 ± 0.055 ***	0.374 ± 0.068 **	0.472 ± 0.075	0.139 ± 0.010
NC Males PD21	2.286 ± 0.086	0.376 ± 0.042	0.376 ± 0.059	0.305 ± 0.033	0.146 ± 0.016
HF Males PD21	2.717 ± 0.126 **	0.561 ± 0.043	0.472 ± 0.042	0.528 ± 0.054	0.144 ± 0.012
NC Males PD28	4.872 ± 0.405	1.029 ± 0.135	1.109 ± 0.212	1.993 ± 0.405	0.249 ± 0.025
HF Males PD28	4.566 ± 0.178	1.472 ± 0.065	1.879 ± 0.281	2.183 ± 0.386	0.226 ± 0.026
NC Males PD60	6.377 ± 0.326	1.216 ± 0.182	2.599 ± 0.381	2.324 ± 0.375	0.269 ± 0.025
HF Males PD60	7.218 ± 0.465 *	1.864 ± 0.152 **	3.162 ± 0.422	3.642 ± 0.503 **	0.284 ± 0.026

Data represent mean ± SEM of 10 NC females, 10 NC males, 9 HF females, and 9 HF males per time-point. * $p < 0.05$, ** $p < 0.01$; *** $p < 0.001$ two-way ANOVA with Sidak's multiple comparison test comparing animals submitted to normal chow (NC) diet and animals submitted to high-fat diet (HF) of the same age. PD21—postnatal day 21; PD28—postnatal day 28; PD60—postnatal day 60.

3.3. *Effect of Overnutrition during Pregnancy and Lactation on Behavior Phenotype in the Offspring*

In order to assess the impact of overnutrition during pregnancy and lactation on the offspring, a series of behavioral tests were conducted on the offspring at PD60. These tests were categorized into three main groups, presented in the following sections: Section 3.3.1. Anxiety and Stress; Section 3.3.2. Memory and Learning; and Section 3.3.3. Food/Drink Behavior.

3.3.1. Anxiety and Stress

In this battery of tests (Figure 3), we incorporated the OF, EPM, and the LDB tests, which are all designed to measure anxiety and stress levels in the offspring [22].

Regarding the OF results, significant differences ($p < 0.05$) were evident among the female offspring groups concerning the distance covered within the inner zone. This suggests that the female offspring from mothers on an HF diet spent notably less time in the inner zone compared to those from the control mothers. However, no significant differences were observed among the other groups for the remaining three parameters: total distance, total immobility/total test time, and immobility within the inner zone/total test time (Figure 3A).

Concerning the results of the elevated plus maze (EPM) test, there were no significant differences detected between the groups (NC vs. HF) for the five parameters assessed: ratio of open arm/open + closed arm, ratio of closed arm/open + closed arm, number of poops, number of pees, and the frequency of alternations between arms (Figure 3B).

Finally, when it comes to the outcomes of the light–dark box test, there were no significant differences observed between the groups (NC vs. HF) in the three parameters analyzed: the proportion of time spent in the light zone relative to the total test time, the proportion of time spent in the dark zone relative to the total test time, and the number of alternations (Figure 3C).

3.3.2. Memory and Learning

For this set of assessments (Figure 4), we used the Y maze, the novel object recognition (NOR), and the block test, designed to assess memory and learning [22].

The Y maze test was used to evaluate spatial learning and memory capacity in the offspring of both the NC and HF animals. There were no significant differences observed in the four parameters analyzed: the time the animals spent in the new arm, the ratio of time spent in the novel arm to the sum of time spent in both the novel and home arms, the number of entries into the novel arm, and alternative behaviors. The last parameter calculates the ratio of the number of alternations to the possible number (total number of arm entries minus two) (Figure 4A).

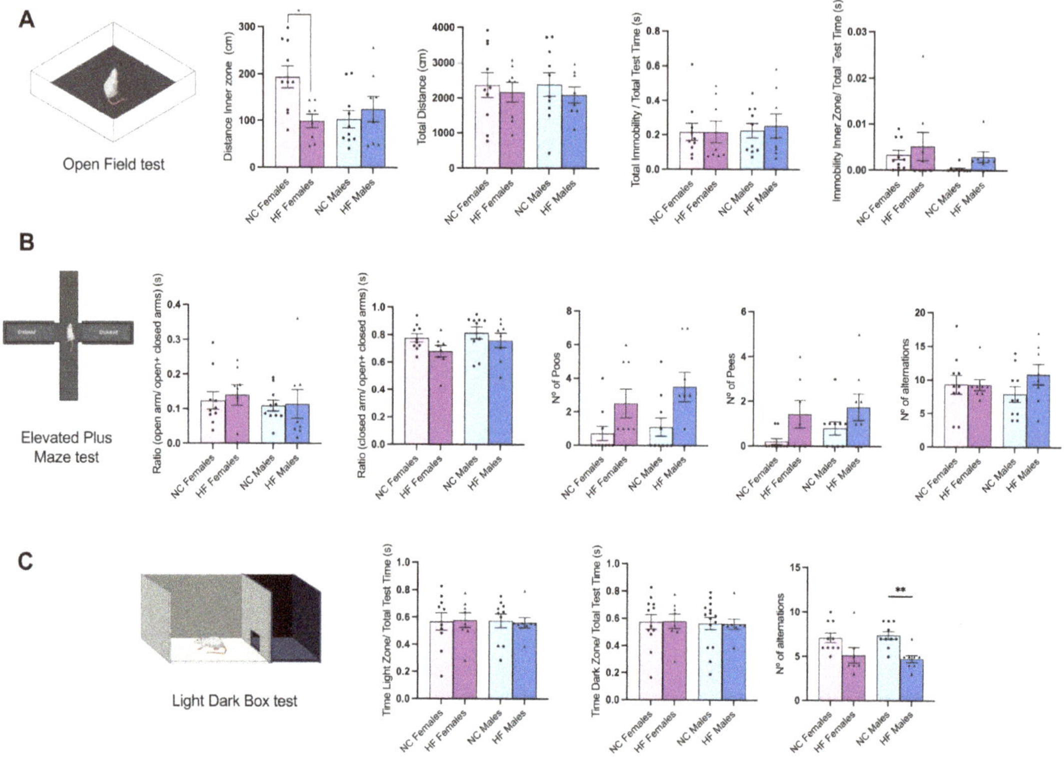

Figure 3. Effect of overnutrition during pregnancy and lactation on anxiety and stress, evaluated using the open field test (**A**), elevated plus maze test (**B**), and light–dark box test (**C**), in the offspring at postnatal day 60. (**A**) From the left to the right, the panel shows a schematic representation of the open field maze, the distance in the inner zone (cm), total distance covered (cm), total immobility (s), and the immobility in the inner zone (s); (**B**) from left to right, the panel shows a schematic representation of the elevated plus maze, the time spent in the open arms in comparison to the time spent in all arms (s), the time spent in the closed arms in comparison to the time spent in all arms (s), the number of poops in the maze, the number of pees, and the number of alternations between the open arm and the closed arms. Data represent mean ± SEM of 10 NC females, 10 NC males, 8 HF females, and 8 HF males. Black circles correspond to offspring born form mothers submitted to NC diet and black triangles to offspring born form mothers submitted to HF diet. Two-way ANOVA with Bonferroni and Sidak's multiple comparison tests were used; * $p < 0.05$ and ** $p < 0.01$ comparing NC animals.

Regarding the NOR results, there were no significant differences observed between the NC and HF groups in the three parameters measured: the time required to identify the new object, the ratio of time spent interacting with the novel object over the total time spent interacting with both the novel and familiar objects, and the ratio of time spent interacting with the familiar object over the total time spent interacting with both the novel and familiar objects (Figure 4B).

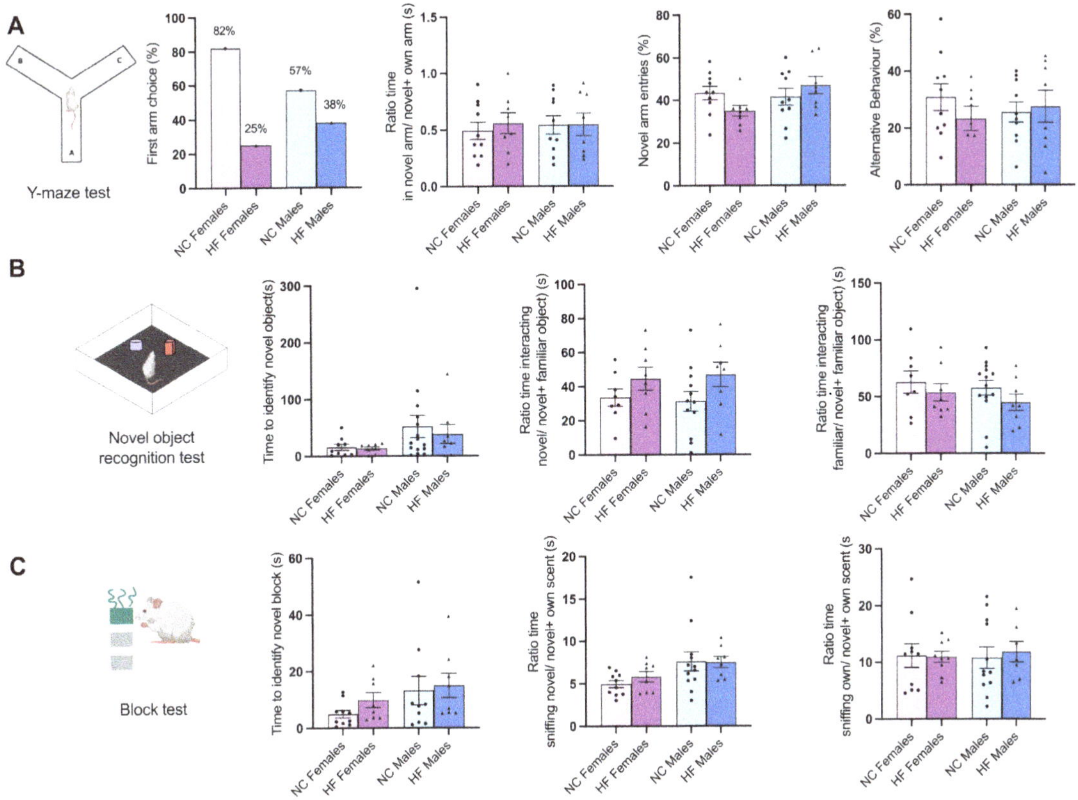

Figure 4. Effect of overnutrition during pregnancy and lactation on memory and learning, evaluated using the Y maze test (**A**), novel object recognition test (**B**), and block test (**C**), in the offspring at postnatal day 60. (**A**) From the left to the right, panel shows a schematic representation of the Y maze, the first arm choice (%), the ratio between the novel arm vs. the time in both the novel and familiar arm, the % of novel arm entries, and the alternative behaviors (%). (**B**) From the left to the right, the panel presents a schematic representation of novel object recognition apparatus, the time taken to identify the novel object (s), the ratio of the time the animals spent interacting with the novel object vs. the total interaction time (s), and the time the animals spent interacting with the familiar object vs. the total interaction time (s). (**C**) From left to right, the panel presents a schematic representation of the block test, the time taken to identify the novel block (s), the time the animals spent sniffing the novel block vs. the total time sniffing all the blocks, and the time the animals spent sniffing the familiar blocks vs. the total time sniffing all of the blocks. Data represent mean ± SEM of 10 NC females, 10 NC males, 8 HF females, and 8 HF males. Black circles correspond to offspring born form mothers submitted to NC diet and black triangles to offspring born form mothers submitted to HF diet. Two-way ANOVA with Bonferroni and Sidak's multiple comparison tests were conducted.

Finally, the animals' olfactory working memories were assessed using the block test that assesses the ability of the animals to discriminate between familiar and novel, innocuous scents [21]. The results are presented in Figure 3C. No significant differences were observed between the experimental groups for the three parameters analyzed: the ratio of time spent sniffing the novel block over the total time spent sniffing both the novel and familiar blocks, the ratio of time spent sniffing the familiar block over the total time spent sniffing both the novel and familiar blocks, and the time required to identify the new block.

However, it is worth noting that there is a noticeable trend in the last parameter, where the offspring of the HF animals tend to take more time to complete this task.

3.3.3. Food/Drink Behavior

To assess the competition for food and drink, the respective two tests were used: food competition and water/sucrose competition (Figure 5).

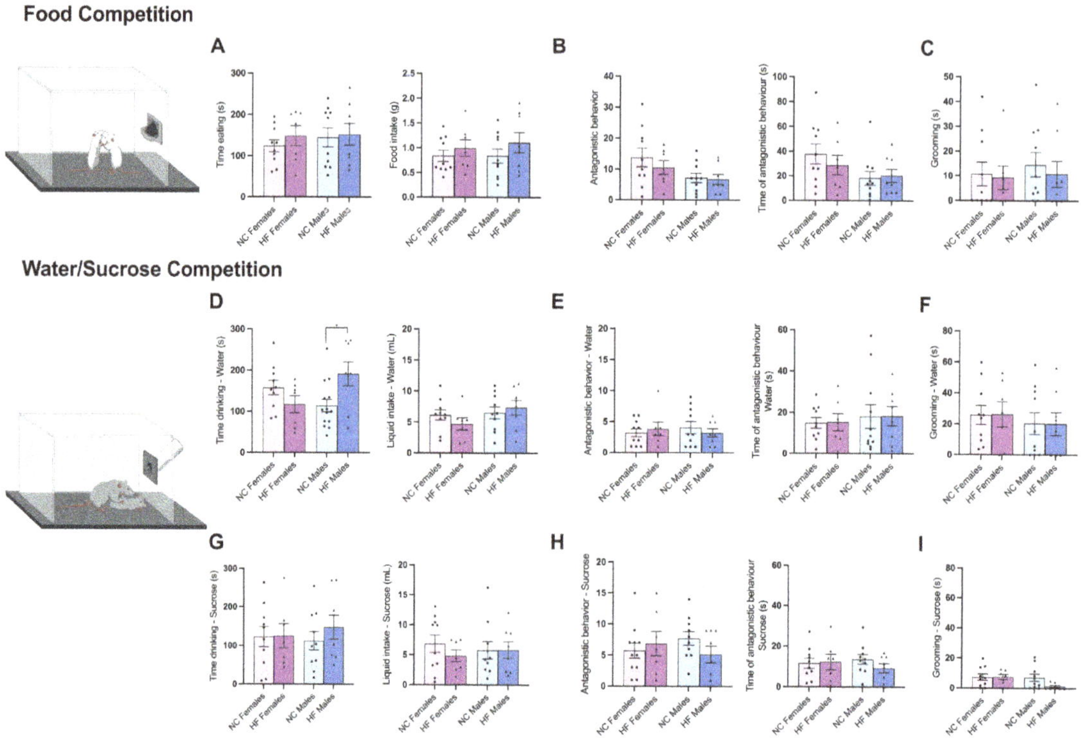

Figure 5. Effect of overnutrition during pregnancy and lactation on food behavior, evaluated using food competition test (**A–C**) and water/sucrose competition test (**D–I**). (**A**) From the left to the right, the panel shows a schematic representation of the food competition test, the time spent eating (s) and the food intake (g) (**A**), the antagonistic behavior and the time spent displaying antagonistic behavior (s) (**B**), and the time of grooming (s) (**C**). (**D–F**) From the left to the right, the panel represents a schematic representation of water/sucrose competition, time drinking (s) and the liquid intake (mL) (**D**), the antagonistic behavior and the time spent displaying antagonistic behavior (s) (**E**), and time of grooming (s) (**F**) in response to water competition. (**G–I**) From the left to the right, the panel shows the time drinking (s) and the liquid intake (mL) (**G**), the antagonistic behavior and the time spent displaying antagonistic behavior (s) (**H**), and time of grooming (s) (**I**) in the presence of sucrose. Data represent mean ± SEM of 10 NC females, 10 NC males, 8 HF females, and 8 HF males. Black circles correspond to offspring born form mothers submitted to NC diet and black triangles to offspring born form mothers submitted to HF diet. Two-way ANOVA with Bonferroni and Sidak's multiple comparison test were conducted; * $p < 0.05$ comparing NC animals.

The food competition test was used to assess both the competition for food, the voracity of feeding, as well as the dominance in the offspring in both the NC and HF animals. No significant differences were observed in any of the five analyzed parameters: time spent eating, food intake, antagonistic behavior, time spent displaying antagonistic behavior, and grooming. However, a trend can be observed for two parameters: the time spent eating and food intake. The HF animals tend to feed for a longer period and consume a higher amount of food compared to the NC animals.

In these tests, antagonistic behaviors indicating dominance over the other were considered (see methods in Sections 2.3.7 and 2.3.8).

Like the previous test, the water/sucrose competition test is used to determine if there is a preference for drinking between sucrose and water. To carry this out, the amount of liquid ingested, the speed at which animals consume it, as well as the dominance of this resource among NC and HF animals were measured. When comparing significant differences within the same trial (when the liquid is the same), it was clear that, when they competed for water, males born from the HF mothers drank water for a longer time, although without significant differences in the amount of liquid intake. Note, also, that no differences were observed in the females in the time spent drinking water or the amount of water intake. Moreover, the animals, when exposed to water, did not exhibit significant differences in antagonistic behavior, the time spent displaying antagonistic behavior, and grooming. When the animals competed for sucrose, no significant differences were observed among the five parameters previously mentioned. When comparing the water trial vs. the sucrose trial, it seems that the animals spent more time consuming water than sucrose, exhibited a greater quantity of antagonistic behaviors when exposed to water compared to sucrose, and also engaged in a higher amount of grooming during the water trial.

3.4. Effect of Overnutrition during Pregnancy and Lactation on Hypothalamic, Hippocampal, and Prefrontal Markers of Synaptic Transmission, Metabolic Signaling, and Inflammation

To assess the impact of excessive nutrition during pregnancy and lactation in some areas of the brain related with stress and anxiety—the hypothalamus, hippocampus, and prefrontal cortex—a Western blot technique was employed to examine the levels of some proteins and receptors involved in synaptic dynamics (Section 3.4.1), metabolism (Section 3.4.2), and inflammation (Section 3.4.3).

3.4.1. Protein Markers of Synaptic Transmission on the Hypothalamus, Hippocampus, and Prefrontal Cortex

Stress, anxiety, and cognition have been associated with altered synaptic transmission and plasticity [23,24]. To investigate synaptic dynamics in the three designated areas, we assessed three distinct proteins: the glutamate vesicular transporter (vGLUT), the synaptosome-associated protein 25 kDa (SNAP-25), which contributes to the SNARE complex involved in exocytosis, and the postsynaptic density protein 95 (PSD-95), a member of the membrane-associated guanylate kinase family that, with PSD-93, is recruited to the NMDA receptor and potassium channel clusters.

In the hypothalamus, development induced increases of 37.5% and 44.7% ($p < 0.001$) in the levels of vGLUT, and the females showed a non-significant increase at PD28 and PD60. The male HF offspring did not exhibit significant differences compared to the NC offspring, but the female offspring at PD21 showed a significant increase of 57.5% that was attenuated with age (Figure 6A). Both the males and females showed increases in the SNAP-25 levels from PD21 onwards in the NC animals, with this being statistically significant only for the females at PD60 (increase of 99.3%). Moreover, both the males and females born from HF mothers exhibited altered levels of SNAP-25 at the hypothalamus, with the males always exhibiting higher levels of protein that were significant at PD28, and with the females exhibiting significant higher levels at PD21 that were maintained at PD28 and PD60 (Figure 6B). In relation to PSD-95, it was clear that the males exhibited a tendency to progressively increase the levels of this protein with age, which is an effect that was not observed in the females. Overnutrition during pregnancy and lactation did not change these results (Figure 6C).

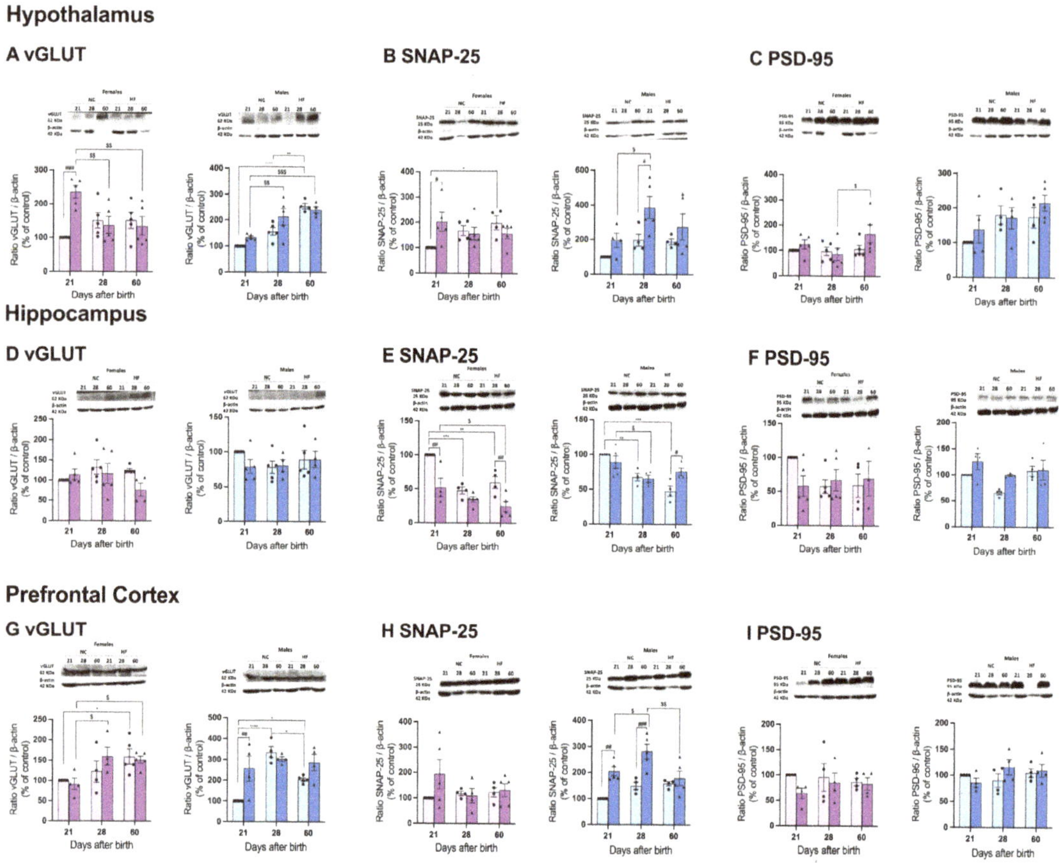

Figure 6. Effect of overnutrition during pregnancy and lactation on the levels of proteins involved in synaptic transmission on the hypothalamus (**A–C**), hippocampus (**D–F**), and prefrontal cortex (**G–I**). From the left to the right, graphs show the mean levels of the ratio between vGLUT and β-actin, the ratio between SNAP-25 and β-actin, and the ratio between PSD-95 and β-actin in males and females, respectively at the hypothalamus (**A–C**), hippocampus (**D–F**), and prefrontal cortex (**G–I**). On the top of the graphs are representative Western blot membranes for the proteins of interest and the respective loading control. Data represent mean ± SEM of 4–5 NC females, 4–5 NC males, 4–5 HF females, and 4–5 HF males for each time-point. Black circles correspond to offspring born form mothers submitted to NC diet and black triangles to offspring born form mothers submitted to HF diet. Two-way ANOVA with Tukey's and Sidak's multiple comparison tests were conducted; * $p < 0.05$, ** $p < 0.01$, *** $p < 0.001$ and **** $p < 0.0001$ comparing NC animals with different ages; # $p < 0.05$, ## $p < 0.01$, and ### $p < 0.001$ comparing NC vs. HF animals with the same age; and $ $p < 0.05$, $$ $p < 0.01$ and $$$ $p < 0.001$ comparing HF animals with different ages.

In the hippocampus, no significant differences were observed regarding age, diet, and sex for vGLUT (Figure 6D) or PSD-95 (Figure 6F). However, notable differences were evident in SNAP-25 in the males and females. Note that there was a clear decrease in the levels of SNAP-25 between PD21, PD28, and PD60 in the NC animals (males: decreases of 32.3% at PD28 and 52.6% at PD60; females: decreases of 52.2% at PD28 and 39.9% at PD60). Interestingly, the overfeeding of mothers during pregnancy and lactation did not change the levels of SNAP-25 in the males at PD21 and PD28, but it increased its levels by 37.5% at

PD60. Moreover, females born from HF mothers showed significantly increased levels at PD21 and PD60 of 47.7% and 59.8%, respectively.

The prefrontal cortex development in males and females produced significant changes in the levels of vGLUT, but not on the other proteins studied, where no differences were observed between PD21, PD28, and PD60 (Figure 6G–I). Note that the vGLUT levels were significantly higher at PD28 (increase of 70.3%) and PD60 (increase of 51.1%) in the males and at PD60 (increase of 37.4%%) in the females. Interestingly, only the male offspring from the HF mothers showed differences at PD21 that were attenuated with age (Figure 6G). The SNAP-25 levels were increased in the males and females born from HF mothers, an effect that was attenuated with age, but there was more marked in the males. Note that at PD21 and PD28, the males born from HF mothers showed high levels of SNAP-25 that were attenuated at PD60. Interestingly, this increase, although non-significant, was only seen in the females at PD21 (increase by 51.1%%) (Figure 6H). An HF diet during pregnancy and lactation did not change the levels of PSD-95 in the males nor females (Figure 6I).

3.4.2. Protein Markers of Metabolism on the Hypothalamus, Hippocampus, and Prefrontal Cortex

Figure 7 depicts the impact of hypercaloric diets consumed by mothers in some metabolic markers—insulin receptor and AMPK—in specific brain areas of the offspring. We can observe that in the hypothalamus, development significantly and progressively increased the levels of the ratio of IR-P/IR-T in the males and females born from NC mothers (males at PD28, 31.9% increase; males at PD60, 43.0% increase; females at PD28, 42.8% increase; females at PD60, 44.4% increase), suggesting a higher phosphorylation of the IR with development. Interestingly, while no changes were observed in the levels of this ratio at PD21 and PD28 in the males born from HF mothers, a significant reduction was seen at PD60 (49.3% decrease). In contrast, in the females, an HF diet in mothers promoted a higher ratio of IR-P/IR-T at PD21 (40.2% increase) that was attenuated with age (Figure 7A). Panel B shows the ratio of AMPK-P/AMPK-T, a metabolic sensor. Age increased the levels of the ratio of AMPK-P/AMPK-T in both the females and males, with these only being significant in the males (Figure 7B), suggesting a higher phosphorylation of AMPK with development. HF diet consumption in mothers did not change this ratio in neither sex (Figure 7B). In the hippocampus, the levels of the ratio of IR-P/IR-T followed, more or less, the same trend than in the hypothalamus, with the phosphorylation of IR increasing by 49.8% at PD60 in the males and increasing progressively with development in the females, although without statistical significance (Figure 7C).

HF diet intake by mothers did not significantly alter the ratio of IR-P/IR-T in the offspring in neither sex (Figure 7C). Interestingly, we observed sexual dimorphism in the ratio of AMPK-P/AMPK-T at the hippocampus, with the females in the NC group showing a progressively and statistically significant increase in the phosphorylation of AMPK at PD28 (36.4%) and PD60 (36.3%), but no effect on the males was observed (Figure 7D). An HF diet in mothers significantly changed the ratio of AMPK-P/AMPK-T in the offspring in both sexes (Figure 7D). Figure 7E and F depict the levels of the same proteins in the prefrontal cortex. Note that there was a progressive decrease in the ratio of IR-P/IR-T with development in the offspring from NC mothers, which was more accentuated and statistically significant in the males (PD28, 38.8% decrease, and PD60, 55.4% decrease compared with PD21) than in the females (Figure 7E). Also note that HF diet intake in mothers promoted a decrease of 48.6% in the phosphorylation of IR at PD21 in the males that was not altered with age. In females, no differences were observed (Figure 7E). In contrast, the ratio of AMPK-P/AMPK-T statistically increased in the offspring of the NC mothers from PD21 onwards, both in the males (PD28m 13.6% increase; PD60, 15.4% increase) and females (PD28, 20.2% increase; PD60, 18.3% increase). HF diet intake in mothers induced a higher phosphorylation of AMPK in the offspring at PD21 in both the males (14.1% increase) and females (17.5% increase) that was maintained with age (Figure 7F).

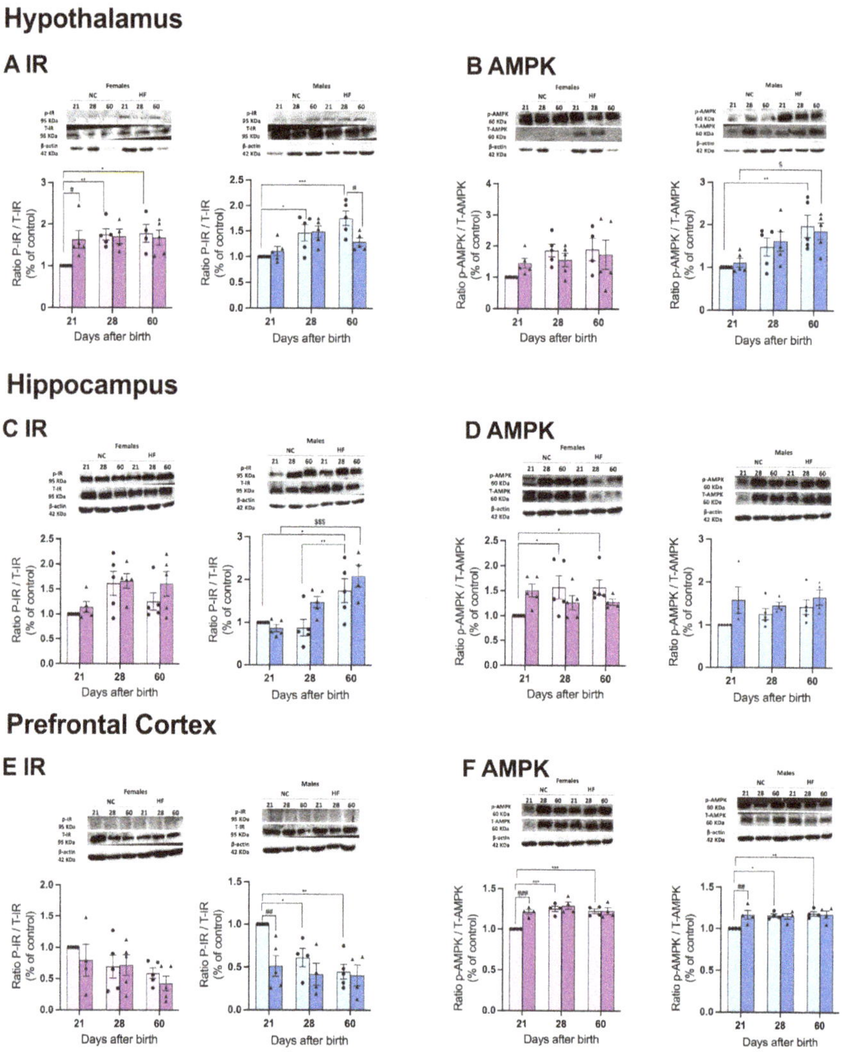

Figure 7. Effect of overnutrition during pregnancy and lactation on the levels of insulin receptor and AMPK on the hypothalamus (**A**,**B**), hippocampus (**C**,**D**), and prefrontal cortex (**E**,**F**). From the left to the right, graphs represent mean levels of the ratio between phosphorylated insulin receptor/total insulin receptor (P-IR/T-IR) and β-actin, and the ratio between phosphorylated AMPK and total AMPK (p-AMPK/T-AMPK) and β-actin in males and females, respectively at the hypothalamus (**A**,**B**), hippocampus (**C**,**D**), and prefrontal cortex (**E**,**F**). On the top of the graphs are representative Western blot membranes for the proteins of interest and the respective loading control. Data represent mean ± SEM of 4–5 NC females, 4–5 NC males, 4–5 HF females, and 4–5 HF males for each time-point. Black circles correspond to offspring born form mothers submitted to NC diet and black triangles to offspring born form mothers submitted to HF diet. Two-way ANOVA with Tukey's and Sidak's multiple comparison tests were conducted; * $p < 0.05$, ** $p < 0.01$ and *** $p < 0.001$ comparing NC animals with different ages; $^{\#}$ $p < 0.05$, $^{\#\#}$ $p < 0.01$, and $^{\#\#\#}$ $p < 0.001$ comparing NC vs. HF animals with the same age; and $^{\$}$ $p < 0.05$, $^{\$\$\$}$ $p < 0.001$ comparing HF animals with different ages.

3.4.3. Protein Markers of Inflammation on the Hypothalamus, Hippocampus, and Prefrontal Cortex

Inflammation has been implicated in several mood and anxiety disorders as well as in cognitive decline [25,26]. To explore the inflammation as being involved in the effects of overnutrition during pregnancy and lactation in stress and anxiety in the offspring, we evaluated the levels of three inflammatory markers: GFAP, whose expression was shown to be increased in brain inflammatory states [26]; TNF-α, a pro-inflammatory cytokine expressed in many brain pathologies and associated with neuronal loss [27]; and IL6-R, the receptor for IL6, a major cytokine that has anti- and pro-inflammatory effects on the brain and has been linked with neurodegenerative diseases and with chronic stress contributing to neurobehavioral complications [28].

In the hypothalamus, no significant changes were observed in the GFAP levels in males concerning diet or age. However, although the GFAP levels in the females did not change with age in the NC offspring, they exhibited a 38.5% significant increase from PD21 to PD28 in the HF offspring (Figure 8A). Surprisingly, the levels of TNF-α decreased from PD21 to PD28 and PD60 both in the males (PD28, 32.7%; PD60, 26.2%) and females (PD28, 36.4%; PD60, 52.4%) in the NC offspring. Overnutrition during pregnancy and lactation in mothers did not alter the levels of TNF-α in the hypothalamus in the offspring (Figure 8B). Considering IL6-R there was an increase in the levels of this protein with development in both the males (PD28, 36.4%; PD60, 91.8%) and females (PD28 32.5%) in the NC offspring. Overnutrition during pregnancy and lactation in the mothers did not significantly alter the levels of IL6-R in the hypothalamus in the offspring (Figure 8C).

In the hippocampus, when examining GFAP, no significant differences were observed concerning age and diet for neither sex in the offspring (Figure 8D). In the analysis of the TNF-α, it became apparent that diet had a notable impact on the males. While there was a decrease in TNF with age in the male NC offspring, there was an increase in the levels of IL6-R in the HF offspring at PD28 and PD60 when compared to the NC group (35.8% and 51.2%, respectively), a trend that was not observed in the females (Figure 8E). When comparing the IL6-R levels in the males, an HF diet in the mothers increased the levels by 35.8% at PD21, which is an effect that was attenuated at PD28 and PD60. In females, age did not change the IL6-R levels in the NC offspring, but HF diet intake in mothers increased the IL6-R levels significantly by 28.1% at PD60 (Figure 8F).

In the prefrontal cortex, the GFAP levels increased significantly with age in the males from the NC group (PD28, 39.6% and PD60, 32.5%), which is an effect that was absent in the females. Overnutrition during pregnancy and lactation in mothers did not significantly alter the levels of GFAP in the prefrontal cortex in the offspring (Figure 8G). TNF-α decreased with age in both the males and females born from NC mothers; however, the decrease was only statistically significant in the males (PD28, 41.5%; PD60, 47.5% in comparison with PD21). HF diet intake in mothers did not change the TNF-α levels in the males nor females (Figure 8H). In the analysis of the IL6-R levels, there was a 29.9% increase in the IL6-R levels at PD60 in the male offspring from NC mothers, while in the females, no alterations were observed with age. Overnutrition during pregnancy and lactation in mothers did not significantly alter the levels of IL6-R in the prefrontal cortex in the offspring (Figure 8I).

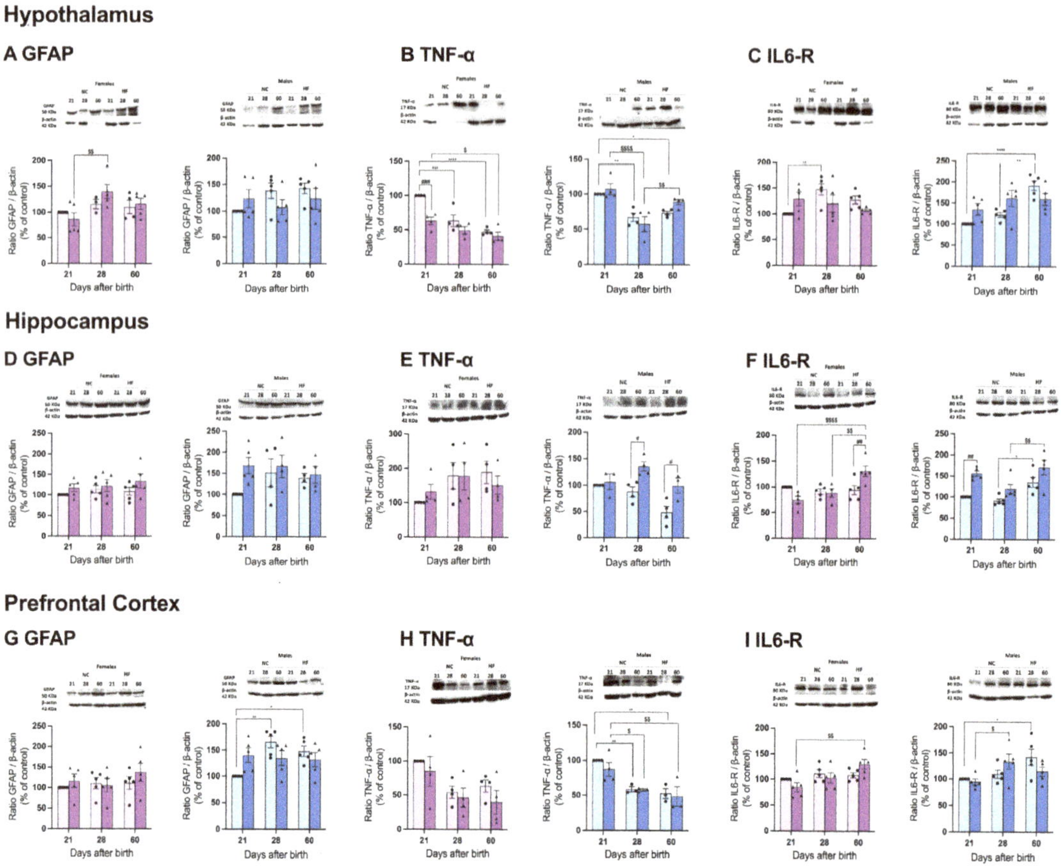

Figure 8. Effect of overnutrition during pregnancy and lactation on the levels of proteins involved in inflammation on the hypothalamus (**A–C**), hippocampus (**D–F**), and prefrontal cortex (**G–I**). From the left to the right, graphs show mean levels of the ratio between GFAP and β-actin, the ratio between TNF-α and β-actin, and the ratio between IL6-R in males and females, respectively at the hypothalamus (**A**,**B**), hippocampus (**C**,**D**), and prefrontal cortex (**E**,**F**). On the top of the graphs are representative Western blot membranes for the proteins of interest and the respective loading control. Data represent mean ± SEM of 4–5 NC females, 4–5 NC males, 4–5 HF females, and 4–5 HF males for each time-point. Black circles correspond to offspring born form mothers submitted to NC diet and black triangles to offspring born form mothers submitted to HF diet. Two-way ANOVA with Tukey's and Sidak's multiple comparison tests were conducted; * $p < 0.05$, ** $p < 0.01$, *** $p < 0.001$ and **** $p < 0.0001$ comparing NC animals with different ages; # $p < 0.05$, ## $p < 0.01$, and ### $p < 0.001$ comparing NC vs. HF animals with the same age; and \$ $p < 0.05$, \$\$ $p < 0.01$, and \$\$\$\$ $p < 0.0001$ comparing HF animals with different ages.

4. Discussion

In this study, we demonstrated that (1) female offspring born from overfeeding mothers during pregnancy and lactation showed increased weight gain and decreased glucose tolerance that attenuated with age; (2) offspring males born from overfeeding mothers exhibited increased weight gain that worsened with age, while glucose tolerance remained unchanged; (3) offspring from HF mothers exhibited increased levels of anxiety and stress during behavioral tests, displaying decreased predisposition for curiosity compared to the control group; and (4) offspring born from overfeeding mothers exhibited alterations in exocytotic capacity in the hypothalamus, hippocampus, and prefrontal cortex and in some inflammatory markers in the hippocampus that are different in males and females. As a whole, we demonstrated that maternal HF diet feeding during pregnancy and lactation induces dysmetabolism in the offspring accompanied by heightened stress and anxiety and by alterations in synaptic dynamics and neuroinflammation. Moreover, we found that there was sexual dimorphism in metabolic traits but not in behavior phenotypes.

4.1. Effect of Overnutrition during Pregnancy and Lactation on Metabolic Function in the Offspring

Herein, we showed, as expected, that throughout offspring development, distinctions arose concerning sex, diet, and age. As a pup develops, the body weight undergoes fluctuations depending on age, consistently showing a tendency to increase, and is largely influenced by dietary factors. Sexual dimorphism becomes apparent from 21 days onwards, with males generally exhibiting larger sizes and consequently greater weights than females. Over time, particularly in males, there is an escalating weight disparity influenced by diet. Conversely, in females, who commence hormone production at day 28 [29], the impact of diet on weight variation gradually diminishes. The estrogen hormone plays a pivotal role in regulating food intake and body weight, extending its influence to the modulation of insulin receptor abundance and ultimately being responsible for these differences between genders in overnutrition-dependent body weight gain [30].

Sexual dimorphism observed in the weight was also evident in the basal glycemia levels, with female offspring exhibiting changes in basal glycemia, while male offspring do not. This increased basal glycemia in female offspring at PD21 and PD28, compared to NC animals, suggests a significant impact of maternal diet during early development that could be attributed to a phenomenon known as developmental programming [31], where maternal influences during critical periods of fetal and neonatal development shape the long-term health of the offspring. Maternal overnutrition induces changes in the intrauterine environment, influencing the release of hormones and leading to epigenetic modifications [32]. These epigenetic changes can subsequently impact gene expression, potentially contributing to the observed differences in basal glycemia [33]. Interestingly, at PD60, the basal glycemia in female offspring from HF mothers did not differ significantly from the NC female offspring. This stabilization or resolution of metabolic effects over time suggests potential adaptive mechanisms or compensatory changes in response to early-life exposures and a possible contribution of hormonal effects [34]. Hormonal and genetic differences between male and female offspring may contribute to the observed variations in metabolic outcomes, and in this case, in basal glycemia [35].

Sexual dimorphism was also observed in insulin sensitivity. At PD21, the male offspring from HF mothers demonstrated a significant increase in the area under the curve of the insulin tolerance test, indicating decreased insulin sensitivity. This early effect suggests that maternal overnutrition during critical developmental periods may induce alterations in insulin signaling pathways, contributing to reduced insulin sensitivity in male offspring during early postnatal life. Surprisingly, at PD28, neither HF maternal feeding nor sex seemed to significantly impact the glycemia profiles during the ITT. This lack of effect at PD28 suggests potential adaptive responses or the normalization of insulin sensitivity in the offspring, possibly involving compensatory mechanisms that counteract the initial impact of maternal overnutrition [36]. By PD60, the male offspring, irrespective of maternal diet, displayed a significant increase in the area under the curve of the ipITT. This effect at PD60 suggests that the effects of age on insulin sensitivity that are clearly described in adulthood are already seen at early ages [37]. Regarding the female offspring, a different pattern emerged. The glycemia during ipITT indicated a significant increase in insulin sensitivity from PD21 to PD28 and a subsequent decrease from PD28 to PD60 in both the NC and HF female descendants. These dynamic changes suggest a time-dependent modulation of insulin sensitivity in females. Hormonal fluctuations, metabolic adaptation, or other sex-specific factors may contribute to these variations in insulin sensitivity during different developmental stages [37]. Moreover, the absence of effects of maternal dysmetabolism on insulin sensitivity with development contrasts with the documented impact of maternal obesity on insulin resistance later in life in animals [38] and humans [39]. However, this lack of effect on insulin sensitivity with development might be attributable to the duration of the exposure of mothers to the high-fat diet and their level of metabolic dysfunction.

The glycemia profiles during the OGTT revealed that at each PD evaluated, neither the maternal diet nor offspring sex exerted a significant impact on glucose tolerance. This suggests a robustness or resilience in glucose tolerance to variations in maternal nutrition during the assessed developmental stages, indicating that the offspring can maintain a relatively stable glucose tolerance despite differences in the maternal diet or offspring sex [40]. However, a notable exception emerged in the female offspring at PD21. The female descendants of HF mothers exhibited a significant decrease in glucose tolerance compared to their counterparts from NC mothers. Interestingly, this effect in the female offspring was attenuated during development. The area under the curve of the OGTT decreased in the HF female offspring from PD21 to PD28 and remained relatively stable from PD28 to PD60. We have to take into account that females start hormonal production at PD28 [29]; therefore, at this age, there must be a potential adaptive response or compensatory mechanism in HF female offspring as they mature, indicating a dynamic regulation of glucose tolerance, with hormonal fluctuations in female development potentially having an important impact [40]. Conversely, the male offspring did not exhibit alterations in glucose tolerance during development based on the diet fed to the mothers. This lack of effect in males underscores a potential sex-specific response to maternal diet in the context of glucose tolerance during the assessed developmental period. Moreover, a recent meta-analysis showed that the exposure to maternal hyperglycemia during pregnancy might be associated with offspring obesity and abnormal glucose tolerance, although the association depends on the duration and intensity of intrauterine exposure to hyperglycemia [41], which agrees with the contrasting effects of our data with some of the published literature.

4.2. *Effect of Overnutrition during Pregnancy and Lactation on Behavior and CNS Functions in the Offspring*

In the present manuscript, we show that maternal exposure to an HF diet is correlated with a propensity for stress and anxiety-like behavior in the offspring without alterations in memory and learning and on food behavior. Our study demonstrates that offspring from HF mothers exhibited diminished exploratory behaviors and reduced curiosity, reflecting the same outcomes observed in previous studies involving offspring from HF mothers [42]. This behavioral trend is consistent with the existing literature that shows that rats subjected to an HF diet exhibit prolonged dwelling times in the shadowy corners of mazes and an increased frequency of defecation, aligning with the patterns reported by other researchers [43] and with human studies that demonstrated that prenatal maternal obesity is associated with offspring anxiety disorders, and that these associations may be long-lasting [44]. Moreover, it was shown that the severity of diabetes during pregnancy may increase the offspring's vulnerability to depression/anxiety during childhood and adolescence [42], which is in agreement with our data presented herein.

Trying to unveil the underlying neurobiological mechanisms by which maternal overnutrition induces stress and anxiety in the offspring, we focused on the synaptic dynamics and transmission, on metabolism, and on neuroinflammation. In the evaluation of synaptic dynamics, notably, neuroprotection was observed in the females, which could be primarily attributed to the presence of hormones, such as estrogen, which serve crucial roles in preserving and maintaining neuronal health [45]. This neuroprotective effect contributes to the manifestation of sexual dimorphism, particularly evident in the prefrontal cortex concerning markers like SNAP-25, a marker for exocytosis, and PSD-95, a marker for postsynaptic glutamatergic transmission. Additionally, the increase in vGLUT in the early days of life, which is also linked to glutamatergic transmission, may be associated with age and, once again, is influenced by the presence of hormones, especially in females, as noted in the hypothalamus and hippocampus [44]. Regarding the impact of overnutrition in mothers in the offspring, this is mainly observed at the exocytosis level, with HF-exposed pups displaying an increased rate of exocytosis in the three regions studied—the hypothalamus, hippocampus, and prefrontal cortex. However, we can also see a sexual dimorphism in synaptic transmission since the increase in the exocytotic marker, SNAP-25, seen in males was not observed in females neither in the hypothalamus nor prefrontal cortex, and it was even decreased at PD60 in the hippocampus in females. Conversely, in the post-synaptic region, no discernible changes were evident with respect to either diet or age on the glutamatergic marker. While there is some evidence reporting altered glutamatergic signaling in the amygdala in the offspring from obese mothers [45,46], in the three regions studied in the present manuscript, overnutrition in mothers clearly modifies markers of glutamatergic signaling, suggesting that the effects of a high fat diet involve neurotransmitters other than glutamate. This agrees with some demonstrations of decreased levels of GABAergic and serotonergic neurotransmitters in the whole brain [47]. Moreover, since hypothalamic neurotransmitters like POMC and NPY are critical for energy balance and feeding [48] (it was found that a POMC-originated circuit regulates stress [49], the ablation of POMC neurons led to anxiety-like behavior [50], and NPY knockouts are related to anxiety [51]), other neurotransmitters will probably be keys in driving this stress/anxiety phenotype in the offspring of dysmetabolic mothers. As a whole, our results clearly suggest a selective influence of dietary factors on specific aspects of synaptic function, underscoring the intricate interplay between nutrition, gender-related hormonal influences, and synaptic mechanisms.

Concerning metabolism molecular pathways, an intriguing pattern emerges as males exhibit an age-related increase in the activation of the insulin signaling pathway, while females maintain a consistent level throughout the early days of development. Notably, in the prefrontal cortex of males, there is a decline in insulin receptor phosphorylation with age. This trend suggests a plausible scenario wherein the initial higher activation of these receptors to participate in the high-rate metabolism at birth diminishes as the pup matures [46–48]. Interestingly, no significant effect of HF diets in mothers was appreciated in the insulin signaling cascade in the offspring except for the statistically significant decrease in activation at PD60 in the hypothalamus in males. Note that, at the hypothalamus, insulin suppresses food intake, is involved in glucose and fat metabolism regulation [52,53], and controls sympathetic activity [54]. Therefore, a decreased activation of this pathway at PD60 can anticipate the late development of cardiometabolic complications. Regarding AMP-activated protein kinase (AMPK), the activation of this protein experiences an increase across various brain regions and age groups. This versatile protein not only facilitates the conversion of AMP into ATP but also responds to nutrients, metabolites, and hormones involved in energy balance [55], and it is also involved in the regulation of growth and reprogramming metabolism [56], which could explain the increase during development. Herein, we observed slight increases in AMPK signaling at PD21 in the offspring of HF mothers, both in males and females, that were attenuated with age. Within the hormones that modulate AMPK-dependent metabolic control, there is insulin, leptin, ghrelin, estrogens, etc. Therefore, hormonal maturation is a crucial step that will change AMPK pathway activation in the control of food intake and energy metabolism. Moreover, the distinct hormonal maturation between sexes will make AMPK modulation different in males and females due to the roles of sex-specific hormones that were previously described [54]. Given its multifaceted role in various pathways, further investigations are warranted to pinpoint the specific pathways activated by AMPK and its specific role in the contexts of brain function [49,50] and the transgenerationality of dysmetabolism.

When the inflammatory markers were analyzed, we showed a general trend where the females exhibited lower levels of these markers compared to the males. This distinction is notably apparent from the age of 28 days onward, coinciding with the onset of hormone production in females, which may act as a protective mechanism against external stressors, thereby mitigating the risk of elevated inflammation [57]. Note also that in contrast, males,

26. Won, E.; Kim, Y.-K. Neuroinflammation-Associated Alterations of the Brain as Potential Neural Biomarkers in Anxiety Disorders. *Int. J. Mol. Sci.* **2020**, *21*, 6546. [CrossRef] [PubMed]
27. Neniskyte, U.; Vilalta, A.; Brown, G.C. Tumour necrosis factor alpha-induced neuronal loss is mediated by microglial phagocytosis. *FEBS Lett.* **2014**, *588*, 2952–2956. [CrossRef]
28. McGregor, B.A.; Schommer, J.; Guo, K.; Raihan, M.O.; Ghribi, O.; Hur, J.; Porter, J.E. Corrigendum to 'Alpha-Synuclein-induced DNA Methylation and Gene Expression in Microglia. *Neuroscience* **2021**, *468*, 186–198. [CrossRef]
29. Boberg, J.; Li, T.; Christiansen, S.; Draskau, M.K.; Damdimopoulou, P.; Svingen, T.; Johansson, H.K.L. Comparison of female rat reproductive effects of pubertal versus adult exposure to known endocrine disruptors. *Front. Endocrinol.* **2023**, *14*, 1126485. [CrossRef]
30. Sanchez-Garrido, M.A.; Ruiz-Pino, F.; Pozo-Salas, A.I.; Castellano, J.M.; Vazquez, M.J.; Luque, R.M.; Tena-Sempere, M. Early overnutrition sensitizes the growth hormone axis to the impact of diet-induced obesity via sex-divergent mechanisms. *Sci. Rep.* **2020**, *10*, 13898. [CrossRef]
31. Şanlı, E.; Kabaran, S. Maternal Obesity, Maternal Overnutrition and Fetal Programming: Effects of Epigenetic Mechanisms on the Development of Metabolic Disorders. *Curr. Genom.* **2019**, *20*, 419–427. [CrossRef] [PubMed]
32. Bianco, M.E.; Josefson, J.L. Hyperglycemia During Pregnancy and Long-Term Offspring Outcomes. *Curr. Diabetes Rep.* **2019**, *19*, 143. [CrossRef] [PubMed]
33. Rosen, E.D.; Kaestner, K.H.; Natarajan, R.; Patti, M.-E.; Sallari, R.; Sander, M.; Susztak, K. Epigenetics and Epigenomics: Implications for Diabetes and Obesity. *Diabetes* **2018**, *67*, 1923–1931. [CrossRef] [PubMed]
34. Shou, J.; Chen, P.-J.; Xiao, W.-H. Mechanism of increased risk of insulin resistance in aging skeletal muscle. *Diabetol. Metab. Syndr.* **2020**, *12*, 14. [CrossRef]
35. Nicholas, L.M.; Nagao, M.; Kusinski, L.C.; Fernandez-Twinn, D.S.; Eliasson, L.; Ozanne, S.E. Exposure to maternal obesity programs sex differences in pancreatic islets of the offspring in mice. *Diabetologia* **2020**, *63*, 324–337. [CrossRef]
36. Eitmann, S.; Németh, D.; Hegyi, P.; Szakács, Z.; Garami, A.; Balaskó, M.; Solymár, M.; Erőss, B.; Kovács, E.; Pétervári, E. Maternal overnutrition impairs offspring's insulin sensitivity: A systematic review and meta-analysis. *Matern. Child Nutr.* **2020**, *16*, e13031. [CrossRef]
37. Melo, B.F.; Sacramento, J.F.; Capucho, A.M.; Sampaio-Pires, D.; Prego, C.S.; Conde, S.V. Long-Term Hypercaloric Diet Consumption Exacerbates Age-Induced Dysmetabolism and Carotid Body Dysfunction: Beneficial Effects of CSN Denervation. *Front. Physiol.* **2022**, *13*, 889660. [CrossRef]
38. Desai, M.; Han, G.; Mossayebi, E.; Beall, M.H.; Ross, M.G. 552: Programmed insulin resistance of offspring of obese mothers. *Am. J. Obstet. Gynecol.* **2017**, *216*, S325–S326. [CrossRef]
39. Mingrone, G.; Manco, M.; Mora, M.E.V.; Guidone, C.; Iaconelli, A.; Gniuli, D.; Leccesi, L.; Chiellini, C.; Ghirlanda, G. Influence of Maternal Obesity on Insulin Sensitivity and Secretion in Offspring. *Diabetes Care* **2008**, *31*, 1872–1876. [CrossRef]
40. van den Beld, A.W.; Kaufman, J.-M.; Zillikens, M.C.; Lamberts, S.W.J.; Egan, J.M.; Van Der Lely, A.J. The physiology of endocrine systems with ageing. *Lancet Diabetes Endocrinol.* **2018**, *6*, 647–658. [CrossRef]
41. Kawasaki, M.; Arata, N.; Miyazaki, C.; Mori, R.; Kikuchi, T.; Ogawa, Y.; Ota, E. Obesity and abnormal glucose tolerance in offspring of diabetic mothers: A systematic review and meta-analysis. *PLoS ONE* **2018**, *13*, e0190676. [CrossRef] [PubMed]
42. Xiang, A.; Carter, S.A.; Lin, J.C.; Chow, T.; Martinez, M.P.; Alves, J.M.; Page, K.A.; Mcconnell, R.; Negriff, S. 251-OR: Maternal Diabetes during Pregnancy and Incidence of Depression and Anxiety in Offspring from Childhood to Young Adulthood. *Diabetes* **2022**, *71* (Suppl. S1), 251-OR. [CrossRef]
43. Aslani, S.; Vieira, N.; Marques, F.; Costa, P.S.; Sousa, N.; Palha, J.A. The effect of high-fat diet on rat's mood, feeding behavior and response to stress. *Transl. Psychiatry* **2015**, *5*, e684. [CrossRef] [PubMed]
44. Daniels, R.W.; Miller, B.R.; DiAntonio, A. Increased vesicular glutamate transporter expression causes excitotoxic neurodegeneration. *Neurobiol. Dis.* **2011**, *41*, 415–420. [CrossRef] [PubMed]
45. Sohrabji, F. Guarding the blood–brain barrier: A role for estrogen in the etiology of neurodegenerative disease. *Gene Expr.* **2006**, *13*, 311–319. [CrossRef] [PubMed]
46. Glendining, K.A.; Fisher, L.C.; Jasoni, C.L. Maternal high fat diet alters offspring epigenetic regulators, amygdala glutamatergic profile and anxiety. *Psychoneuroendocrinology* **2018**, *96*, 132–141. [CrossRef] [PubMed]
47. Erbas, O.; Erdogan, M.A.; Khalilnezhad, A.; Gürkan, F.T.; Yiğittürk, G.; Meral, A.; Taskiran, D. Neurobehavioral effects of long-term maternal fructose intake in rat offspring. *Int. J. Dev. Neurosci.* **2018**, *69*, 68–79. [CrossRef]
48. Qi, Y.; Lee, N.J.; Ip, C.K.; Enriquez, R.; Tasan, R.; Zhang, L.; Herzog, H. Agrp-negative arcuate NPY neurons drive feeding under positive energy balance via altering leptin responsiveness in POMC neurons. *Cell Metab.* **2023**, *35*, 979–995.e7. [CrossRef]
49. Qu, N.; He, Y.; Wang, C.; Xu, P.; Yang, Y.; Cai, X.; Liu, H.; Yu, K.; Pei, Z.; Hyseni, I.; et al. A POMC-originated circuit regulates stress-induced hypophagia, depression, and anhedonia. *Mol. Psychiatry* **2020**, *25*, 1006–1021. [CrossRef]
50. Greenman, Y.; Kuperman, Y.; Drori, Y.; Asa, S.L.; Navon, I.; Forkosh, O.; Gil, S.; Stern, N.; Chen, A. Postnatal Ablation of POMC Neurons Induces an Obese Phenotype Characterized by Decreased Food Intake and Enhanced Anxiety-like Behavior. *Mol. Endocrinol.* **2013**, *27*, 1091–1102. [CrossRef]
51. Reichmann, F.; Holzer, P. Neuropeptide Y: A stressful review. *Neuropeptides* **2016**, *55*, 99–109. [CrossRef] [PubMed]
52. Ono, H. Molecular Mechanisms of Hypothalamic Insulin Resistance. *Int. J. Mol. Sci.* **2019**, *20*, 1317. [CrossRef] [PubMed]

53. Esler, M.; Rumantir, M.; Wiesner, G.; Kaye, D.; Hastings, J.; Lambert, G. Sympathetic nervous system and insulin resistance: From obesity to diabetes. *Am. J. Hypertens.* **2001**, *14*, S304–S309. [CrossRef] [PubMed]
54. Wang, B.; Cheng, K.K.-Y. Hypothalamic AMPK as a Mediator of Hormonal Regulation of Energy Balance. *Int. J. Mol. Sci.* **2018**, *19*, 3552. [CrossRef] [PubMed]
55. Kola, B. Role of AMP-Activated Protein Kinase in the Control of Appetite. *J. Neuroendocr.* **2008**, *20*, 942–951. [CrossRef] [PubMed]
56. Mihaylova, M.M.; Shaw, R.J. The AMPK signalling pathway coordinates cell growth, autophagy and metabolism. *Nature* **2011**, *13*, 1016–1023. [CrossRef] [PubMed]
57. Liu, Y.-Z.; Wang, Y.-X.; Jiang, C.-L. Inflammation: The Common Pathway of Stress-Related Diseases. *Front. Hum. Neurosci.* **2017**, *11*, 316. [CrossRef] [PubMed]
58. Burch, K.E.; McCracken, K.; Buck, D.J.; Davis, R.L.; Sloan, D.K.; Curtis, K.S. Relationship Between Circulating Metabolic Hormones and Their Central Receptors During Ovariectomy-Induced Weight Gain in Rats. *Front. Physiol.* **2022**, *12*, 800266. [CrossRef]
59. Fagundes, C.P.; Bennett, J.M.; Derry, H.M.; Kiecolt-Glaser, J.K. Relationships and Inflammation across the Lifespan: Social Developmental Pathways to Disease. *Soc. Personal. Psychol. Compass.* **2011**, *5*, 891–903. [CrossRef]
60. Humberg, A.; Fortmann, I.; Siller, B.; Kopp, M.V.; Herting, E.; Göpel, W.; Härtel, C. Preterm birth and sustained inflammation: Consequences for the neonate Network, German Center for Lung Research and Priming Immunity at the beginning of life (PRIMAL) Consortium. *Semin. Immunopathol.* **2020**, *42*, 451–468. [CrossRef]
61. Schultz, C.; Temming, P.; Bucsky, P.; Göpel, W.; Strunk, T.; Härtel, C. Immature anti-inflammatory response in neonates. *Clin. Exp. Immunol.* **2003**, *135*, 130–136. [CrossRef] [PubMed]
62. Sasaki, A.; de Vega, W.; Sivanathan, S.; St-Cyr, S.; McGowan, P. Maternal high-fat diet alters anxiety behavior and glucocorticoid signaling in adolescent offspring. *Neuroscience* **2014**, *272*, 92–101. [CrossRef] [PubMed]
63. Martin, H.; Mounsey, R.; Sathe, K.; Mustafa, S.; Nelson, M.; Evans, R.; Teismann, P. A peroxisome proliferator-activated receptor-δ agonist provides neuroprotection in the 1-methyl-4-phenyl-1,2,3,6-tetrahydropyridine model of Parkinson's disease. *Neuroscience* **2013**, *240*, 191–203. [CrossRef] [PubMed]

nutrients

MDPI

Article

Salt-Intake-Related Behavior Varies between Sexes and Is Strongly Associated with Daily Salt Consumption in Obese Patients at High Risk for MASLD

Bianca Heller [1], Florian P. Reiter [1], Hans Benno Leicht [1], Cornelia Fiessler [2], Ina Bergheim [3], Peter U. Heuschmann [2], Andreas Geier [1] and Monika Rau [1,*]

1 Division of Hepatology, Department of Internal Medicine II, University Hospital Würzburg, 97080 Würzburg, Germany
2 Institute of Clinical Epidemiology and Biometry, Julius-Maximilians-University of Würzburg, 97080 Würzburg, Germany
3 Department of Nutritional Sciences, Molecular Nutritional Science, University of Vienna, 1040 Vienna, Austria
* Correspondence: rau_m@ukw.de

Abstract: Background: Metabolic dysfunction-associated steatotic liver disease (MASLD) imposes a significant burden on Westernized regions. The Western diet, high in salt intake, significantly contributes to disease development. However, there are a lack of data on salt literacy and salt intake among MASLD patients in Germany. Our study aims to analyze daily salt intake and salt-intake-related behavior in MASLD patients. Methods: 234 MASLD patients were prospectively included. Daily salt intake and salt-intake-related behavior were assessed via a food frequency questionnaire (FFQ—DEGS) and a salt questionnaire (SINU). Statistical analyses were performed using SPSS. Results: Mean daily salt intake was higher in men than in women (7.3 ± 5 g/d vs. 5.3 ± 4 g/d; $p < 0.001$). There was significant agreement between increased daily salt intake (>6 g/d) and the behavioral salt index (SI) ($p < 0.001$). Men exhibited higher SI scores compared to women, indicating lower awareness of salt in everyday life. Multivariate analysis identified specific salt-intake-related behaviors impacting daily salt consumption. Conclusions: Our study reveals a strong link between daily salt intake and salt-intake-related behavior, highlighting sex-specific differences in an MASLD cohort. To enhance patient care in high-cardiovascular-risk populations, specific behavioral approaches may be considered, including salt awareness, to improve adherence to lifestyle changes, particularly in male patients.

Keywords: MASLD; steatotic liver disease; salt consumption; salt-intake-related behavior

1. Introduction

Western-type diets (WDs) with a high consumption of calorically rich, (ultra-) processed foods combined with chronic overnutrition and a sedentary lifestyle evoke a state of chronic metabolic inflammation [1], which contributes to diseases, such as type 2 diabetes mellitus (T2DM) as well as metabolic dysfunction-associated steatotic liver disease (MASLD). Recently, a consensus statement on a new nomenclature for fatty liver disease was published, and the term nonalcoholic fatty liver disease (NAFLD) is now replaced by MASLD [2]. MASLD is the most frequent chronic liver disease, with a global prevalence in the general population of about 25%, with a significant increase expected in the next few years [3,4].

A WD is characterized by an excessive intake of highly processed foods, including red meat, sugary drinks, snacks, cakes, and biscuits [5]. The salt content in processed foods can be more than 100-times higher in comparison to similar home-made meals, and an estimated 75% of sodium intake comes from processed or restaurant-prepared food [6].

So far, the high salt content in the WD has been mainly studied in the pathogenesis of cardiovascular diseases [7]. Here, the deleterious effect of a high-salt diet (HSD) is driven by arterial hypertension and associated with increased morbidity and mortality on a global scale [7]. The consumption of major foods and nutrients across 195 countries was recently analyzed for the Global Burden of Disease Study [8]. In this study, the worldwide sodium consumption greatly exceeded the recommended threshold, reaching 6 g of sodium per day (equivalent to 15 g of salt (NaCl) per day and 86% greater than the optimal amount) [8]. Furthermore, this analysis showed that, in 2017, more than half of diet-related deaths were attributable to a high intake of sodium [8]. Regional variations were evident, with only the Caribbean and African regions showing sodium intake within the recommended levels [8].

The consumption of ultra-processed foods (UPFs) has increased worldwide in the last few years. In the European Prospective Investigation into Cancer and Nutrition (EPIC) study, the mean energy intake of highly processed foods varied between 61% (Spain) and 79% (Germany) [9]. The short-time effect of UPF consumption was studied in a cross-over trial with 20 volunteers, who received either ultra-processed or unprocessed diets for 14 days each [10]. Diets were matched for calories, sugar, fat, fiber, and macronutrients but provided ad libitum. The UPF group consumed an additional 500 kcal/day, and BMI increased accordingly compared to the unprocessed group. There are several hypotheses regarding the possible mechanistic links between UPF consumption and the incidence of chronic diseases [11]. One of these hypotheses describes the poorer nutritional quality of UPFs, as they are characterized by a higher content of saturated fat, added-sugar energy density, and salt, along with a lower fiber and vitamin content [11]. In a cross-sectional study of 789 volunteers, a high consumption of UPFs was further associated to metabolic syndrome and, in MASLD patients, with fibrosis [12]. In a prospective Chinese study cohort, the consumption of ultra-processed foods was associated with a higher risk for MASLD after adjusting for various variables including total energy intake [13]. Another prospective study in China revealed that a high daily salt intake was associated with a higher risk of developing MASLD in the future [14]. High salt intake is a risk factor for metabolic diseases, including obesity, hypertension, dyslipidemia, and T2DM [15,16]. In preclinical MASLD models, a high-salt diet decreased antioxidant defenses and increased inflammation and fibrosis in the liver of animals [17,18]. The DASH (Dietary Approaches to Stop Hypertension) diet is characterized, amongst other things, by a sodium reduction. This diet was analyzed in a randomized clinical trial over 8 weeks in MASLD patients and showed beneficial effects on serum markers and body weight [19]. Reductions in UPF consumption and salt intake could be a strategy for the prevention of obesity and associated diseases such as MASLD.

To date, no data exist about daily salt intake or the salt-intake-related behavior of MASLD patients in Germany. The term "food literacy" describes that knowledge, skills, and behaviors are required to be put into practice for healthy diets [20]. A survey on knowledge and salt-intake-related behavior in an Italian cohort showed that the study population had a decent level of knowledge about salt but less satisfactory behavior [21]. A further gender-specific difference is observed with higher regular sodium consumption in men compared to women [22]. The current first-line treatment for MASLD patients is lifestyle change, with weight reduction and the recommendation of a Mediterranean diet [23]. But, only a few patients show long-term adherence to these recommendations [24,25].

The main focus of this study was to prospectively evaluate salt-intake-related behavior in a well-characterized cohort of MASLD patients and analyze its association with their daily salt intake. Gender-specific differences in salt-related behavior were further analyzed, as regular salt consumption is reported to be higher in men. Specific salt related behavior could be further addressed in a personalized manner of nutrition recommendations for MASLD patients to facilitate long-term adherence.

2. Materials and Methods

2.1. Cohort Characteristics

In this prospective single-center study, 234 patients were included after providing informed consent from 04/2021–01/2022. The study protocol was implemented in accordance with Good Clinical Practice Guidelines and the Declaration of Helsinki. The study was reviewed and approved by local Ethics Committee (AZ-188/17-mk). All patients were >18 years old. MASLD was diagnosed either via controlled attenuation parameter (CAP) measurement or liver histology. Every MASLD patient included in the study had at least one cardiometabolic risk factor, as per the defined criteria [2]. Patients with significant alcohol consumption (women > 20 g/day and men > 30 g/day) and other concomitant chronic liver diseases were excluded. Thus, 107 patients had histologically diagnosed MASLD, and 15 patients underwent liver biopsy without histological diagnosis of MASLD but with a clinical diagnosis of MASLD. Further, 90 patients were included before undergoing bariatric surgery.

2.2. Food Frequency Questionnaire

To analyze the usual food consumption of the patient cohort, a self-administered and semi-quantitative food frequency questionnaire (FFQ) based on the German Health Examination Survey for Adults 2008–2011 (DEGS) with slight modifications was used [26]. Specifically, in addition to the frequency and portion size of 53 ordinary German food products, long-term eating behavior was assessed by extending the assessment time. Furthermore, three food products, including pretzels, rolls, or pretzels, canned cold fish, and smoked salmon, were added for more accurate salt detection. In addition, the FFQ was supplemented with the question of how often the patient prepares a meal from convenience foods. Seven questions of the DEGS-FFQ concerning low-fat products, deep-fried meals, or beverage consumption were removed as they did not fit the purpose of this study.

2.3. Daily Salt Intake

To calculate daily salt intake, the daily consumption of each food in grams per day was determined using a Healthy Eating Index developed specifically for the DEGS study (HEI-DEGS1) [27]. Average consumption of one food group was calculated on the basis of consumption frequency in relation to opportunities per 28 days and portion size of each product. The procedure for missing responses was adapted to that of the HEI-DEGS1: If both details for a question (frequency of consumption and portion size) were missing, this food was considered as "never consumed". If the portion size information (frequency of consumption available) was the only missing data, it was substituted with the average information of the corresponding food for both sexes. The salt content of the individual food products and dishes was taken from a food table. For this study, "Die Nährwerttabelle" was used [28]. It contains the most frequently consumed foods and dishes in Germany, based on consumption studies. Each salt content per 100 g of ready-to-eat food is listed in milligrams. To calculate the salt content of each consumed food product in grams per day, the following equation was used:

$$\text{daily salt intake by one food product [g/d]} = (\text{daily consumption of the food product [g/d]} \times \text{it's amount NaCl per 100 g} \times 10^{-3}\,[\text{g}])/100\,[\text{g}]$$

Beverages were excluded for this analysis as they do not account for a relevant contribution to salt consumption. To obtain the daily salt intake of each patient, the calculated salt amounts of the individual food products were added up.

2.4. Salt-Intake-Related behavior

Salt-intake-related behavior of MASLD patients was assessed by adapting questions from the Italian Salt Literacy Questionnaire of the SINU (Società Italiana di Nutrizione Umana) [21]. The questions are self-administered multiple-choice questions about behavior and knowledge related to salt intake. In this study, 14 SINU questions were used, from which a salt index (SI) was calculated. The number of response options was standardized to 3-point Likert scale for this study. Points were assigned from 0 to 2. All questions used and their scores are listed in Table A1. For the calculation of the SI, all scores of the individual salt behavior questions were added up. Finally, a score between 0 and 28 points could be achieved. The lower the SI, the higher the awareness of salt in the patient's everyday life. In addition to the SINU questions, the "Würzburg Salt Questionnaire" comprised 5 extra questions that evaluated the frequency of meal preparation using basic ingredients/fresh foods or convenience foods for each participant. Additionally, the questionnaire inquired about the presence of a saltshaker during meals and the frequency of restaurant or cafeteria visits before and during the COVID-19 pandemic. The corresponding response options and scores are shown in Table A2.

2.5. Statistical Analysis

To analyze the association between salt-related behavior and the daily salt consumption, the statistical software IBM® SPSS® Statistics, Version 28.0.0.0 (190) for Windows (IBM Corp.) was used. The Kruskal–Wallis test was performed to analyze differences between two or more groups of the salt behavior questions. Subsequently, pairwise comparisons were calculated using the Mann–Whitney-U-test. Finally, to find out which salt behavior questions have the greatest influence on daily salt intake, a multiple linear regression with backward selection was performed as the requirements for this multivariate analysis were met. Pearson's chi-square test and Fisher–Freeman–Halton exact test were used for the gender-specific analysis of the data. The listed results include medians with interquartile ranges (IQRs), mean values with standard deviations, frequencies, and p-values of asymptotic significances. A significance level of <0.05 was set. Deviating n-numbers in the results are due to missing items.

3. Results

3.1. Patient Characteristics

In this prospective study, 234 MASLD patients were included, 56.4% women and 43.6% men. The median age was 55 ± 17 years, and BMI was 32.8 ± 9.9 kg/m^2. MASLD patients had a high prevalence of arterial hypertension (58.6%), T2DM (37.7%), and dyslipidemia (23.0%); 107 patients had histologically confirmed MASLD (82 with MASH and 25 with MAFL); and 127 patients received a clinical diagnosis via CAP measurement. Median serum ALT was slightly elevated, with 36 ± 35 U/L. Further laboratory findings are depicted in Table 1, including the median CAP value of 320 ± 90 dB/m and stiffness of 6.0 ± 3.7 kPa.

Table 1. Patient characteristics.

Cohort	
Sex m/f	102/132
Age in years (median ± IQR)	55 ± 17
BMI in kg/m^2 (median ± IQR)	32.8 ± 9.9
Disesases	
Arterial Hypertension % (n)	58.6 (116)
Diabetes Mellitus Type II % (n)	37.7 (75)
Hyperlipidemia % (n)	23.0 (46)
Hypertriglyceridemia % (n)	5.5 (11)
Coronary Heart Disease % (n)	3.5 (7)
Clinical findings in MASLD patients	
Diagnosis histological/clinical	107/127
ALT in U/L (median ± IQR)	35.5 ± 35
AST in U/L (median ± IQR)	30.0 ± 18
GGT in U/L (median ± IQR)	44.1 ± 66
CAP in dB/m (median ± IQR)	320 ± 90
Stiffness in kPa (median ± IQR)	6.0 ± 3.7

3.2. Lower Salt Intake and Higher Salt Awareness

The calculated daily salt intake in this cohort of MASLD patients was 6.0 ± 3.8 g/d (median ± IQR). The daily salt intake was subsequently analyzed together with specific salt-related behavior. The mean salt index (SI) reflecting the salt-related behavior as the sum of the 14 single questions was 16 ± 3.5 points (median ± IQR) (range from 0 = high awareness to 28 = low awareness). To examine SI in relation to daily salt intake, the cohort was divided into two groups. The threshold between the two groups was selected based on the median daily salt intake in our cohort of 6 g/d. As shown in Figure 1, patients with lower daily salt intake also had significantly lower SI values ($p < 0.001$, *t*-test) and, thus, showed higher awareness of salt in daily life. Daily salt intake in the group of patients with ≤6 g/d ranged from 1.1 g/d to 6 g/d (mean ± SD: 4.2 ± 1.2 g/d), while in the group of patients with >6 g/d, it ranged from 6 g/d to 28 g/d (mean ± SD: 9.7 ± 4.3 g/d).

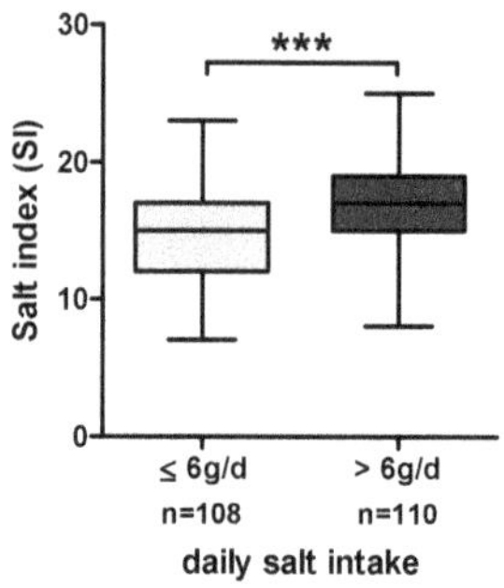

Figure 1. Boxplots of SI in the total cohort of MASLD patients stratified by daily salt intake (≤6 g/d and >6 g/d). *** = $p \leq 0.001$.

Given that daily salt intake can vary across one's lifetime, we conducted an analysis of daily salt intake in relation to age. No correlation was detected between daily salt intake and age (Spearman's Rho = −0.037, $p = 0.577$). However, a minor negative correlation between age and SI, signifying increasing salt awareness with age, was evident (Spearman's Rho = −0.189, $p = 0.005$). To investigate the age-dependent effects further, we divided the cohort into two groups based on the median age of 55 years within our cohort. Patients aged > 55 years exhibited a lower median SI of 16 compared to their younger counterparts (≤55 years, median SI = 17) (Appendix A Figure A1).

3.3. Specific Salt-Intake-Related Behavioral Patterns and Low Salt Intake

Further analysis of the single questions for the salt index (SI) addressing specific salt-intake-related behavior was performed. Figure 2 presents the mean ± SD of responses on the Likert scale for each question related to salt-intake-related behavior. A gender-specific analysis of these data is shown in Figure A2. To perform a statistical analysis, a Kruskal–Wallis test was used, comparing daily salt intake with the responses to the questions about salt-intake-related behavior. The following behavioral patterns were significantly associated with daily salt intake: considering salt information on nutrition tables ($p = 0.004$), no addition of salt during cooking ($p = 0.009$), using spices instead of salt ($p = 0.022$), checking salt content on nutrition tables ($p = 0.023$), avoiding eating out ($p = 0.018$), avoiding consumption of salty products ($p = 0.028$), and salty taste of food out of home ($p = 0.030$). In contrast, no significant association with daily salt intake was observed for the following behavioral questions: estimated daily salt intake ($p = 0.183$), salting during meal preparation ($p = 0.415$), addition of salt at the table ($p = 0.457$), buying lower-salt alternative products ($p = 0.233$), thirsting after meals ($p = 0.711$), using iodized salt ($p = 0.754$), and asking for low-salt alternatives ($p = 0.253$). The analysis of the "Würzburg Salt Questionnaire" is depicted in Figure 3, showing the mean ± SD of responses on the Likert scale for each question. To conduct the statistical analysis, a Kruskal–Wallis test was once again used to compare daily salt intake with the responses to the "Würzburg Salt Questionnaire". The statistical analysis revealed a significant association between daily salt intake and the consumption of convenience foods ($p = 0.001$). This association was also seen with eating away from home, e.g., in restaurants or cafeterias, before ($p = 0.001$) and during the COVID-19 pandemic ($p < 0.001$). There was no significant association with patients' salt intake when asked about the frequency of cooking with whole/fresh foods ($p = 0.638$) and when asked about the saltshaker at the table during a meal ($p = 0.437$). The summary of the results of the "Würzburg Salt Questionnaire" can be found in Table A2.

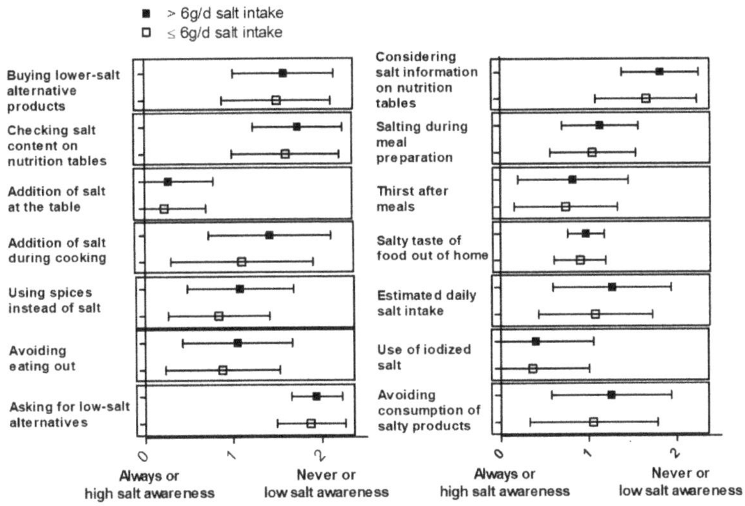

Figure 2. Mean ± SD are shown for each salt-intake-related behavior question for the group with high (>6 g/d, $n = 114$) and low ($\leq$6 g/d, $n = 120$) daily salt intake.

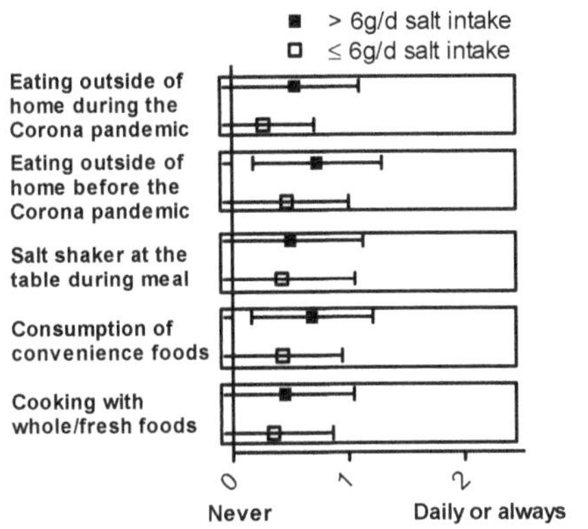

Figure 3. Mean ± SD are depicted for each item of the "Würzburg salt questionnaire" for the group with high (>6 g/d, $n = 114$) and low ($\leq$6 g/d, $n = 120$) daily salt intake.

3.4. *Multivariate Analysis—The Influence of Salt-Specific Salt-intake-related behavior on Salt Intake*

After the analysis of single salt questions in relation to daily salt intake, a multivariate model was created to determine which salt behavior questions have the greatest influence on salt intake. The salt-intake-related behavior questions and the "Würzburg salt questionnaire" were analyzed together. For our research question, a multiple linear regression model including 208 patients' data with backward selection was used, which was statistically significant ((F(6, 201) = 7.59, $p < 0.001$), with an R^2 of 0.185 (corrected $R^2 = 0.160$)). All 19 salt questions were included in the analysis, and after selection (pout(.10)), as appropriate, the final model contained six salt questions: consumption of convenience foods, considering salt information on nutrition tables, salty taste of food out of home, use of iodized salt, salting during meal preparation, and eating outside of home (restaurants or cafeterias) during the COVID-19 pandemic. Four questions represented significant factors influencing the prediction of daily salt consumption in grams per day: consumption of convenience foods (b = 0.867; CI 0.183–1.551; $p = 0.013$), considering salt information on nutrition tables (b = 0.977; CI 0.247–1.707; $p = 0.009$), salting during meal preparation (b = 0.504; CI 0.007–1.001; $p = 0.047$), and eating outside of home (restaurants or canteens) during the COVID-19 pandemic (b = 1.017; CI 0.298–1.735; $p = 0.006$). The questions about the salty taste of food out of home ($p = 0.084$) and use of iodized salt ($p = 0.097$) did not contribute significantly to the model (Table 2).

Table 2. Results of the multiple linear regression, and the influence of salt questions on salt consumption.

	Dependent Variable: Daily Salt Intake [g/d]							
Coefficients	b	SE	β	T	95% CI		*p*	S
					LL	UL		
(Constant)	1.837	0.972		1.890	−0.079	3.753	0.060	
Consumption of convenience foods	0.867	0.347	0.166	2.499	0.183	1.551	0.013	*
Considering salt information on nutrition tables	0.977	0.370	0.173	2.639	0.247	1.707	0.009	**
Salty taste of food out of home	1.228	0.707	0.115	1.736	−0.167	2.622	0.084	
Use of iodized salt	0.485	0.291	0.112	1.669	−0.088	1.059	0.097	
Salting during meal preparation	0.504	0.252	0.135	1.999	0.007	1.001	0.047	*
Eating outside of home (restaurants or canteens) during the COVID-19 pandemic	1.017	0.364	0.190	2.791	0.298	1.735	0.006	**

b = regression coefficient, SE = standard error, β = standardized coefficient, T = T-value, 95% KI = Confidence interval with lower and upper limit, *p* = significance value, S = significance for * $p \leq 0.05$, ** $p \leq 0.01$.

3.5. Gender-Specific Differences in Salt Consumption and Salt-Intake-Related Behavior

Our study focused on gender-specific differences in salt consumption and salt-intake-related behavior. In this cohort, slightly more women than men were included (56.4% women and 43.6% men). Women had significantly higher BMI compared to men (35.0 ± 12.1 kg/m^2 vs. 30.5 ± 8.2 kg/m^2, $p = 0.001$). Non-invasive liver stiffness measurements via Fibroscan and liver fat quantification via CAP were higher in men (men: 6.8 ± 4.0 kPa vs. women: 5.7 ± 4.2 kPa, $p = 0.019$; men: 329 ± 93 dB/m vs. women: 313 ± 100 dB/m vs., $p = 0.039$).

Furthermore, men had higher daily salt intake 7.3 ± 4.8 g/d compared to women (5.3 ± 3.5 g/d, ($p < 0.001$); Figure 4A). According to the daily salt intake, men showed higher SI scores compared to women, reflecting a lower awareness of salt-intake-related behavior in everyday life ($p < 0.001$) (Figure 4B). A comprehensive analysis of gender-specific salt-intake-related behavior in conjunction with daily salt intake revealed a significant difference, indicating higher SI (meaning lower salt awareness) in men in both the high (>6 g/d) and low (≤6 g/d) daily salt intake groups compared to women (Figure 4C). Only in women was the difference regarding SI between high and low daily salt intake groups statistically significant (Figure 4C).

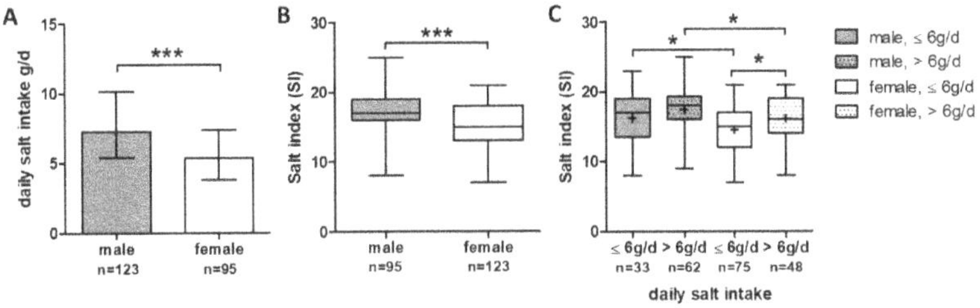

Figure 4. Gender-specific differences in daily salt intake and salt-intake-related behavior: (**A**) gender-specific daily salt intake shown as bars with median ± IQR. (**B**) Boxplots of salt index in men and women; (**C**) boxplots of salt index in men and women with low and high daily salt intake (≤6 g/d, >6 g/d). *** $p \leq 0.001$, * $p < 0.05$.

In our cohort, women showed different salt-intake-related behavior than men. Specifically, women reported that they read the salt content on nutrition tables more often ($p = 0.002$), used spices instead of salt ($p = 0.004$), and avoided eating out ($p = 0.002$), whereas men more frequently used salt in cooking ($p = 0.005$) or salt at the table ($p = 0.016$). Regarding the Würzburg salt questions, women more often reported that they cooked with whole/fresh foods ($p = 0.005$), whereas men showed a higher frequency in the preparation of convenience foods ($p < 0.001$) and the frequency of restaurant visits before ($p < 0.001$) and during ($p < 0.001$) the COVID-19 pandemic. No gender difference could be found when asked about the saltshaker at the table during a meal.

4. Discussion

The Western diet is characterized by its high salt content and the prevalence of highly processed foods. The consumption of such highly processed foods and adherence to a Western-type diet are significant risk factors contributing to the pathogenesis of MASLD. Growing evidence indicates that patients with MASLD are at substantial risk for the development of hypertension, coronary heart disease, and cardiac arrhythmias, which clinically result in increased cardiovascular morbidity and mortality [29]. High salt intake is associated with hypertension and risk for cardiovascular diseases, but its role in the prevention or treatment of MASLD is not known. As lifestyle changes are the primary therapeutic goal in MASLD patients and adherence to these goals is difficult to achieve, nutritional behavior has a major impact on the treatment of these patients.

In this study, daily salt intake together with salt-intake-related behavior were analyzed in a cohort of obese MASLD patients. Increased daily salt intake was linked to reduced awareness of salt consumption in everyday life. In a multivariate linear regression model, eating away from home, considering salt information on nutrition tables, consumption of convenience foods, and salting during meal preparation had the highest impact on daily salt intake. Sex-specific differences were observed, with higher daily salt intake in men compared to women, together with lower salt-intake-related behavior in men.

Daily salt intake with 6.0 g/d in our cohort (men 7.3 ± 4.8 g/d compared to women 5.3 ± 3.5 g/d) was lower than estimated in the general German Health Interview and Examination Survey for Adults (DEGS; first wave DEGS1 2008–2011), with 10.4 g/d in men and 9.2 g/d in women in the age group of 50–59 years [30]. In the DEGS study, daily salt intake exceeded 10 g/day for almost 50% of men and 40% of women [31]. However, it is important to note that the assessment of daily salt intake differed between the DEGS study and our cohort. In the DEGS study, salt intake was estimated using urinary sodium excretion (spot urine), and it was assumed that renal sodium excretion was equivalent to salt intake. In contrast, our study calculated sodium intake based on a comprehensive FFQ, which may have led to observed differences due to the potential under-reporting of salt intake within our patient cohort. These differences are in line with the significant lower estimated daily salt intake of 8.7 g/d for men and 6.3 g/d for women in a German national consumption study (Nationale Verzehrstudie II) based on diet history interviews or 24 h dietary recalls [32]. The daily salt intake observed within our MASLD patient cohort surpassed the WHO-recommended cut off (<5 g/day) [33]. The German Nutrition Society (Deutsche Gesellschaft für Ernährung e.V., DGE) offers an approximate value for salt intake in Germany of up to 6 g of table salt per day [30]. The median salt consumption identified in our study was 6 g/d, which is the upper limit of DGE's suggested salt intake, but men are still above this limit compared to women.

Specific salt-intake-related behaviors were found to correlate with the detected daily salt intake in our cohort. The individual salt index (SI) was calculated based on the behavioral questions. Men exhibited higher daily salt intake and SI scores compared to women, indicating a greater consumption of salt and lower awareness of salt in men. In the original study using the SINU salt questionnaire, women outperformed men in both knowledge and behavior categories related to salt, highlighting the gender-specific aspect of nutritional behavior [21]. When examining specific salt-intake-related behavioral questions, we observed a significant association with lower daily salt intake for the following aspects: considering salt information on nutrition tables, salting during meal preparation, using spices instead of salt, checking salt content on nutrition tables, avoiding eating out, avoiding consumption of salty foods, and salty taste of food out of home. In the Italian analysis of behavioral questions of the SINU salt questionnaire, the majority of participants reported reducing or avoiding the consumption of salty foods. Additionally, they mentioned refraining from adding salt while cooking or eating [21]. These statements align with the specific salt-intake-related behaviors observed in our cohort. Salt awareness was also assessed in another study using a 68-item, self-administered questionnaire in a population of 141 individuals in Switzerland, specifically among individuals in a workplace setting [34]. In a multiple linear regression model exploring the relationship between health literacy, food literacy, and the variable 'salt content impacting food/menu choice' with salt intake, only the variable 'salt content impacting food/menu choice' showed a significant association with salt intake, while health literacy and food literacy did not demonstrate significant associations. Participants indicating that salt content influences their decision to purchase/choose a food or menu item had a 1.5 g lower daily salt intake [34]. In our cohort, 23 percent of patients reported that they paid attention to the salt content on nutrition tables, and this statement was also found to be significant in the multivariate analysis of this study. In a Spanish cross-sectional analysis among children and their parents, "Checking sodium content on food labels and the use of table salt by the children or father" was associated with a lower preference for salty foods. In this study, the majority of families (59%) stated that they never look at the sodium content on food labels [35]. Awareness of salt content on food and nutrition labels appears to be a crucial indicator of high salt-intake-related behavior, but it is only practiced by a minority of the analyzed populations.

In our cohort, we observed sex-specific differences not only in daily salt intake but also in salt-intake-related behavior, with men showing higher daily salt intake and lower salt awareness compared to women. Women are reported to have a 9–11% lower salt intake worldwide, which is suggested to be associated with their lower calorie intake [36]. In NHANES data, an age–gender interaction for adding table salt suggests that younger women (<30 years) tend to add salt more frequently than men, but after age 30, men more often add salt compared to women [36]. In our cohort, we also observed an age-dependent effect on salt-related behavior, indicating higher salt awareness with increasing age. However, this effect was not reflected in daily salt intake. In addition to the awareness of salt-intake-related behavior, an Australian survey identified another factor contributing to the emergence of the gender difference. This survey, which included 530 participants, revealed that female participants had lower levels of salt taste beliefs, and these beliefs had the strongest positive association with salt use [37]. Sex-specific differences in salt-intake-related behavior are observed worldwide and seem to be associated with age and salt tase beliefs.

There are certain limitations in our study that should be acknowledged. The daily salt intake was calculated using a detailed FFQ, but no measurements of sodium in 24 h urine collections were taken. As a result, intake-related bias must be considered when interpreting and comparing our data. Our analysis is based on a cohort of MASLD patients without a comparative analysis of a matched cohort of healthy obese people without steatotic liver disease. Consequently, the presented findings concentrate on variations in salt-related behavior and gender-specific distinctions within the MASLD patient cohort, without drawing comparisons to healthy controls. Additionally, our patient cohort originates from a tertiary center with specific patient selection. Larger multicentric cohort studies are suggested to validate these findings.

WHO Member States have agreed to reduce the global population's intake of salt by a relative 30% by 2025, as reducing salt intake has been identified as one of the most cost-effective measures countries can take to improve population health outcomes [33]. The WHO's country support package for the European region aims to accelerate salt reduction efforts and encompasses specific objectives, including raising population awareness about salt intake, empowering the public with actionable knowledge, and providing targeted assistance for reducing salt consumption [38]. To attain these objectives, it is essential to explore personalized and gender-specific nutritional approaches that can effectively drive behavioral changes in patients.

To conclude, our study demonstrates a robust association between daily salt intake and salt-intake-related behavior, highlighting sex-specific differences within an MASLD cohort of obese patients. Sex-specific behavioral nutrition patterns should be validated and analyzed in future studies to be considered for individualized nutritional treatment, which remains the first-line approach for managing MASLD patients. Given that cardiovascular diseases are the leading cause of mortality within this patient population, it is crucial to utilize all available strategies to promote adherence to lifestyle modifications. Therefore, it is essential to explore various avenues for improving patient care, including the incorporation of tailored behavioral approaches to enhance adherence to these lifestyle changes.

Author Contributions: Conceptualization, B.H. and M.R.; data curation, F.P.R., H.B.L., A.G. and M.R.; methodology, P.U.H. and M.R.; formal analysis, B.H., C.F. and M.R.; investigation, B.H. and M.R.; visualization, B.H. and M.R.; supervision, I.B. and A.G.; resources A.G.; project administration M.R.; writing—original draft preparation, B.H. and M.R.; writing—review and editing, all authors. All authors have read and agreed to the published version of the manuscript.

Funding: B.H. received a scholarship by the Graduate School of Life Sciences of the University of Würzburg.

Institutional Review Board Statement: The study was conducted in accordance with the Declaration of Helsinki, and approved by the Ethics Committee of the University of Würzburg (AZ-188/17-mk and date of approval: 10/05/2019).

Informed Consent Statement: Informed consent was obtained from all subjects involved in the study.

Data Availability Statement: The data presented in this study are available on request from the corresponding author. The data are not publicly available due to confidentiality reasons.

Conflicts of Interest: The authors declare no conflict of interest.

Appendix A

Table A1. Salt-related behavior questions, their scores, and the frequencies of the answer options. Deviating *n*-numbers in the results are due to missing items.

Salt-Intake-Related Behavior Questions	**Score**	*N* (%)	*N* (%)
Daily salt intake		≤6 g/d	>6 g/d
Does the information about salt quantity reported on nutrition tables affect your food choices when grocery shopping?		*N* = 120	*N* = 113
- Never	2	85 (70.8)	94 (83.2)
- Sometimes	1	29 (24.2)	17 (15.0)
- Always	0	6 (5.0)	2 (1.8)
When you cook/eat a meal, you add salt:		*N* = 119	*N* = 114
- both while cooking and at the table	2	17 (14.3)	19 (16.7)
- only during preparation/only at the table	1	91 (76.5)	91 (79.8)
- neither during preparation nor at the table	0	11 (9.2)	4 (3.5)
Are you thirsty after meals?		*N* = 120	*N* = 114
- Always	2	9 (7.5)	14 (12.3)
- Sometimes	1	71 (59.2)	66 (57.9)
- Never	0	40 (33.3)	34 (29.8)
When eating away from home to me food tastes:		*N* = 117	*N* = 112
- Flavorless	2	0 (0)	1 (0.9)
- Normal	1	106 (90.6)	107 (95.5)
- Salty	0	11 (9.4)	4 (3.6)
In conclusion, in your opinion, how much is your daily salt intake? Taking into account the foods you buy and the ones you personally cook (including the salt added during preparation and/or after cooking)?		*N* = 118	*N* = 113
- more than 20 g - between 10 and 20 g	2	29 (24.6)	44 (38.9)
- between 5 and 10 g	1	69 (58.5)	55 (48.7)
- less than 5 g	0	20 (16.9)	14 (12.4)
Do you usually use iodized salt for seasoning or cooking?		*N* = 119	*N* = 114
- Never	2	11 (9.2)	11 (9.6)
- Sometimes	1	21 (17.7)	23 (20.2)
- Always	0	87 (73.1)	80 (70.2)
I avoid eating, or I reduce the consumption of products rich in salt		*N* = 119	*N* = 114
- Never	2	34 (28.6)	44 (38.6)
- Sometimes	1	57 (47.9)	55 (48.2)
- Always	0	28 (23.5)	15 (13.2)
I buy alternative products, with a low salt content		*N* = 118	*N* = 114
- Never	2	65 (55.1)	69 (60.5)
- Sometimes	1	46 (39.0)	41 (36.0)
- Always	0	7 (5.9)	4 (3.5)

Table A1. *Cont.*

Salt-Intake-Related Behavior Questions	Score	*N* (%)	*N* (%)
I read the salt content on the nutrition tables		*N* = 120	*N* = 114
- Never	2	78 (65.0)	86 (75.4)
- Sometimes	1	35 (29.2)	25 (21.9)
- Always	0	7 (5.8)	3 (2.6)
Salt at the table		*N* = 116	*N* = 113
- I always add salt at the table.	2	2 (1.7)	3 (2.7)
- I add salt at the table sometimes.	1	23 (19.8)	25 (22.1)
- I add no, or very little salt at the table.	0	91 (78.4)	85 (75.2)
Salt while cooking		*N* = 118	*N* = 114
- I always add salt while cooking.	2	44 (37.3)	60 (52.6)
- I add salt while cooking sometimes.	1	42 (35.6)	41 (36.0)
- I don't add salt while cooking (or only very little)	0	32 (27.1)	13 (11.4)
I use spices instead of salt in cuisine		*N* = 118	*N* = 113
- Never	2	11 (9.3)	25 (22.1)
- Sometimes	1	77 (65.6)	72 (63.7)
- Always	0	30 (25.4)	16 (14.2)
I avoid going out to eat out too often		*N* = 118	*N* = 114
- Never	2	18 (15.3)	23 (20.2)
- Sometimes	1	68 (57.6)	69 (60.5)
- Always	0	32 (27.1)	22 (19.3)
If I eat out, I ask for low-salt options		*N* = 117	*N* = 113
- Never	2	104 (88.9)	105 (92.9)
- Sometimes	1	11 (9.4)	7 (6.2)
- Always	0	2 (1.7)	1 (0.9)

Table A2. "Würzburg Salt Questionnaire", their scores, and the frequencies of the answer options. Deviating *n*-numbers in the results are due to missing items.

Würzburg Salt Questionnaire	Score	*N* (%)	*N* (%)
Daily salt intake		≤6 g/d	>6 g/d
How often a week do you prepare a hot meal (lunch or dinner) yourself from basic ingredients/fresh foods?		*N* = 120	*N* = 114
- Daily - 5–6 times per week	0	80 (66.7)	69 (60.5)
- 3–4 times per week - 1–2 times per week	1	38 (31.6)	39 (34.2)
- Never	2	2 (1.7)	6 (5.3)

Table A2. *Cont.*

Würzburg Salt Questionnaire	Score	*N* (%)	*N* (%)
How often a week do you prepare a hot meal (lunch or dinner) yourself from convenience foods?		*N* = 119	*N* = 114
- Daily - 5–6 times per week	2	1 (0.8)	3 (2.6)
- 3–4 times per week - 1–2 times per week	1	49 (41.2)	72 (63.2)
- Never	0	69 (58.0)	39 (34.2)
Do you have a saltshaker at your table when you eat a meal?		N = 120	N = 114
- Always	2	9 (7.5)	8 (7.0)
- Sometimes	1	33 (27.5)	41 (36.0)
- Never	0	78 (65.0)	65 (57.0)
How many times a week did you eat away from home (e.g., restaurant, canteen) before the COVID-19 pandemic?		N = 119	N = 113
- Daily - 5–6 times per week	2	2 (1.7)	6 (5.3)
- 3–4 times per week - 1–2 times per week	1	52 (43.7)	71 (62.8)
- Never	0	65 (54.6)	36 (31.9)
How many times a week now, during the COVID-19 pandemic, do you eat away from home (e.g., delivery/pickup, restaurant/canteen)?		N = 119	N = 113
- Daily - 5–6 times per week	2	0 (0)	2 (1.8)
- 3–4 times per week - 1–2 times per week	1	32 (26.9)	56 (49.5)
- Never	0	87 (73.1)	55 (48.7)

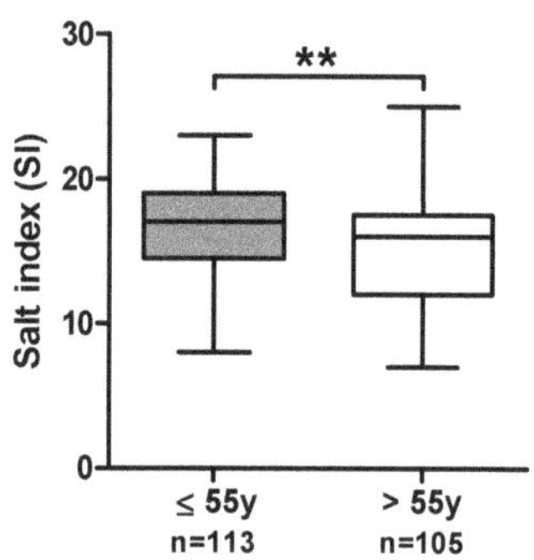

Figure A1. Boxplots of SI in the total cohort of MASLD patients stratified by age (≤55 years and >55 years). ** = $p < 0.01$.

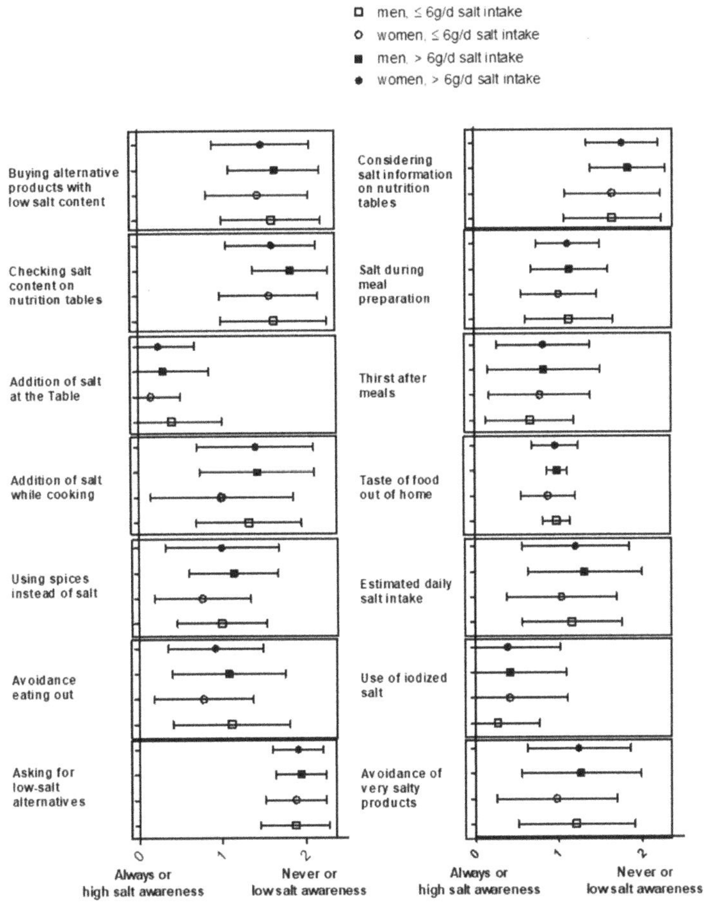

Figure A2. Gender-specific analysis of salt-related behavior questions. Mean ± SD are shown for each salt-intake-related behavior question for the group with high (>6 g/d, n = 114) and low (≤6 g/d, n = 120) daily salt intake and further separated by gender.

References

1. Christ, A.; Lauterbach, M.; Latz, E. Western Diet and the Immune System: An Inflammatory Connection. *Immunity* **2019**, *51*, 794–811. [CrossRef]
2. Rinella, M.E.; Lazarus, J.V.; Ratziu, V.; Francque, S.M.; Sanyal, A.J.; Kanwal, F.; Romero, D.; Abdelmalek, M.F.; Anstee, Q.M.; Arab, J.P.; et al. A multi-society Delphi consensus statement on new fatty liver disease nomenclature. *J. Hepatol.* **2023**. [CrossRef]
3. Estes, C.; Anstee, Q.M.; Arias-Loste, M.T.; Bantel, H.; Bellentani, S.; Caballeria, J.; Colombo, M.; Craxi, A.; Crespo, J.; Day, C.P.; et al. Modeling NAFLD disease burden in China, France, Germany, Italy, Japan, Spain, United Kingdom, and United States for the period 2016–2030. *J. Hepatol.* **2018**, *69*, 896–904. [CrossRef]
4. Younossi, Z.M.; Koenig, A.B.; Abdelatif, D.; Fazel, Y.; Henry, L.; Wymer, M. Global epidemiology of nonalcoholic fatty liver disease-Meta-analytic assessment of prevalence, incidence, and outcomes. *Hepatology* **2016**, *64*, 73–84. [CrossRef]
5. Berná, G.; Romero-Gomez, M. The role of nutrition in non-alcoholic fatty liver disease: Pathophysiology and management. *Liver Int.* **2020**, *40* (Suppl. 1), 102–108. [CrossRef]
6. Brown, I.J.; Tzoulaki, I.; Candeias, V.; Elliott, P. Salt intakes around the world: Implications for public health. *Int. J. Epidemiol.* **2009**, *38*, 791–813. [CrossRef]
7. Mozaffarian, D.; Fahimi, S.; Singh, G.M.; Micha, R.; Khatibzadeh, S.; Engell, R.E.; Lim, S.; Danaei, G.; Ezzati, M.; Powles, J. Global sodium consumption and death from cardiovascular causes. *N. Engl. J. Med.* **2014**, *371*, 624–634. [CrossRef]
8. GBD 2017 Diet Collaborators. Health effects of dietary risks in 195 countries, 1990–2017: A systematic analysis for the Global Burden of Disease Study 2017. *Lancet* **2019**, *393*, 1958–1972. [CrossRef]

9. Slimani, N.; Deharveng, G.; Southgate, D.A.; Biessy, C.; Chajès, V.; van Bakel, M.M.; Boutron-Ruault, M.C.; McTaggart, A.; Grioni, S.; Verkaik-Kloosterman, J.; et al. Contribution of highly industrially processed foods to the nutrient intakes and patterns of middle-aged populations in the European Prospective Investigation into Cancer and Nutrition study. *Eur. J. Clin. Nutr.* **2009**, *63* (Suppl. 4), S206–S225. [CrossRef]
10. Hall, K.D.; Ayuketah, A.; Brychta, R.; Cai, H.; Cassimatis, T.; Chen, K.Y.; Chung, S.T.; Costa, E.; Courville, A.; Darcey, V.; et al. Ultra-Processed Diets Cause Excess Calorie Intake and Weight Gain: An Inpatient Randomized Controlled Trial of Ad Libitum Food Intake. *Cell Metab.* **2020**, *32*, 690. [CrossRef]
11. Srour, B.; Kordahi, M.C.; Bonazzi, E.; Deschasaux-Tanguy, M.; Touvier, M.; Chassaing, B. Ultra-processed foods and human health: From epidemiological evidence to mechanistic insights. *Lancet Gastroenterol. Hepatol.* **2022**, *7*, 1128–1140. [CrossRef] [PubMed]
12. Ivancovsky-Wajcman, D.; Fliss-Isakov, N.; Webb, M.; Bentov, I.; Shibolet, O.; Kariv, R.; Zelber-Sagi, S. Ultra-processed food is associated with features of metabolic syndrome and non-alcoholic fatty liver disease. *Liver Int.* **2021**, *41*, 2635–2645. [CrossRef]
13. Zhang, S.; Gan, S.; Zhang, Q.; Liu, L.; Meng, G.; Yao, Z.; Wu, H.; Gu, Y.; Wang, Y.; Zhang, T.; et al. Ultra-processed food consumption and the risk of non-alcoholic fatty liver disease in the Tianjin Chronic Low-grade Systemic Inflammation and Health Cohort Study. *Int. J. Epidemiol.* **2022**, *51*, 237–249. [CrossRef] [PubMed]
14. Shen, X.; Jin, C.; Wu, Y.; Zhang, Y.; Wang, X.; Huang, W.; Li, J.; Wu, S.; Gao, X. Prospective study of perceived dietary salt intake and the risk of non-alcoholic fatty liver disease. *J. Hum. Nutr. Diet.* **2019**, *32*, 802–809. [CrossRef] [PubMed]
15. Takahashi, F.; Hashimoto, Y.; Kaji, A.; Sakai, R.; Kawate, Y.; Okamura, T.; Kitagawa, N.; Okada, H.; Nakanishi, N.; Majima, S.; et al. The Association of Salt Intake and Non-alcoholic Fatty Liver Disease in People with Type 2 Diabetes: A Cross-Sectional Study. *Front. Nutr.* **2022**, *9*, 943790. [CrossRef]
16. Yi, S.S.; Kansagra, S.M. Associations of sodium intake with obesity, body mass index, waist circumference, and weight. *Am. J. Prev. Med.* **2014**, *46*, e53–e55. [CrossRef]
17. Dornas, W.C.; de Lima, W.G.; dos Santos, R.C.; Guerra, J.F.; de Souza, M.O.; Silva, M.; Souza e Silva, L.; Diniz, M.F.; Silva, M.E. High dietary salt decreases antioxidant defenses in the liver of fructose-fed insulin-resistant rats. *J. Nutr. Biochem.* **2013**, *24*, 2016–2022. [CrossRef]
18. Uetake, Y.; Ikeda, H.; Irie, R.; Tejima, K.; Matsui, H.; Ogura, S.; Wang, H.; Mu, S.; Hirohama, D.; Ando, K.; et al. High-salt in addition to high-fat diet may enhance inflammation and fibrosis in liver steatosis induced by oxidative stress and dyslipidemia in mice. *Lipids Health Dis.* **2015**, *14*, 6. [CrossRef]
19. Razavi Zade, M.; Telkabadi, M.H.; Bahmani, F.; Salehi, B.; Farshbaf, S.; Asemi, Z. The effects of DASH diet on weight loss and metabolic status in adults with non-alcoholic fatty liver disease: A randomized clinical trial. *Liver Int.* **2016**, *36*, 563–571. [CrossRef]
20. Vidgen, H.A.; Gallegos, D. Defining food literacy and its components. *Appetite* **2014**, *76*, 50–59. [CrossRef]
21. Iaccarino Idelson, P.; D'Elia, L.; Cairella, G.; Sabino, P.; Scalfi, L.; Fabbri, A.; Galletti, F.; Garbagnati, F.; Lionetti, L.; Paolella, G.; et al. Salt and Health: Survey on Knowledge and Salt Intake Related Behaviour in Italy. *Nutrients* **2020**, *12*, 279. [CrossRef]
22. Jackson, S.L.; King, S.M.; Zhao, L.; Cogswell, M.E. Prevalence of Excess Sodium Intake in the United States—NHANES, 2009–2012. *MMWR Morb. Mortal. Wkly. Rep.* **2016**, *64*, 1393–1397. [CrossRef] [PubMed]
23. European Association for the Study of the Liver (EASL); European Association for the Study of Diabetes (EASD); European Association for the Study of Obesity (EASO). EASL-EASD-EASO Clinical Practice Guidelines for the management of non-alcoholic fatty liver disease. *J. Hepatol.* **2016**, *64*, 1388–1402. [CrossRef]
24. Promrat, K.; Kleiner, D.E.; Niemeier, H.M.; Jackvony, E.; Kearns, M.; Wands, J.R.; Fava, J.L.; Wing, R.R. Randomized controlled trial testing the effects of weight loss on nonalcoholic steatohepatitis. *Hepatology* **2010**, *51*, 121–129. [CrossRef] [PubMed]
25. Tsofliou, F.; Vlachos, D.; Hughes, C.; Appleton, K.M. Barriers and Facilitators Associated with the Adoption of and Adherence to a Mediterranean Style Diet in Adults: A Systematic Review of Published Observational and Qualitative Studies. *Nutrients* **2022**, *14*, 314. [CrossRef] [PubMed]
26. Haftenberger, M.; Heuer, T.; Heidemann, C.; Kube, F.; Krems, C.; Mensink, G.B. Relative validation of a food frequency questionnaire for national health and nutrition monitoring. *Nutr. J.* **2010**, *9*, 36. [CrossRef] [PubMed]
27. Kuhn, D.-A. *Entwicklung eines Index zur Bewertung der Ernährungsqualität in der Studie zur Gesundheit Erwachsener in Deutschland (DEGS1)*; Robert Koch-Institut: Berlin, Germany, 2018.
28. Heseker, H.; Heseker, B. *Die Nährwerttabelle*; Neuer Umschau Buchverlag: Neustadt, Germany, 2019; Volume 6, p. 144.
29. Kasper, P.; Martin, A.; Lang, S.; Kütting, F.; Goeser, T.; Demir, M.; Steffen, H.M. NAFLD and cardiovascular diseases: A clinical review. *Clin. Res. Cardiol.* **2021**, *110*, 921–937. [CrossRef]
30. Strohm, D.B.H.; Leschik-Bonnet, E.; Heseker, H.; Arens-Aze-vêdo, U.; Bechthold, A.; Knorpp, L.; Kroke, A. for the German Nutrition Society (DGE). Salt intake in Germany, health consequences, and resulting recommendations for action. A scientific statement from the German Nutrition Society (DGE). *Ernahrungs Umschau* **2016**, *63*, 62–70.
31. Johner, S.A.; Thamm, M.; Schmitz, R.; Remer, T. Current daily salt intake in Germany: Biomarker-based analysis of the representative DEGS study. *Eur. J. Nutr.* **2015**, *54*, 1109–1115. [CrossRef]
32. Max Rubner-Institut. *Ergebnisbericht Teil 2, Nationale Verzehrstudie II*; Max Rubner-Institut: Karlsruhe, Germany, 2008.
33. WHO. Available online: https://www.who.int/news-room/fact-sheets/detail/salt-reduction (accessed on 29 July 2023).

34. Luta, X.; Hayoz, S.; Gréa Krause, C.; Sommerhalder, K.; Roos, E.; Strazzullo, P.; Beer-Borst, S. The relationship of health/food literacy and salt awareness to daily sodium and potassium intake among a workplace population in Switzerland. *Nutr. Metab. Cardiovasc. Dis.* **2018**, *28*, 270–277. [CrossRef]
35. Cuadrado-Soto, E.; Peral-Suarez, Á.; Rodríguez-Rodríguez, E.; Aparicio, A.; Andrés, P.; Ortega, R.M.; López-Sobaler, A.M. The association of parents' behaviors related to salt with 24 h urinary sodium excretion of their children: A Spanish cross-sectional study. *PLoS ONE* **2019**, *14*, e0227035. [CrossRef] [PubMed]
36. Santollo, J.; Daniels, D.; Leshem, M.; Schulkin, J. Sex Differences in Salt Appetite: Perspectives from Animal Models and Human Studies. *Nutrients* **2023**, *15*, 208. [CrossRef] [PubMed]
37. Sarmugam, R.; Worsley, A.; Wang, W. An examination of the mediating role of salt knowledge and beliefs on the relationship between socio-demographic factors and discretionary salt use: A cross-sectional study. *Int. J. Behav. Nutr. Phys. Act.* **2013**, *10*, 25. [CrossRef] [PubMed]
38. WHO. Accelerating Salt Reduction in Europe. A Country Support Package to Reduce Population Salt Intake in the WHO European Region. 2020. Available online: https://apps.who.int/iris/handle/10665/340028 (accessed on 31 August 2023).

nutrients

MDPI

Article

The Microbiome, Epigenome, and Diet in Adults with Obesity during Behavioral Weight Loss

Emily B. Hill [1,†], Iain R. Konigsberg [2,†], Diana Ir [3], Daniel N. Frank [3], Purevsuren Jambal [1], Elizabeth M. Litkowski [2,4,5], Ethan M. Lange [2,6], Leslie A. Lange [2,5], Danielle M. Ostendorf [7,8], Jared J. Scorsone [8], Liza Wayland [7,8], Kristen Bing [8], Paul S. MacLean [7], Edward L. Melanson [7,9], Daniel H. Bessesen [7], Victoria A. Catenacci [7,8], Maggie A. Stanislawski [2,*,‡] and Sarah J. Borengasser [1,*,‡]

1 Section of Nutrition, Department of Pediatrics, University of Colorado Anschutz Medical Campus, Aurora, CO 80045, USA; emily.b.hill@cuanschutz.edu (E.B.H.)
2 Department of Biomedical Informatics, University of Colorado Anschutz Medical Campus, Aurora, CO 80045, USA; iain.konigsberg@cuanschutz.edu (I.R.K.)
3 Division of Infectious Diseases, Department of Medicine, University of Colorado Anschutz Medical Campus, Aurora, CO 80045, USA
4 Department of Epidemiology, University of Colorado School of Public Health, Aurora, CO 80045, USA
5 Eastern Colorado Veterans Affairs Geriatric Research, Education, and Clinical Center, Aurora, CO 80045, USA
6 Department of Biostatistics and Informatics, University of Colorado School of Public Health, Aurora, CO 80045, USA
7 Division of Endocrinology, Metabolism, and Diabetes, Department of Medicine, University of Colorado Anschutz Medical Campus, Aurora, CO 80045, USA
8 Anschutz Health and Wellness Center, Department of Medicine, University of Colorado Anschutz Medical Campus, Aurora, CO 80045, USA
9 Division of Geriatric Medicine, Department of Medicine, University of Colorado Anschutz Medical Campus, Aurora, CO 80045, USA
* Correspondence: maggie.stanislawski@cuanschutz.edu (M.A.S.); sarah.borengasser@cuanschutz.edu (S.J.B.)
† These authors contributed equally to the manuscript and share first authorship.
‡ These authors contributed equally to the manuscript and share senior authorship.

Abstract: Obesity has been linked to the gut microbiome, epigenome, and diet, yet these factors have not been studied together during obesity treatment. Our objective was to evaluate associations among gut microbiota (MB), DNA methylation (DNAme), and diet prior to and during a behavioral weight loss intervention. Adults (n = 47, age 40.9 ± 9.7 years, body mass index (BMI) 33.5 ± 4.5 kg/m^2, 77% female) with data collected at baseline (BL) and 3 months (3 m) were included. Fecal MB was assessed via 16S sequencing and whole blood DNAme via the Infinium EPIC array. Food group and nutrient intakes and Healthy Eating Index (HEI) scores were calculated from 7-day diet records. Linear models were used to test for the effect of taxa relative abundance on DNAme and diet cross-sectionally at each time point, adjusting for confounders and a false discovery rate of 5%. Mean weight loss was 6.2 ± 3.9% at 3 m. At BL, one MB taxon, *Ruminiclostridium*, was associated with DNAme of the genes *COL20A1* (r = 0.651, p = 0.029), *COL18A1* (r = 0.578, p = 0.044), and *NT5E* (r = 0.365, p = 0.043). At 3 m, there were 14 unique MB:DNAme associations, such as *Akkermansia* with DNAme of *GUSB* (r = −0.585, p = 0.003), *CRYL1* (r = −0.419, p = 0.007), *C9* (r = −0.439, p = 0.019), and *GMDS* (r = −0.559, p = 0.046). Among taxa associated with DNAme, no significant relationships were seen with dietary intakes of relevant nutrients, food groups, or HEI scores. Our findings indicate that microbes linked to mucin degradation, short-chain fatty acid production, and body weight are associated with DNAme of phenotypically relevant genes. These relationships offer an initial understanding of the possible routes by which alterations in gut MB may influence metabolism during weight loss.

Keywords: DNA methylation; epigenetics; gut microbiome; diet; lifestyle; obesity

1. Introduction

Modifiable factors, such as diet and physical activity patterns, influence both prevention and treatment of obesity and its sequalae. However, underlying mechanisms by which lifestyle affects weight status are complex. Potential etiologic pathways in obesity development may include changes in gene expression due to environmentally induced epigenetic modifications that lead to reprogramming of endocrine or metabolic regulator circuits [1,2]. DNA methylation (DNAme) profiles also change in response to lifestyle interventions, suggesting that alterations in nutrition or exercise may favorably modulate gene activity to improve phenotypes during obesity treatment [3]. In addition, the gut microbiota (MB) is involved in several physiological functions that maintain metabolic homeostasis, including mechanisms involving energy balance, inflammation, and appetite regulation [4]. Compositional and functional changes in the MB through environmental or lifestyle exposures, such as diet may, therefore, impact the development of obesity and cardiometabolic disease [5–9]. There is evidence to suggest that the gut MB may contribute towards weight loss, indicating that the gut MB may serve as useful therapeutic target for obesity management [10,11].

While the gut MB and epigenome may independently act to influence weight and other indicators of health in response to lifestyle, little is known about the interplay between them [12]. It is possible that microbes may themselves contribute to weight regulation through direct interaction with host cells to influence metabolism or through microbial signaling that influences other metabolic factors [13]. It has also been proposed that circulating metabolites may not only lead to beneficial metabolic effects but could also provide a link between the gut MB and epigenetic changes [14–17]. However, this relationship is poorly understood, with few primary research articles investigating the relationship between these two measures and limited mechanistic understanding of microbiota-host epigenome interactions [15,16].

Though previous literature suggests both the gut MB and epigenome are responsive to changes in lifestyle, these factors have not been considered together in the context of a behavioral weight loss intervention. Thus, the primary objective of the present study was to evaluate associations among the gut MB and DNAme prior to and during a behavioral weight loss intervention, with a secondary objective to evaluate the relationship between these measures, as well as a targeted plasma metabolomics panel and diet. We hypothesized that DNAme of genes within metabolically relevant pathways would be associated with MB composition and that these taxa would be responsive to changes in dietary intakes within 3 months of initiating a behavioral weight loss intervention. Evaluation of these relationships will aid in hypothesis generation for follow-up studies investigating how the gut MB influences metabolism during weight loss.

2. Materials and Methods

2.1. Participants and Study Design

Participants in this ancillary study were healthy individuals with overweight or obesity (age 18–55 years, BMI 27–45 kg/m^2) enrolled in a randomized controlled trial (NCT03411356) comparing weight loss generated by daily caloric restriction (DCR) or intermittent fasting (IMF) during a one-year behavioral weight loss intervention. All participants were provided comprehensive behavioral support, encouraged to reduce calorie intake by ~34% per week from baseline estimated energy requirements, and increase physical activity to 300 min of moderate-to-vigorous activity per week. Stool samples, whole blood, clinical, and dietary intake data were collected at in-person assessments, and individuals from both arms of the first two of five cohorts of enrolled participants (starting April 2018 through February 2019) who had complete biospecimen collection at baseline and an early intervention time point of 3 months and agreed to optional ancillary studies were included in the present study. All study procedures were reviewed and approved by the Colorado Multiple Institutional Review Board, and individuals provided written, informed consent. Additional details regarding study procedures have been previously published [11,18,19]. Forty-seven participants provided viable stool and blood samples after data quality control (Figure 1).

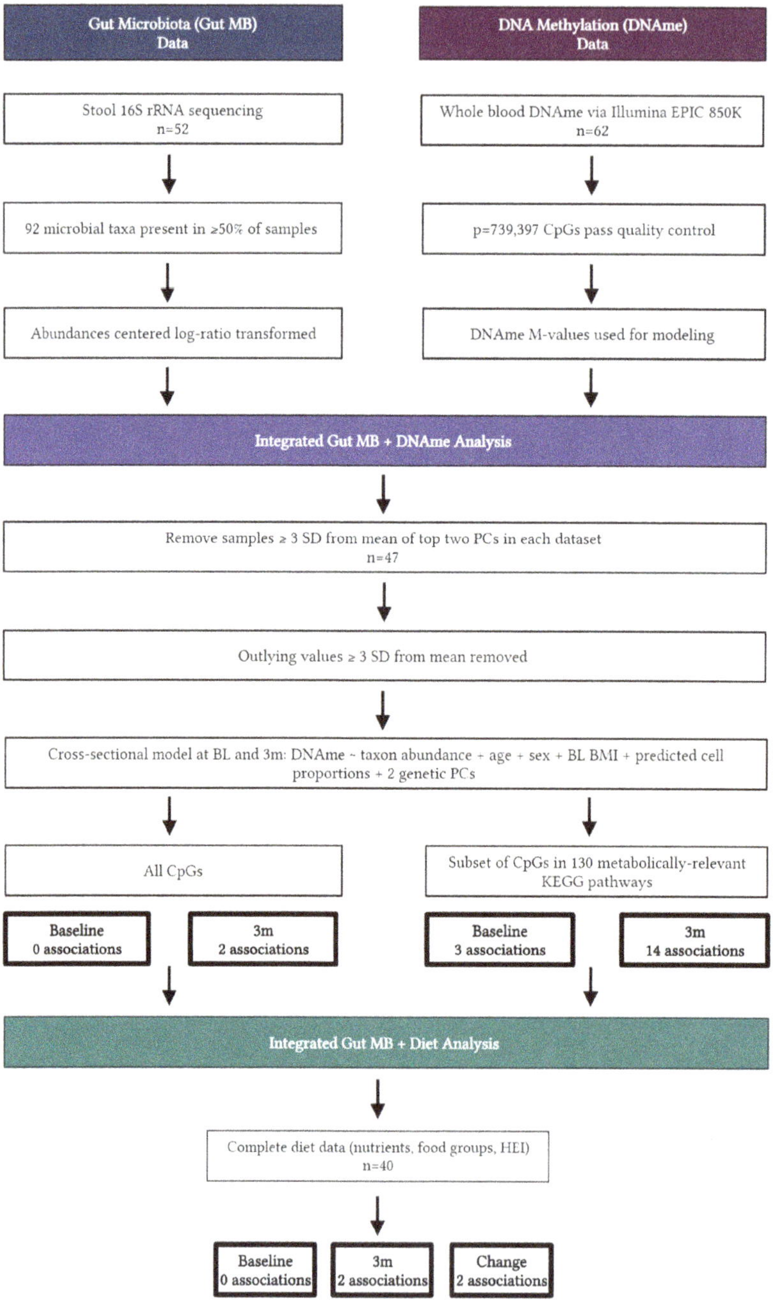

Figure 1. Flow diagram displaying data analysis for n = 47 individuals, created with BioRender.com.

2.2. Data Collection

2.2.1. Clinical Assessments and Anthropometrics

Individuals attended assessment visits at baseline and 3 months. Prior to visits, participants were asked to fast for at least 12 h. Anthropometry measures, including weight and waist circumference, clinical assessments, including systolic blood pressure and diastolic blood pressure, blood collection and processing, and cardiometabolic assays were completed according to standard protocols by trained research personnel at the University of Colorado Anschutz Health and Wellness Center (AHWC) and the University of Colorado Clinical and Translational Research Center (CTRC). Weight was obtained using a digital scale accurate to 0.1 kg, and height measured with a fixed stadiometer to the nearest 1 mm at baseline only. Waist circumference was measured in duplicate to the nearest 1 mm using a tape measure at the top of the iliac crest, and the average was recorded. Blood pressure was measured using a manual sphygmomanometer using the average of two seated values taken after five minutes of rest. Participants self-collected stool at home using the Alpco EasySampler® Stool Collection Kit (ALPCO Diagnostics, Salem, NH, USA) and provided samples to study staff at clinic visits. Briefly, ~1–2 g of stool were collected, sterilely transferred to provided collection tubes, and stored at −20 °C in home freezers prior to transfer to the clinic on ice packs within one week of collection at baseline and 3 months. Whole blood and stool samples were stored at −80 °C until subsequent analysis for DNAme and gut MB sequencing, respectively.

2.2.2. DNA Methylation and Data Pre-Processing

Genomic DNA was isolated from whole blood samples using the commercially available PureLink Genomic DNA Kit (Invitrogen, Carlsbad, CA, USA). Quality was assessed via a NanoDrop 2000/2000c spectrophotometer (ThermoFisher Scientific, Grand Island, NY, USA) and concentrations were measured by a Qubit Fluorometer 2.0 (Invitrogen, Carlsbad, CA, USA). DNAme was assessed via the Illumina Infinium Human Methylation EPIC BeadChip Array (EPIC 850K) at the University of Colorado Anschutz Medical Campus Genomics and Microarray Core. DNA samples were bisulfite converted using the EZ DNA Methylation Kit (Zymo Research, Irvine, CA, USA) and used as inputs for the EPIC 850K array. Positive and negative controls from each conversion assay were included, and data were visualized using GenomeStudio Software (Illumina, Inc., San Diego, CA, USA).

Array data were pre-processed and normalized using the SeSAMe R package [20]. After removing cross-hybridizing probes, probes overlying SNPs with minor allele frequency >1% in the general population [21], and probes that failed across multiple samples with an average detection p-value > 0.05, a total of 739,397 cytosine–phosphate–guanine dinucleotide (CpG) probes remained for analysis. Cell type frequencies (percentage of CD8 T cells, CD4 T cells, NK cells, B cells, monocytes, and neutrophils) were estimated using estimateCellCounts with the IlluminaHumanMethylationEPIC reference panel in the minfi R package [22].

All CpGs from the 850K array that passed quality control were included in subsequent analyses. In addition, a sensitivity analysis using only CpGs within genes mapped to 130 Kyoto Encyclopedia of Genes and Genomes (KEGG) pathways involved in metabolism (e.g., lipid and carbohydrate metabolism), organismal systems (e.g., digestive, endocrine, and immune systems), and human diseases (e.g., metabolic disease, cardiovascular disease) and, therefore, deemed highly relevant to overweight and obesity, was also completed (Table S1). This resulted in 82,889 CpG probes included in the sensitivity analyses. Array fluorescent intensities were transformed into M-values for statistical analysis [23].

2.2.3. Microbiome Sequencing and Pre-Processing

Stool samples were thawed from −80 °C on ice and homogenized using the Roche MagNA Lyser (Roche, Inc., Basel, Switzerland) and DNA was extracted using the QiaAmp PowerFecal DNA kit (Qiagen, Hilden, Germany). To assess bacterial composition, amplification and sequence analysis of 16S rRNA genes was conducted as previously described [24,25]. Amplicons were generated using primers targeting base pairs within the V3V4 variable region of the 16S rRNA gene. PCR products were normalized using a SequalPrep™ kit (Invitrogen, Carlsbad, CA, USA), pooled, lyophilized, and then purified and concentrated using the DNA Clean and Concentrator Kit (Zymo, Irvine, CA, USA). The Qubit Fluorometer 2.0 was used to quantify pooled amplicons prior to sequencing. In brief, the amplicon pool was diluted to 4 nM and denatured using 0.2 N NaOH at room temperature. Denatured DNA was diluted to 15 pM and spiked with 25% Illumina PhiX control DNA prior to performing paired-end sequencing using the MiSeq platform (version 2.4) with a 600-cycle version 3 reagent kit.

All reads were quality filtered and trimmed to a uniform length using the average position of the first low-quality base pair among all samples using Qiime2 software version 2019.10 [26]. The data were de-noised using DADA2 under default parameters [27], and quality-filtered sequences were inserted into the SILVA 12.8 taxonomic database [28] using the SATé-enabled phylogenetic placement (SEPP) technique [29]. Analyses were standardized at 3407 sequences per sample. Microbe abundances were center log-ratio transformed, and after quality control to remove taxa present in <50% of samples, 92 genus-level taxa were retained for analysis.

2.2.4. Dietary Intake Assessment and Data Pre-Processing

Participants completed 7-day diet records at baseline and 3 months. Individuals were asked to record all foods and beverages consumed each day, including detail on brand names, preparation, cooking methods, and portion sizes in household measurements. Diet records were analyzed by trained registered dietitian nutritionists (RDNs) at the Colorado Clinical and Translational Sciences Institute (CCTSI) Nutrition Core using the Nutrition Data System for Research (NDSR) nutrient analysis software versions 2018 and 2019 to align with year of data collection, developed by the Nutrition Coordinating Center (NCC) at the University of Minnesota, Minneapolis, MD, USA.

Daily intake of energy, macronutrients, fiber, micronutrients, and NCC food group serving counts were exported from NDSR to calculate mean intakes over seven days at each time point. Individual food subgroups (e.g., citrus juice, fruit excluding citrus fruit) were further collapsed into food groups (e.g., fruit) based upon NCC food group serving count subgroup IDs. Nutrient and food group data were used to calculate Healthy Eating Index (HEI) total and component scores using all seven days of intake for each time point using published methods for the simple HEI scoring algorithm and publicly available SAS code [30]. Briefly, the HEI measures adherence to the US Dietary Guidelines for Americans, and total scores represent energy adjusted intakes of key food groups for adequacy (9 components) and moderation (4 components). Total scores range from 0–100, with higher scores reflective of greater adherence to dietary guidelines and higher diet quality [30].

For subsequent modeling, dietary variables were retained only for those with nonzero values among >50% participants at both baseline and 3 months. This resulted in 169 nutrients, 29 individual food subgroups, 20 collapsed food groups, and all measures of diet quality (HEI total and 13 HEI component scores) for analysis. Dietary variables included in the analyses are presented in Table S2.

2.2.5. Plasma Targeted Metabolomics

Plasma samples obtained at clinic visits were assessed for a targeted panel of betaine, choline, carnitine, and trimethylamine oxide (TMAO) using liquid chromatography/mass spectroscopy (LC/MS) by the Mayo Clinic Metabolomics Core using previously published methods [31,32]. Plasma was spiked with D9-isotopes of internal standards prior to a cold methanol crash to remove proteins. The supernatant was dried and resuspended in running buffer for separation using a Grace Altima HP HILIC column (150 mm × 2.1 mm × 5 μm) on a Cohesive TX2 LC system (Franklin, MA). Samples were then introduced into a Sciex 6500 triple quadrupole mass spectrometer (Framingham, MA, USA) via electrospray ionization. Data acquisition was performed using selective ion monitoring mode, and concentrations of each analyte were determined using an 8-point standard curve for each respective analyte.

2.3. Statistical Analysis

Changes in indicators of health from baseline to 3 months were assessed via paired *t*-tests. Due to heteroscedasticity, laboratory measures were log-transformed prior to analysis. A two-step analysis process was used to assess associations between gut MB, DNAme, and dietary intakes (Figure 1). First, we tested the effect of MB taxa on DNAme cross-sectionally at baseline and 3 months. After removing samples with values ≥3 standard deviations from the mean for each feature, a linear model was fitted to test for the effect of individual MB taxa abundance on DNAme of all CpGs passing quality control while adjusting for age, sex, baseline BMI, predicted cell proportions, and two genetic principal components. This same analysis was completed for a subset of CpGs present in genes within metabolically relevant KEGG pathways as described above. Next, a linear model was fitted to test for the association between significant taxa from MB:DNAme analyses and dietary intake for each type of diet data separately (nutrients, food groups, diet quality) cross-sectionally at baseline and 3 months, as well as longitudinally over this time period. DNAme probes, MB taxa, and diet variables identified as significant in these analyses were further assessed for associations with plasma metabolites cross-sectionally at baseline and 3 months while adjusting for relevant covariates. All analyses were adjusted for multiple testing using the Benjamini–Hochberg false discovery rate (FDR) with a threshold for significance of 0.05 [33].

3. Results

3.1. Participant Demographic and Clinical Characteristics

Participants in this study (n = 47) were primarily non-Hispanic White females with a mean age of 40.9 ± 9.7 years (Table 1). At baseline, mean BMI was 33.5 ± 4.5 kg/m^2, and individuals achieved significant weight loss on average, with the mean BMI decreasing by 2.1 ± 1.4 kg/m^2 and waist circumference by nearly 9 cm (−8.5 ± 6.0 cm) at 3 months (both $p < 0.001$). Similar improvements were noted for plasma indicators of cardiometabolic health, including significant reductions in total cholesterol, triglycerides, glucose, and insulin from baseline to 3 months.

Table 1. Participant demographic characteristics at baseline and 3 months (n = 47).

Characteristic		Baseline Mean ± SD or % (n)	3 Months Mean ± SD or % (n)	Change (p Value)
Age, years		40.9 ± 9.7	-	-
Sex	Male	23 (11)	-	-
	Female	77 (36)	-	-
Race	White	89 (42)	-	-
	Black	6 (3)	-	-
	Multiracial	4 (2)	-	-
Ethnicity	Hispanic	9 (19)	-	-
	Non-Hispanic	81 (38)	-	-
Income	<$25,000 USD	11 (5)	-	-
	$25,000–$45,000 USD	4 (2)	-	-
	$45,001–$70,000 USD	23 (11)	-	-
	$70,001–$110,000 USD	25 (12)	-	-
	>$110,000 USD	36 (17)	-	-
Education	Some college	11 (5)	-	-
	Four-year degree	45 (21)	-	-
	Master's degree	34 (16)	-	-
	Doctorate degree	11 (5)	-	-

Table 1. *Cont.*

Characteristic	Baseline Mean ± SD or % (n)	3 Months Mean ± SD or % (n)	Change (p Value)
Weight, kg	96.1 ± 16.1	90.2 ± 15.3	−6.0 ± 3.9 (<0.001)
Body mass index, kg/m^2	33.5 ± 4.5	31.5 ± 4.3	−2.1 ± 1.4 (<0.001)
Waist circumference, cm	109.4 ± 10.3	100.9 ± 10.5	−8.5 ± 6.0 (<0.001)
Systolic blood pressure, mmHg	117 ± 14	114 ± 12	−3 ± 12 (0.104)
Diastolic blood pressure, mmHg	74 ± 8	76 ± 9	2 ± 12 (0.173)
Total cholesterol, mg/dL [a]	180 ± 34	165 ± 30	−15 ± 26 (<0.001)
High-density lipoprotein (HDL) cholesterol, mg/dL [a]	48 ± 12	47 ± 12	−1 ± 6 (0.288)
Triglycerides, mg/dL [a]	136 ± 79	107 ± 56	−29 ± 61 (0.002)
Glucose, mg/dL [a]	93 ± 11	88 ± 8	−5 ± 11 (0.002)
Insulin, uIU/mL [a]	12 ± 8	7 ± 5	−4 ± 6 (<0.001)

[a] log-transformed prior to analysis, n = 45 due to missing data.

3.2. Integrated Microbiome and DNAme Analysis

Integrated gut MB and DNAme analyses showed several significant cross-sectional associations (Table S3). In the full model containing all CpGs, there were no associations between gut MB and DNAme at baseline. At 3 months, there were two significant associations between the gut MB and DNAme (Figure 2A). These included a moderate inverse association between the abundance of *Lachnospiraceae NK4A136* and DNAme within an intron in Ornithine Transcarbamylase (*OTC*) and a strong positive association between the abundance of *Megasphaera* and DNAme in the promoter region of CCAAT/enhancer-binding protein delta (*CEBPD*).

In the reduced model using only CpGs in metabolically relevant KEGG pathways, three cross-sectional associations were noted at baseline (Figure 2B). Abundance of *Ruminiclostridium* was moderately positively associated with DNAme in the promoter region of 5′-Nucleotidase Ecto (*NT5E*), a gene involved in nicotinate and nicotinamide metabolism, as well as DNAme within intron regions of two collagen proteins involved in protein digestion and absorption, namely Collagen Type XVIII Alpha 1 Chain (*COL18A1*), and Collagen Type XX Alpha 1 Chain (*COL20A1*). Scatterplots displaying the underlying relationships for these three associations are presented in Figure S1. At 3 months, 14 unique MB:DNAme cross-sectional associations were observed (Figure 2B). Among these, abundance of *Akkermansia* was moderately inversely associated with DNAme in four genes: GDP-Mannose 4,6-Dehydratase (*GMDS*), complement 9 (*C9*), Crystallin Lambda 1 (*CRYL1*), and Glucuronidase Beta (*GUSB*). Similarly, *Lachnospiraceae UCG-001* was moderately inversely associated with DNAme in four genes: Nuclear Receptor Subfamily 5 Group A Member 2 (*NR5A2*), Hes Family BHLH Transcription Factor 1 (*HES1*), Piccolo Presynaptic Cytomatrix Protein (*PCLO*), and Lecithin Retinol Acyltransferase (*LRAT*). In the reduced model, *Ruminococcus gnavus* was the only MB taxon positively associated with DNAme at 3 months, demonstrating a significant association between abundance and DNAme in the gene Carbonic Anhydrase 3 (*CA3*). Scatterplots displaying the underlying relationships for these associations are presented in Figure S2.

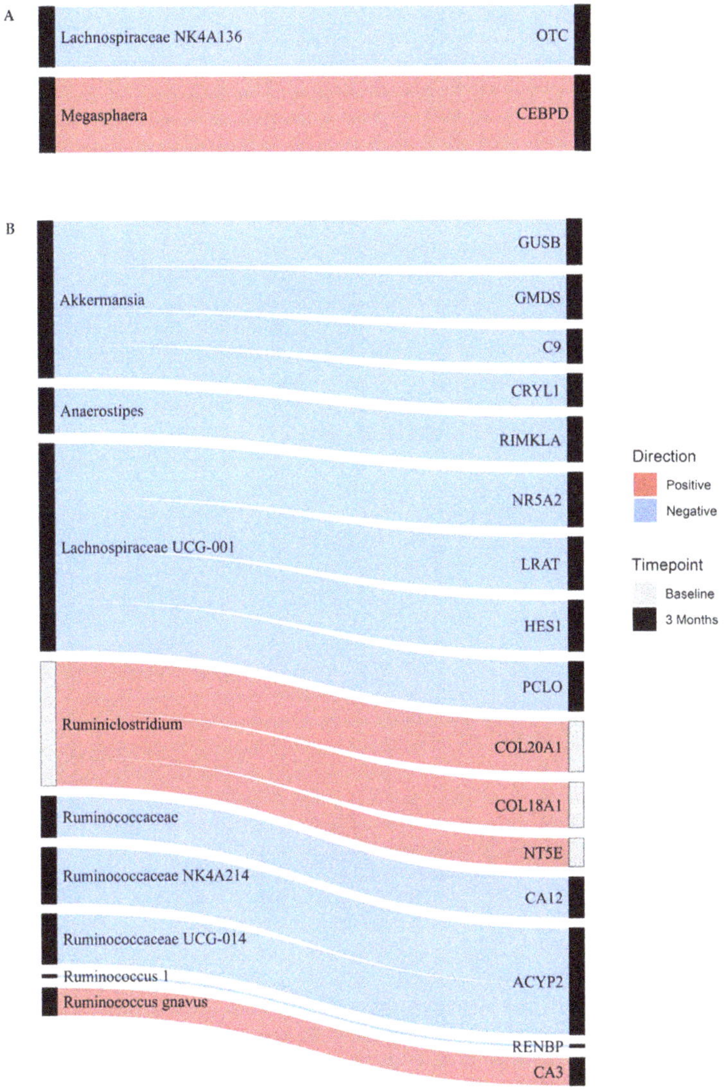

Figure 2. Cross-sectional associations between DNAme and gut microbiota. The Sankey plot displays associations between gut microbiota and DNAme of CpGs within labeled genes from (**A**) full and (**B**) reduced models at baseline (gray nodes) and 3 months (black nodes). Blue indicates an inverse association, whereas red indicates a positive association, while the thickness of each line indicates the strength of the association.

3.3. Integrated Dietary Analysis

Dietary intakes were assessed among those with complete biospecimen collections and corresponding gut MB and DNAme data who also provided 7-day diet records at both baseline and 3 months (n = 40). Participants self-reported decreasing energy intakes by nearly 500 kcal/day (-479 ± 445 kcal/day) from baseline to 3 months ($p < 0.001$, Table 2). The percentage of calories from fat decreased by nearly 4%, while intakes from protein increased by a similar amount. Overall diet quality, as assessed by HEI total score, was also improved at 3 months ($p = 0.022$).

Table 2. Participant dietary intakes at baseline and 3 months (n = 40).

Characteristic	Baseline Mean ± SD	3 Months Mean ± SD	Change (p Value)
Energy (kcal/day)	1764 ± 338	1284 ± 380	−479 ± 445 (<0.001)
Carbohydrate (% kcal)	42 ± 8	42 ± 7	0 ± 6 (0.797)
Fat (% kcal)	39 ± 7	35 ± 5	−4 ± 6 (<0.001)
Protein (% kcal)	17 ± 3	21 ± 4	4 ± 5 (<0.001)
Fiber (g/day)	16 ± 5	14 ± 6	−2 ± 6 (0.015)
Diet quality (total HEI score)	57 ± 12	62 ± 12	4 ± 12 (0.022)

Integrated analysis between the DNAme-associated MB taxa and dietary intake data identified no associations between components of diet quality as measured by HEI. However, at 3 months, *Ruminococcaceae NK4A214* was associated with food-group level intakes of reduced-fat margarines. In change analyses, *Ruminococcus gnavus* was associated with dietary intakes of the nutrient, trans-octadecenoic acid, as well as total trans fatty acids, both of which significantly decreased from baseline to 3 months.

3.4. Targeted Metabolomics Analysis

Plasma concentrations of betaine, choline, carnitine, and TMAO were not significantly associated with gut MB, DNAme, or dietary intakes at baseline or 3 months.

4. Discussion

This study is one of few primary investigations of the association between the gut microbiome and epigenome in humans and the first to evaluate these relationships within the context of a behavioral weight loss intervention. At this 3-month early intervention time point, most individuals had lost a clinically significant amount of weight, showed improvements in several measures of cardiometabolic health, and had improved overall diet quality. Our results suggest several cross-sectional associations between the gut MB and DNAme in metabolically relevant pathways, providing evidence that gut MB composition correlates with epigenetic markers and, thus, could influence mechanisms underlying phenotypes observed before and during a weight loss intervention.

The only microbial taxon associated with the epigenome at baseline was *Ruminiclostridium*, a member of the Firmicutes (now renamed Bacillota) phylum and Ruminococcaceae family. The Ruminococcaceae family has been identified as one of primary differentiating characteristics among different gut MB enterotypes and was noted as a one of few family-level taxa that discriminate lean from obese phenotypes in a preclinical model of fecal transplantation from humans to mice [34–36]. *Ruminiclostridium* may be a beneficial component of the gut MB, as it is involved in secretion of short chain fatty acids (SCFAs) and plays a role in maintenance of intestinal epithelial cells [37]. In our analyses, baseline abundance of this taxon was positively associated with DNAme in three metabolically relevant genes. Expression of *COL18A1* has been associated with human adipocyte differentiation and susceptibility to obesity, and preclinical work demonstrates that knockout of *COL18A1* is associated with a reduction in adiposity in murine models [38,39]. A genome-wide meta-analysis linked the *COL20A1* gene to diabetic kidney disease [40]. The expression of *NT5E* has also been associated with the consumption of a high fat diet in preclinical models and, thus, may be associated with obesity [41]. The positive association between *Ruminiclostridium* abundance and DNAme in these genes indicates that this taxon may be associated with repression of these genes and may, therefore, play a beneficial role in regulation of adiposity among healthy adults with overweight or obesity. Future work should evaluate the association between measures of body fat and *Ruminiclostridium* abundance to test this hypothesis.

Other members of the Ruminococcaceae family (*Ruminococcaceae uncultured, Ruminococcus 1, Ruminococcaceae NK4A214,* and *Ruminococcaceae UCG-014*) were inversely associated with DNAme in the promoter region of several genes involved in pyruvate, amino sugar and nucleotide sugar, and nitrogen metabolism at 3 months. Our previous analysis of multiomic predictors of weight loss in this cohort indicated that baseline *Ruminococcaceae UCG-014* was disadvantageous for weight loss, while *Ruminococcaceae NK4A214* and *Ruminococcus 1* were advantageous for weight loss and change in waist circumference [11,19], suggesting additional follow-up is needed to elucidate the mechanistic underpinnings of these relationships.

In the full model at 3 months, our analyses indicate a positive association between the abundance of *Megasphaera* and DNAme of the *CEBPD* gene, which is one of three CCAAT/enhancer-binding protein genes that have been implicated, along with peroxisome proliferator-activated receptor gamma (*PPAR*γ) genes, as key regulators of adipogenesis [42,43]. Reduced expression of *CEBPD* has been associated with impaired adipocyte development. Furthermore, obesity has been associated with lower differentiation capacity of preadipocytes [44,45]. *Megasphaera*, a member of the Firmicutes (Bacillota) phylum, has been shown to be higher in individuals with obesity when compared to those of normal weight and has been positively correlated with body fat [46,47]. Thus, our data suggest that *Megasphaera* may be associated with epigenetic modification in pathways that could reduce the expression of major regulatory genes and impair appropriate development of adipocytes, leading to lower differentiation capacity and highlighting a potential mechanism by which the gut MB may influence the epigenome in obesity.

At 3 months, *Akkermansia* was inversely associated with DNAme in four metabolically relevant genes. *Akkermansia* has demonstrated benefit in preventing high fat diet-induced obesity as well as in alleviating obesity-related insulin resistance and inflammation [48–50]. The most common *Akkermansia* species, *Akkermansia muciniphilia*, is a mucin-degrading and SCFA-producing bacterium that has been associated with lower body fat mass and improvements in multiple indicators of cardiometabolic health [51–54]. Though *Akkermansia* has been associated with beneficial changes in gene expression related to adipocyte differentiation and inversely correlated to gene variants associated with body mass index [53,55], to our knowledge, no previous studies have investigated the relationship between *Akkermansia* and DNAme in the genes identified in the present study. Thus, future work should characterize functional capacity of this taxon in relation to epigenetic markers and corresponding gene expression, particularly in the context of behavioral weight loss interventions for the management of overweight and obesity.

Integrated analyses evaluating the relationship between the gut MB and epigenome demonstrated inverse cross-sectional associations between the abundance of *Lachnospiraceae* and DNAme in several genes at both baseline and 3 months. This family is within the Firmicutes (Bacillota) phylum and includes a heterogenous group with diverse functions, with some studies suggesting a beneficial effect through mechanisms, such as enhanced SCFA production, and others finding associations with metabolic disease [10,56,57]. Interestingly, research indicates that a greater abundance of some *Lachnospiraceae* species is associated with obesity and altered lipid and glucose metabolism, while others, such as *Lachnospiraceae NK4A136*, have been described as protective against obesity, with potential anti-inflammatory effects and previously reported associations with greater adherence to healthy dietary patterns, such as a Mediterranean diet [56,58,59].

At baseline, a greater abundance of *Lachnospiraceae NK4A136* was correlated with hypo-methylation in the *OTC* gene, which plays a role in mitochondrial metabolism, amino acid homeostasis in the liver and small intestine, and has been associated with insulin response [60]. Decreased *OTC* expression has also been linked to reduced enterocyte mass and function as well as poorer liver function, though it is unclear if these are causes or consequences of underlying disease. At 3 months, *Lachnospiraceae UCG-001* abundance was inversely associated with DNAme in the promoter region of four metabolically relevant genes, including those involved in insulin secretion. While some work indicates that a greater abundance of *Lachnospiraceae UCG-001* may be associated with obesity, other preclinical and human studies suggest this species may be predictive of weight loss among those with overweight/obesity [61,62]. In our previous analyses of associations between gut MB and cardiometabolic health indicators in this cohort, we observed that greater abundance of an unclassified member of *Lachnospiraceae* was associated with less decrease in waist circumference from baseline to 3 months [11,19]. However, as detailed above, findings from our current analyses suggest different *Lachnospiraceae* species may be beneficial in modulating epigenetic changes, suggesting further investigation at the sequence variant level is warranted. These inconsistencies may be a result of using 16S data versus shotgun metagenomics, or they could be due to substantial inter-individual variation in the structure and function of the gut MB, highlighting the individualized role some taxa may play in metabolic regulation.

While dietary patterns have been suggested to be more closely associated with gut MB composition than individual nutrients [63,64], no associations with metrics of overall diet quality, nor expected dietary components, such as fiber, or food groups, such as grains, vegetables, or dairy, were noted when gut MB was evaluated alongside dietary intake data in our cohort. This may be because the gut MB is highly personalized in its response to dietary intake, with few diet–MB interactions conserved across people [65,66]. In addition, food composition databases underlying dietary analysis software capture only a fraction of the components within foods that may affect the gut MB [67]. Previous analyses have also demonstrated that habitual or long-term dietary intake is more closely associated with gut MB composition than intake reported from shortly before MB sampling [36], and, thus, use of food frequency questionnaires or weighted daily intake data may be useful for identifying relationships between gut MB and diet. Of note, however, is that an increased abundance of *Ruminococcus gnavus* was associated with decreases in intakes of trans fatty acids from baseline to 3 months. Previous studies have implicated *Ruminococcus gnavus*, which is a mucin-degrading group, in inflammatory bowel disease and in altered lipid metabolism related to non-alcoholic fatty liver disease (NAFLD) and obesity [46,68,69]. Relatedly, our analysis of multiomic predictors of change in weight and cardiometabolic health in this cohort indicate that this taxon is disadvantageous for reductions in triglycerides [19]. Furthermore, though not observed in this study, this taxon has been inversely associated with long-term diet quality, as assessed by HEI-2015 and positively associated with consumption of a pro-inflammatory diet in a large, multiethnic cohort [70,71], suggesting abundance may be higher among those with suboptimal dietary intake.

The strengths of this study include a rigorously designed weight loss trial with intensive behavioral support and careful monitoring of dietary intake by RDNs using multi-day diet records and comprehensive diet analysis software. In addition, blood and fecal sample collection aligned with other assessments and were analyzed for DNAme and gut MB, respectively, using validated and widely employed methods. Despite these strengths, this study is not without limitations. The participants included in this analysis represent two of five recruitment cohorts for the parent trial, and, thus, the sample size is relatively small and not representative of the general population of those with overweight and obesity in terms of race, ethnicity, socioeconomics, and other factors. Use of short-read 16S rRNA amplicon sequencing for the evaluation of gut MB composition limited our ability to evaluate functional capacity, and assessment of DNAme in whole blood did not allow for comparisons with metabolically active tissues. Furthermore, inclusion of transcriptomics, proteomics, or more thorough characterization of metabolomics in future studies would help to further elucidate relationships with the gut MB. Observations may have been related to unmeasured factors, such as physical activity or change in fat mass at 3 months. Lastly, while current best practice, self-reported dietary intake is prone to innate bias [72]; thus, underreporting, particularly for energy intake, may be likely within this population.

5. Conclusions

Healthy adults with overweight or obesity enrolled in a behavioral weight loss program demonstrate changes in body weight, cardiometabolic health, and dietary intakes after 3 months of intervention. At baseline and this early intervention time point, there are several associations between the gut MB and DNAme in genes known to be involved in KEGG pathways related to obesity and metabolism. Thus, the identified relationships may provide initial insights into potential pathways through which changes in the gut MB may affect metabolism before and during weight loss. Though relationships were not associated with overall dietary patterns or a subset of gut MB-derived metabolites, such as TMAO, future work may consider modeling other dietary features or integration of untargeted metabolomics to further explore the mechanisms by which the gut MB and epigenome interact to influence weight status. Furthermore, this study provides a foundation for the co-evaluation of multiomics in lifestyle interventions to discern associations between modifiable factors that may be useful for identifying baseline predictors of response, drivers of interindividual metabolic variability, or future therapeutic targets.

Supplementary Materials: The following supporting information can be downloaded at: https://www.mdpi.com/article/10.3390/nu15163588/s1, Table S1: 130 metabolically relevant KEGG pathways for inclusion in sensitivity analyses; Table S2: Dietary variables from the Nutrition Data System for Research employed for integrated diet and microbiome analyses; Table S3: Full and reduced model results for integrated DNAme and microbiome analyses; Figure S1: Scatterplots of baseline cross-sectional associations between DNAme in genes within metabolically relevant KEGG pathways and gut microbial taxa abundance; Figure S2: Scatterplots of 3-month cross-sectional associations between DNAme in genes within metabolically relevant KEGG pathways and gut microbial taxa abundance.

Author Contributions: Conceptualization, E.B.H., I.R.K., M.A.S. and S.J.B.; data curation, E.B.H., I.R.K., D.M.O. and K.B.; formal analysis, I.R.K. and E.M.L. (Elizabeth M. Litkowski); funding acquisition V.A.C., M.A.S. and S.J.B.; investigation, D.I., D.N.F., P.J., D.M.O., J.J.S., L.W. and K.B.; methodology, E.B.H. and I.R.K.; project administration, V.A.C., M.A.S. and S.J.B.; resources, V.A.C., M.A.S. and S.J.B.; supervision, D.N.F., E.M.L. (Ethan M. Lange), L.A.L., V.A.C., M.A.S. and S.J.B.; visualization, I.R.K.; writing—original draft preparation, E.B.H. and I.R.K.; writing—review and editing, E.B.H., I.R.K., D.I., D.N.F., P.J., E.M.L. (Elizabeth M. Litkowski), E.M.L. (Ethan M. Lange), L.A.L., D.M.O., J.J.S., L.W., K.B., P.S.M., E.L.M., D.H.B., V.A.C., M.A.S. and S.J.B. All authors have read and agreed to the published version of the manuscript.

Funding: This research was funded by the National Institutes of Health (NIH) K01 HL157658 (M.A.S.), R01 DK111622 (V.A.C.), T32 DK007658 (E.B.H.), and F32 DK122652 (D.M.O.). This work was also made possible by the support of the American Heart Association 18IPA34170317 (S.J.B.). D.I. and D.N.F. were supported, in part, by the University of Colorado GI and Liver Innate Immunity Program. Additionally, the Colorado Nutrition and Obesity Research Center (NORC) P30 DK048520 and the Colorado Clinical and Translational Sciences Institute (CCTSI) NIH/NCATS Colorado CTSA Grant Number UL1 TR002535 provided resources and support related to outcomes measured in this study. The contents of this study are the authors' sole responsibility and do not necessarily represent official NIH views. The funding sources were not involved in the study design; data collection, analysis, or interpretation; or in the writing and submitting of this work.

Institutional Review Board Statement: The study was conducted in accordance with the Declaration of Helsinki and was approved by the Colorado Multiple Institutional Review Board (protocol #17-0369 and protocol #21-4601, respectively).

Informed Consent Statement: Informed consent was obtained from all subjects involved in the study.

Data Availability Statement: The gut microbiome data used in this study are available from the European Bioinformatics Institute (EBI), accession No. PRJEB64902 (ERP150076). Methylation data have been deposited in the Gene Expression Omnibus and are publicly available through accession GSE240184.

Acknowledgments: The authors thank the study participants and the clinical research team at the Anschutz Health and Wellness Center.

Conflicts of Interest: The authors declare no conflict of interest.

References

1. van Dijk, S.J.; Molloy, P.L.; Varinli, H.; Morrison, J.L.; Muhlhausler, B.S.; Buckley, M.; Clark, S.J.; McMillen, I.C.; Noakes, M.; Samaras, K.; et al. Epigenetics and human obesity. *Int. J. Obes.* **2015**, *39*, 85–97. [CrossRef]
2. Ling, C.; Rönn, T. Epigenetics in Human Obesity and Type 2 Diabetes. *Cell Metab.* **2019**, *29*, 1028–1044. [CrossRef] [PubMed]
3. Rohde, K.; Keller, M.; la Cour Poulsen, L.; Blüher, M.; Kovacs, P.; Böttcher, Y. Genetics and epigenetics in obesity. *Metabolism* **2019**, *92*, 37–50. [CrossRef] [PubMed]
4. Meijnikman, A.S.; Gerdes, V.E.; Nieuwdorp, M.; Herrema, H. Evaluating Causality of Gut Microbiota in Obesity and Diabetes in Humans. *Endocr. Rev.* **2018**, *39*, 133–153. [CrossRef]
5. Sonnenburg, J.L.; Backhed, F. Diet-microbiota interactions as moderators of human metabolism. *Nature* **2016**, *535*, 56–64. [CrossRef]
6. Marzullo, P.; Di Renzo, L.; Pugliese, G.; De Siena, M.; Barrea, L.; Muscogiuri, G.; Colao, A.; Savastano, S. From obesity through gut microbiota to cardiovascular diseases: A dangerous journey. *Int. J. Obes. Suppl.* **2020**, *10*, 35–49. [CrossRef]
7. Muscogiuri, G.; Cantone, E.; Cassarano, S.; Tuccinardi, D.; Barrea, L.; Savastano, S.; Colao, A.; on behalf of the Obesity Programs of nutrition, E.R.; Assessment, g. Gut microbiota: A new path to treat obesity. *Int. J. Obes. Suppl.* **2019**, *9*, 10–19. [CrossRef]
8. Aron-Wisnewsky, J.; Warmbrunn, M.V.; Nieuwdorp, M.; Clément, K. Metabolism and Metabolic Disorders and the Microbiome: The Intestinal Microbiota Associated With Obesity, Lipid Metabolism, and Metabolic Health—Pathophysiology and Therapeutic Strategies. *Gastroenterology* **2021**, *160*, 573–599. [CrossRef]
9. Maruvada, P.; Leone, V.; Kaplan, L.M.; Chang, E.B. The Human Microbiome and Obesity: Moving beyond Associations. *Cell Host. Microbe* **2017**, *22*, 589–599. [CrossRef] [PubMed]
10. Seganfredo, F.B.; Blume, C.A.; Moehlecke, M.; Giongo, A.; Casagrande, D.S.; Spolidoro, J.V.N.; Padoin, A.V.; Schaan, B.D.; Mottin, C.C. Weight-loss interventions and gut microbiota changes in overweight and obese patients: A systematic review. *Obes. Rev.* **2017**, *18*, 832–851. [CrossRef]
11. Stanislawski, M.A.; Frank, D.N.; Borengasser, S.J.; Ostendorf, D.M.; Ir, D.; Jambal, P.; Bing, K.; Wayland, L.; Siebert, J.C.; Bessesen, D.H.; et al. The Gut Microbiota during a Behavioral Weight Loss Intervention. *Nutrients* **2021**, *13*, 3248. [CrossRef] [PubMed]
12. Carbonero, F. Human epigenetics and microbiome: The potential for a revolution in both research areas by integrative studies. *Future Sci. OA* **2017**, *3*, Fso207. [CrossRef] [PubMed]
13. Peterson, L.W.; Artis, D. Intestinal epithelial cells: Regulators of barrier function and immune homeostasis. *Nat. Rev. Immunol.* **2014**, *14*, 141–153. [CrossRef] [PubMed]
14. Vallianou, N.; Stratigou, T.; Christodoulatos, G.S.; Dalamaga, M. Understanding the Role of the Gut Microbiome and Microbial Metabolites in Obesity and Obesity-Associated Metabolic Disorders: Current Evidence and Perspectives. *Curr. Obes. Rep.* **2019**, *8*, 317–332. [CrossRef]
15. Bhat, M.I.; Kapila, R. Dietary metabolites derived from gut microbiota: Critical modulators of epigenetic changes in mammals. *Nutr. Rev.* **2017**, *75*, 374–389. [CrossRef] [PubMed]
16. Woo, V.; Alenghat, T. Epigenetic regulation by gut microbiota. *Gut Microbes* **2022**, *14*, 2022407. [CrossRef] [PubMed]
17. Sharma, M.; Li, Y.; Stoll, M.L.; Tollefsbol, T.O. The Epigenetic Connection Between the Gut Microbiome in Obesity and Diabetes. *Front. Genet.* **2019**, *10*, 1329. [CrossRef]
18. Ostendorf, D.M.; Caldwell, A.E.; Zaman, A.; Pan, Z.; Bing, K.; Wayland, L.T.; Creasy, S.A.; Bessesen, D.H.; MacLean, P.; Melanson, E.L.; et al. Comparison of weight loss induced by daily caloric restriction versus intermittent fasting (DRIFT) in individuals with obesity: Study protocol for a 52-week randomized clinical trial. *Trials* **2022**, *23*, 718. [CrossRef] [PubMed]
19. Siebert, J.C.; Stanislawski, M.A.; Zaman, A.; Ostendorf, D.M.; Konigsberg, I.R.; Jambal, P.; Ir, D.; Bing, K.; Wayland, L.; Scorsone, J.J.; et al. Multiomic Predictors of Short-Term Weight Loss and Clinical Outcomes During a Behavioral-Based Weight Loss Intervention. *Obesity* **2021**, *29*, 859–869. [CrossRef]
20. Zhou, W.; Triche, T.J.; Laird, P.W.; Shen, H. SeSAMe: Reducing artifactual detection of DNA methylation by Infinium BeadChips in genomic deletions. *Nucleic Acids Res.* **2018**, *46*, e123. [CrossRef]
21. Zhou, W.; Laird, P.W.; Shen, H. Comprehensive characterization, annotation and innovative use of Infinium DNA methylation BeadChip probes. *Nucleic Acids Res.* **2017**, *45*, e22. [CrossRef]
22. Houseman, E.A.; Accomando, W.P.; Koestler, D.C.; Christensen, B.C.; Marsit, C.J.; Nelson, H.H.; Wiencke, J.K.; Kelsey, K.T. DNA methylation arrays as surrogate measures of cell mixture distribution. *BMC Bioinform.* **2012**, *13*, 86. [CrossRef] [PubMed]
23. Du, P.; Zhang, X.; Huang, C.-C.; Jafari, N.; Kibbe, W.A.; Hou, L.; Lin, S.M. Comparison of Beta-value and M-value methods for quantifying methylation levels by microarray analysis. *BMC Bioinform.* **2010**, *11*, 587. [CrossRef]
24. Hara, N.; Alkanani, A.K.; Ir, D.; Robertson, C.E.; Wagner, B.D.; Frank, D.N.; Zipris, D. Prevention of virus-induced type 1 diabetes with antibiotic therapy. *J. Immunol.* **2012**, *189*, 3805–3814. [CrossRef]

25. Markle, J.G.; Frank, D.N.; Mortin-Toth, S.; Robertson, C.E.; Feazel, L.M.; Rolle-Kampczyk, U.; von Bergen, M.; McCoy, K.D.; Macpherson, A.J.; Danska, J.S. Sex differences in the gut microbiome drive hormone-dependent regulation of autoimmunity. *Science* **2013**, *339*, 1084–1088. [CrossRef] [PubMed]
26. Bolyen, E.; Rideout, J.R.; Dillon, M.R.; Bokulich, N.A.; Abnet, C.C.; Al-Ghalith, G.A.; Alexander, H.; Alm, E.J.; Arumugam, M.; Asnicar, F.; et al. Reproducible, interactive, scalable and extensible microbiome data science using QIIME 2. *Nat. Biotechnol.* **2019**, *37*, 852–857. [CrossRef] [PubMed]
27. Callahan, B.J.; McMurdie, P.J.; Rosen, M.J.; Han, A.W.; Johnson, A.J.A.; Holmes, S.P. DADA2: High-resolution sample inference from Illumina amplicon data. *Nat. Methods* **2016**, *13*, 581–583. [CrossRef]
28. Quast, C.; Pruesse, E.; Yilmaz, P.; Gerken, J.; Schweer, T.; Yarza, P.; Peplies, J.; Glöckner, F.O. The SILVA ribosomal RNA gene database project: Improved data processing and web-based tools. *Nucleic Acids Res.* **2013**, *41*, D590–D596. [CrossRef]
29. Janssen, S.; McDonald, D.; Gonzalez, A.; Navas-Molina, J.A.; Jiang, L.; Xu, Z.Z.; Winker, K.; Kado, D.M.; Orwoll, E.; Manary, M.; et al. Phylogenetic Placement of Exact Amplicon Sequences Improves Associations with Clinical Information. *mSystems* **2018**, *3*, 00021-18. [CrossRef] [PubMed]
30. Krebs-Smith, S.M.; Pannucci, T.E.; Subar, A.F.; Kirkpatrick, S.I.; Lerman, J.L.; Tooze, J.A.; Wilson, M.M.; Reedy, J. Update of the Healthy Eating Index: HEI-2015. *J. Acad. Nutr. Diet.* **2018**, *118*, 1591–1602. [CrossRef] [PubMed]
31. Koeth, R.A.; Wang, Z.; Levison, B.S.; Buffa, J.A.; Org, E.; Sheehy, B.T.; Britt, E.B.; Fu, X.; Wu, Y.; Li, L.; et al. Intestinal microbiota metabolism of L-carnitine, a nutrient in red meat, promotes atherosclerosis. *Nat. Med.* **2013**, *19*, 576–585. [CrossRef]
32. Kirsch, S.H.; Herrmann, W.; Rabagny, Y.; Obeid, R. Quantification of acetylcholine, choline, betaine, and dimethylglycine in human plasma and urine using stable-isotope dilution ultra performance liquid chromatography-tandem mass spectrometry. *J. Chromatogr. B Anal. Technol. Biomed. Life Sci.* **2010**, *878*, 3338–3344. [CrossRef] [PubMed]
33. Benjamini, Y.; Hochberg, Y. Controlling the false discovery rate: A practical and powerful approach to multiple testing. *J. R. Stat. Soc. Ser. B* **1995**, *57*, 289–300. [CrossRef]
34. Arumugam, M.; Raes, J.; Pelletier, E.; Le Paslier, D.; Yamada, T.; Mende, D.R.; Fernandes, G.R.; Tap, J.; Bruls, T.; Batto, J.M.; et al. Enterotypes of the human gut microbiome. *Nature* **2011**, *473*, 174–180. [CrossRef] [PubMed]
35. Ridaura, V.K.; Faith, J.J.; Rey, F.E.; Cheng, J.; Duncan, A.E.; Kau, A.L.; Griffin, N.W.; Lombard, V.; Henrissat, B.; Bain, J.R.; et al. Gut microbiota from twins discordant for obesity modulate metabolism in mice. *Science* **2013**, *341*, 1241214. [CrossRef]
36. Tang, Z.Z.; Chen, G.; Hong, Q.; Huang, S.; Smith, H.M.; Shah, R.D.; Scholz, M.; Ferguson, J.F. Multi-Omic Analysis of the Microbiome and Metabolome in Healthy Subjects Reveals Microbiome-Dependent Relationships Between Diet and Metabolites. *Front. Genet.* **2019**, *10*, 454. [CrossRef]
37. Tan, J.; McKenzie, C.; Potamitis, M.; Thorburn, A.N.; Mackay, C.R.; Macia, L. The role of short-chain fatty acids in health and disease. *Adv. Immunol.* **2014**, *121*, 91–119. [CrossRef]
38. Errera, F.I.; Canani, L.H.; Yeh, E.; Kague, E.; Armelin-Corrêa, L.M.; Suzuki, O.T.; Tschiedel, B.; Silva, M.E.; Sertié, A.L.; Passos-Bueno, M.R. COL18A1 is highly expressed during human adipocyte differentiation and the SNP c.1136C > T in its "frizzled" motif is associated with obesity in diabetes type 2 patients. *Acad. Bras. Cienc.* **2008**, *80*, 167–177. [CrossRef]
39. Aikio, M.; Elamaa, H.; Vicente, D.; Izzi, V.; Kaur, I.; Seppinen, L.; Speedy, H.E.; Kaminska, D.; Kuusisto, S.; Sormunen, R.; et al. Specific collagen XVIII isoforms promote adipose tissue accrual via mechanisms determining adipocyte number and affect fat deposition. *Proc. Natl. Acad. Sci. USA* **2014**, *111*, E3043–E3052. [CrossRef]
40. Sandholm, N.; Cole, J.B.; Nair, V.; Sheng, X.; Liu, H.; Ahlqvist, E.; van Zuydam, N.; Dahlström, E.H.; Fermin, D.; Smyth, L.J.; et al. Genome-wide meta-analysis and omics integration identifies novel genes associated with diabetic kidney disease. *Diabetologia* **2022**, *65*, 1495–1509. [CrossRef]
41. Kieffer, D.A.; Piccolo, B.D.; Marco, M.L.; Kim, E.B.; Goodson, M.L.; Keenan, M.J.; Dunn, T.N.; Knudsen, K.E.B.; Adams, S.H.; Martin, R.J. Obese Mice Fed a Diet Supplemented with Enzyme-Treated Wheat Bran Display Marked Shifts in the Liver Metabolome Concurrent with Altered Gut Bacteria. *J. Nutr.* **2016**, *146*, 2445–2460. [CrossRef] [PubMed]
42. Rosen, E.D.; Walkey, C.J.; Puigserver, P.; Spiegelman, B.M. Transcriptional regulation of adipogenesis. *Genes Dev.* **2000**, *14*, 1293–1307. [CrossRef] [PubMed]
43. Merrett, J.E.; Bo, T.; Psaltis, P.J.; Proud, C.G. Identification of DNA response elements regulating expression of CCAAT/enhancer-binding protein (C/EBP) β and δ and MAP kinase-interacting kinases during early adipogenesis. *Adipocyte* **2020**, *9*, 427–442. [CrossRef] [PubMed]
44. Rossmeislová, L.; Malisová, L.; Kracmerová, J.; Tencerová, M.; Kovácová, Z.; Koc, M.; Siklová-Vítková, M.; Viquerie, N.; Langin, D.; Stich, V. Weight loss improves the adipogenic capacity of human preadipocytes and modulates their secretory profile. *Diabetes* **2013**, *62*, 1990–1995. [CrossRef] [PubMed]
45. Isakson, P.; Hammarstedt, A.; Gustafson, B.; Smith, U. Impaired preadipocyte differentiation in human abdominal obesity: Role of Wnt, tumor necrosis factor-alpha, and inflammation. *Diabetes* **2009**, *58*, 1550–1557. [CrossRef]
46. Palmas, V.; Pisanu, S.; Madau, V.; Casula, E.; Deledda, A.; Cusano, R.; Uva, P.; Vascellari, S.; Loviselli, A.; Manzin, A.; et al. Gut microbiota markers associated with obesity and overweight in Italian adults. *Sci. Rep.* **2021**, *11*, 5532. [CrossRef]
47. Turnbaugh, P.J.; Ley, R.E.; Mahowald, M.A.; Magrini, V.; Mardis, E.R.; Gordon, J.I. An obesity-associated gut microbiome with increased capacity for energy harvest. *Nature* **2006**, *444*, 1027–1031. [CrossRef]
48. Jayachandran, M.; Chung, S.S.M.; Xu, B. A critical review of the relationship between dietary components, the gut microbe Akkermansia muciniphila, and human health. *Crit. Rev. Food Sci. Nutr.* **2020**, *60*, 2265–2276. [CrossRef]

49. Yoon, H.S.; Cho, C.H.; Yun, M.S.; Jang, S.J.; You, H.J.; Kim, J.H.; Han, D.; Cha, K.H.; Moon, S.H.; Lee, K.; et al. Akkermansia muciniphila secretes a glucagon-like peptide-1-inducing protein that improves glucose homeostasis and ameliorates metabolic disease in mice. *Nat. Microbiol.* **2021**, *6*, 563–573. [CrossRef]
50. Collado, M.C.; Derrien, M.; Isolauri, E.; de Vos, W.M.; Salminen, S. Intestinal integrity and Akkermansia muciniphila, a mucin-degrading member of the intestinal microbiota present in infants, adults, and the elderly. *Appl. Environ. Microbiol.* **2007**, *73*, 7767–7770. [CrossRef]
51. Zhao, Q.; Yu, J.; Hao, Y.; Zhou, H.; Hu, Y.; Zhang, C.; Zheng, H.; Wang, X.; Zeng, F.; Hu, J.; et al. Akkermansia muciniphila plays critical roles in host health. *Crit. Rev. Microbiol.* **2023**, *49*, 82–100. [CrossRef]
52. Cani, P.D.; de Vos, W.M. Next-Generation Beneficial Microbes: The Case of Akkermansia muciniphila. *Front. Microbiol.* **2017**, *8*, 1765. [CrossRef] [PubMed]
53. Everard, A.; Belzer, C.; Geurts, L.; Ouwerkerk, J.P.; Druart, C.; Bindels, L.B.; Guiot, Y.; Derrien, M.; Muccioli, G.G.; Delzenne, N.M.; et al. Cross-talk between Akkermansia muciniphila and intestinal epithelium controls diet-induced obesity. *Proc. Natl. Acad. Sci. USA* **2013**, *110*, 9066–9071. [CrossRef] [PubMed]
54. Dao, M.C.; Everard, A.; Aron-Wisnewsky, J.; Sokolovska, N.; Prifti, E.; Verger, E.O.; Kayser, B.D.; Levenez, F.; Chilloux, J.; Hoyles, L.; et al. Akkermansia muciniphila and improved metabolic health during a dietary intervention in obesity: Relationship with gut microbiome richness and ecology. *Gut* **2016**, *65*, 426–436. [CrossRef] [PubMed]
55. Davenport, E.R.; Cusanovich, D.A.; Michelini, K.; Barreiro, L.B.; Ober, C.; Gilad, Y. Genome-Wide Association Studies of the Human Gut Microbiota. *PLoS ONE* **2015**, *10*, e0140301. [CrossRef]
56. Vacca, M.; Celano, G.; Calabrese, F.M.; Portincasa, P.; Gobbetti, M.; De Angelis, M. The Controversial Role of Human Gut Lachnospiraceae. *Microorganisms* **2020**, *8*, 573. [CrossRef]
57. Vital, M.; Karch, A.; Pieper, D.H. Colonic Butyrate-Producing Communities in Humans: An Overview Using Omics Data. *mSystems* **2017**, *2*, 00130-17. [CrossRef]
58. Muralidharan, J.; Moreno-Indias, I.; Bulló, M.; Lopez, J.V.; Corella, D.; Castañer, O.; Vidal, J.; Atzeni, A.; Fernandez-García, J.C.; Torres-Collado, L.; et al. Effect on gut microbiota of a 1-y lifestyle intervention with Mediterranean diet compared with energy-reduced Mediterranean diet and physical activity promotion: PREDIMED-Plus Study. *Am. J. Clin. Nutr.* **2021**, *114*, 1148–1158. [CrossRef]
59. Machado Arruda, S.P.; da Silva, A.A.M.; Kac, G.; Vilela, A.A.F.; Goldani, M.; Bettiol, H.; Barbieri, M.A. Dietary patterns are associated with excess weight and abdominal obesity in a cohort of young Brazilian adults. *Eur. J. Nutr.* **2016**, *55*, 2081–2091. [CrossRef] [PubMed]
60. Couchet, M.; Breuillard, C.; Corne, C.; Rendu, J.; Morio, B.; Schlattner, U.; Moinard, C. Ornithine Transcarbamylase From Structure to Metabolism: An Update. *Front. Physiol.* **2021**, *12*, 748249. [CrossRef]
61. Song, X.; Zhong, L.; Lyu, N.; Liu, F.; Li, B.; Hao, Y.; Xue, Y.; Li, J.; Feng, Y.; Ma, Y.; et al. Inulin Can Alleviate Metabolism Disorders in ob/ob Mice by Partially Restoring Leptin-related Pathways Mediated by Gut Microbiota. *Genom. Proteom. Bioinform.* **2019**, *17*, 64–75. [CrossRef] [PubMed]
62. Atzeni, A.; Galié, S.; Muralidharan, J.; Babio, N.; Tinahones, F.J.; Vioque, J.; Corella, D.; Castañer, O.; Vidal, J.; Moreno-Indias, I.; et al. Gut Microbiota Profile and Changes in Body Weight in Elderly Subjects with Overweight/Obesity and Metabolic Syndrome. *Microorganisms* **2021**, *9*, 346. [CrossRef] [PubMed]
63. Baldeon, A.D.; McDonald, D.; Gonzalez, A.; Knight, R.; Holscher, H.D. Diet Quality and the Fecal Microbiota in Adults in the American Gut Project. *J. Nutr.* **2023**, *153*, 2004–2015. [CrossRef]
64. Cotillard, A.; Cartier-Meheust, A.; Litwin, N.S.; Chaumont, S.; Saccareau, M.; Lejzerowicz, F.; Tap, J.; Koutnikova, H.; Lopez, D.G.; McDonald, D.; et al. A posteriori dietary patterns better explain variations of the gut microbiome than individual markers in the American Gut Project. *Am. J. Clin. Nutr.* **2022**, *115*, 432–443. [CrossRef] [PubMed]
65. Johnson, A.J.; Vangay, P.; Al-Ghalith, G.A.; Hillmann, B.M.; Ward, T.L.; Shields-Cutler, R.R.; Kim, A.D.; Shmagel, A.K.; Syed, A.N.; Walter, J.; et al. Daily Sampling Reveals Personalized Diet-Microbiome Associations in Humans. *Cell Host. Microbe.* **2019**, *25*, 789–802.e785. [CrossRef]
66. Healey, G.R.; Murphy, R.; Brough, L.; Butts, C.A.; Coad, J. Interindividual variability in gut microbiota and host response to dietary interventions. *Nutr. Rev.* **2017**, *75*, 1059–1080. [CrossRef]
67. Scalbert, A.; Brennan, L.; Manach, C.; Andres-Lacueva, C.; Dragsted, L.O.; Draper, J.; Rappaport, S.M.; Van Der Hooft, J.J.; Wishart, D.S. The food metabolome: A window over dietary exposure. *Am. J. Clin. Nutr.* **2014**, *99*, 1286–1308. [CrossRef]
68. Hullar, M.A.J.; Jenkins, I.C.; Randolph, T.W.; Curtis, K.R.; Monroe, K.R.; Ernst, T.; Shepherd, J.A.; Stram, D.O.; Cheng, I.; Kristal, B.S.; et al. Associations of the gut microbiome with hepatic adiposity in the Multiethnic Cohort Adiposity Phenotype Study. *Gut Microbes* **2021**, *13*, 1965463. [CrossRef]
69. Henke, M.T.; Kenny, D.J.; Cassilly, C.D.; Vlamakis, H.; Xavier, R.J.; Clardy, J. Ruminococcus gnavus, a member of the human gut microbiome associated with Crohn's disease, produces an inflammatory polysaccharide. *Proc. Natl. Acad. Sci. USA* **2019**, *116*, 12672–12677. [CrossRef]
70. Ma, E.; Maskarinec, G.; Lim, U.; Boushey, C.J.; Wilkens, L.R.; Setiawan, V.W.; Le Marchand, L.; Randolph, T.W.; Jenkins, I.C.; Curtis, K.R.; et al. Long-term association between diet quality and characteristics of the gut microbiome in the multiethnic cohort study. *Br. J. Nutr.* **2022**, *128*, 93–102. [CrossRef]

71. Lozano, C.P.; Wilkens, L.R.; Shvetsov, Y.B.; Maskarinec, G.; Park, S.-Y.; Shepherd, J.A.; Boushey, C.J.; Hebert, J.R.; Wirth, M.D.; Ernst, T.; et al. Associations of the Dietary Inflammatory Index with total adiposity and ectopic fat through the gut microbiota, LPS, and C-reactive protein in the Multiethnic Cohort–Adiposity Phenotype Study. *Am. J. Clin. Nutr.* **2021**, *115*, 1344–1356. [CrossRef] [PubMed]
72. Ravelli, M.N.; Schoeller, D.A. Traditional Self-Reported Dietary Instruments Are Prone to Inaccuracies and New Approaches Are Needed. *Front. Nutr.* **2020**, *7*, 90. [CrossRef] [PubMed]

nutrients

MDPI

Article

Biochemical Changes Induced by the Administration of *Cannabis sativa* Seeds in Diabetic Wistar Rats

Camelia Munteanu [1], Mihaela Mihai [2], Francisc Dulf [3], Andreea Ona [1,*], Leon Muntean [1], Floricuța Ranga [4], Camelia Urdă [5], Daria Pop [6], Tania Mihaiescu [3], Sorin Marian Mârza [7,*] and Ionel Papuc [7]

1 Department of Plant Culture, Faculty of Agriculture, University of Agricultural Sciences and Veterinary Medicine Cluj-Napoca, Calea Mănăştur 3-5, 400372 Cluj-Napoca, Romania; camelia.munteanu@usamvcluj.ro (C.M.); leon.muntean@usamvcluj.ro (L.M.)

2 Department of Transversal Competencies, University of Agricultural Sciences and Veterinary Medicine Cluj-Napoca, Calea Mănăştur 3-5, 400372 Cluj-Napoca, Romania; mihaela.mihai@usamvcluj.ro

3 Department of Environmental and Plant Protection, Faculty of Agriculture, University of Agricultural Sciences and Veterinary Medicine Cluj-Napoca, Calea Mănăştur 3-5, 400372 Cluj-Napoca, Romania; francisc.dulf@usamvcluj.ro (F.D.); tania.mihaiescu@usamvcluj.ro (T.M.)

4 Department of Food Science, University of Agricultural Sciences and Veterinary Medicine Cluj-Napoca, Calea Mănăştur 3-5, 400372 Cluj-Napoca, Romania; floricutza_ro@yahoo.com

5 Agricultural Research Development Station Turda, 27 Agriculturii Street, 401100 Turda, Romania; camelia.urda@scdaturda.ro

6 Clinic of Obstetrics and Gynecology II "Dominic Stanca", University of Medicine and Pharmacy "Iuliu Hațieganu" Cluj-Napoca, Victor Babeș 8, 400347 Cluj-Napoca, Romania; drdariamariapop@gmail.com

7 Faculty of Veterinary Medicine, University of Agricultural Sciences and Veterinary Medicine Cluj-Napoca, Calea Mănăştur 3-5, 400372 Cluj-Napoca, Romania; ionel.papuc@usamvcluj.ro

* Correspondence: andreea.ona@usamvcluj.ro (A.O.); sorinmarza@yahoo.com (S.M.M.)

Abstract: The present pilot study investigates the blood biochemical changes induced by hemp seeds in rats with diabetes. The composition of industrial hemp seeds, antioxidant activity, identification and quantification of phenols and fatty acids from hemp oil were determined. The Wistar adult rats used in the experiment were divided into three groups ($n = 6$) and kept under standard conditions. Group one, the control group (individuals without diabetes), and group two (diabetic individuals) received water and normal food ad libitum, while the third group, also including diabetic individuals, received specific food (hemp seeds) and water ad libitum. Subsequent blood biochemical parameters were determined. Hemp seeds had higher phenol (14 compounds), flavonoids and PUFA contents compared to other plants seeds. In addition, the antioxidant activity in *Cannabis sativa* was also increased. Moreover, the ratio between *n*-6 and *n*-3 was 4.41, ideal for different diseases. Additionally, all biochemical parameters showed significant changes following the treatment. It was shown that high doses of hemp seeds decreased diabetes-induced biochemical damage in rats most probably due to the high content of active compounds. In order to use these seeds in humans, it is essential to find out which hemp compounds are particularly responsible for these effects. Moreover, for the objective investigation of their effects, longer-term studies are needed.

Keywords: cannabis; hydroxybenzoic acid; phenolic activity; fatty acids; glycemia

1. Introduction

More than a third of the world's population is overweight or obese and therefore at risk of developing diabetes mellitus type 2 (T2DM) [1]. This effect has been accentuated since the beginning of the pandemic period, due to the lack of physical activity and the increase in food intake. The disease involves progressive worsening of insulin resistance and a compensatory increase in its secretion. While T2DM is clearly a multifactorial and complex disorder, there is no doubt that obesity-induced insulin resistance accelerates the depletion of pancreatic β cells in the Langerhans Islands and therefore the onset of

T2DM [2]. One of the ways to prevent obesity and its associated metabolic disorders is the consumption of plant-based foods with known low caloric values [3]. Moreover, it is of the utmost relevance to note that these products provide a good source of bioactive compounds with beneficial health effects [4]. In this regard, the optimum fiber content, high protein concentration, flavonoid concentration, antioxidant activity, phenol compounds and PUFA omega-6:PUFA omega-3 optimal ratio from industrial hemp seeds illustrate the perfect match [5]. In recent decades, cannabis was banned as it was used for recreational purposes, without a clear distinction regarding the tetrahydrocannabinol (THC) content, which is the psychoactive compound in the plant. The term 'hemp' refers to varieties of *Cannabis sativa* that have been grown for industrial purposes and characterized by lower concentrations of THC. According to the Integrated Taxonomic Information System, cannabis belongs to the class Magnoliopsida, order Rosales, family Cannabaceae, genus Cannabis, which has a single species *Cannabis sativa* with the subspecies *sativa* and *indica* [6]. Hemp is a herbaceous and anemophilous plant (Figure 1) native to Central Asia. Due to its long history of cultivation, it is considered one of the oldest cultivated plants. The species was introduced to Europe during the Bronze Age as a cultivated domestic agricultural plant [7]. In the beginning, hemp was cultivated to produce fiber for clothes and seeds for eating, but due to their active compounds, hemp plants are used nowadays in multiple areas besides the earlier ones: in agriculture as a phytoremediation agent of polluted soils, in cosmetics, and as healthy foods, to decrease LDL cholesterol and blood pressure [8–10]. From a nutritional perspective, hemp seeds are rich in healthy nutrients such as minerals (zinc, sodium, magnesium, calcium and iron), antioxidants and proteins [11]. Moreover, hemp oil is rich in arginine, alleviating the disrupted metabolism in obesity, regulating blood pressure, and inhibiting the symptoms of type 2 diabetes [12]. In addition, all the essential amino acids can be found in hemp oil [13] alongside high levels of omega-3 fatty acid, α-linolenic acid (ALA), as well as omega-6 fatty linoleic acid (LA). The latter is considered essential because the mammalian body cannot synthesize it and serves as a precursor for the omega-3 group [14]. The ratio between ALA and LA from seeds seems to be perfect for health [15]. This has antioxidant effects on the cardiovascular system overall. Unfortunately, however, hemp oil also contains some trans-fatty acids which are correlated with atherosclerosis due to their role in blocking other PUFA synthesis [16,17]. Additionally, hemp seeds contain gamma-linolenic acid (GLA), with significant antioxidant effects such as the improvement of diabetic complications via anti-inflammatory mechanisms [18]. Other therapeutic antioxidant compounds in hemp seeds are polyphenols [19] and more than 20 flavonoids, of which the most predominant are prenylated flavonoids [20]. Lignanamides and phenol amides have also been observed [21,22]. Their administration is correlated with hypoglycemic effects [23]. Furthermore, in rats with diabetes induced by streptozotocin, hydroxycinnamic acid derivatives, it has been found that hemp seeds significantly decreased blood glucose concentration [24]. Commonly, flavonoids have overall beneficial effects on health due to their involvement in the decrease in the metabolic complications shown in type 2 diabetes [25]. These changes may be the results of amylase inhibition, aiming at the reduction of carbohydrate absorption [25]. Moreover, proteins from the seeds are used for their hypoglycemic effects [26]. For the reasons listed above, the present pilot research investigates the biochemical changes induced by the administration of hemp seed compounds in rats with diabetes.

Figure 1. (Original) *Cannabis sativa*, 'Zenit' variety.

2. Materials and Methods

2.1. *Cannabis sativa* 'Zenit' Variety Hemp Seed Composition

Hemp seeds were collected from plants grown in the Botanical Garden of UASVM Cluj-Napoca. The plants belonged to the 'Zenit' variety of *Cannabis sativa* and were used to determine the concentration of lipids, proteins, total phenols, total flavonoid antioxidants activity and quantification of phenols (Figure 1).

2.2. Chemical Analysis

The Kjeldahl method was used for protein content, the Soxhlet method was used for determinations of fat content, while the calcination method served for determining both the dry matter and ash contents [27].

Protein content was determined via the Kjeldahl method: mineralization was performed in a Turbotherm TT 265 unit connected to a Turbosorg Tur/K scrubber. A Vapodest 30S unit was used for distillation. All devices were purchased from Gerhardt, Koenigswinter, Germany. Finally, a classic titration was performed, with a 25.00 mL class A burette [27].

Fat content was determined via the Soxhlet method, using a 6-position unit Det Gras N6p (JP Selecta, Barcelona, Spain). Acetone was used as an extraction solvent [27].

The ash content was determined via heating at 550 °C for 5 h in a Nabertherm B180 calcination furnace (Nabertherm GmbH, Lilienthal, Germany) [27].

The dry matter content was determined via heating at 105 °C for 6 h in a forced air oven SLW 53 (POL-EKO-Apparatus, Wodzisław Śląski, Poland).

Total phenols and flavonoids were determined using spectrophotometry VIS/UV/VIS Spectrometer T80+. In this case, the evaluation of the total content of phenols and flavonoids was related to a standard curve of gallic acid. This was dissolved in ethanol. Subsequently, the concentration of total phenols and flavonoids was expressed in mg gallic acid equivalents (GAE) per gram of dried extract [27].

2.2.1. Antioxidant Activity of Hemp Extracts (Methanolic and Ethanolic Extracts)

The antioxidant activity of the samples was determined using the DPPH (1,1-diphenyl-2-picrylhydrazyl) free radical scavenging capacity technique developed by Brand-Williams et al. [28]. To determine sample antioxidant responses, 35 μL of previously methanolic and ethanolic extracted samples mixed with 250 μL of DPPH solution were prepared in triplicate. The reaction solution was incubated for 30 min at room temperature in the darkness before measuring absorbance at 515 nm using a multi-mode plate reader (BioTek, Winuschi, VT, USA). The findings were presented as micromol Trolox equivalents (μmol TE)/100 g sample [28].

2.2.2. Extraction of Phenolic Compounds

To this end, 0.3 g of the crushed sample were extracted with 3 mL of methanol acidified with 1% HCl of 37% concentration, via vortexing for 1 min at the Heidolph Reax top vortex, sonicating for 30 min in the Elmasonic E 15 H sonication bath and centrifuging at 10,000 rpm for 10 min and at 24 °C in the Eppendorf AG 5804 centrifuge. In parallel, the extraction was carried out with 70% ethanol, under the conditions mentioned above. The supernatant was filtered through a 0.45 μm Chromafil Xtra nylon filter and 20 μL were injected into the HPLC system [29].

2.2.3. Fatty Acid Analyses

Hemp oil extraction was used in this experiment to determine the fatty acids in the seeds. The chemicals used for the preparation of fatty acid methyl esters (FAMEs) were purchased from Sigma-Aldrich (Steinheim, Germany), while the FAMEs standard (37 component FAME Mix, SUPELCO) was from Supelco (Bellefonte, PA, USA). The results obtained were checked against the dry matter.

Fatty acid methyl esters (FAMEs) of the total lipids (0.2 g oil) were produced via acid-catalyzed transesterification using 1% sulphuric acid in methanol [30,31].

The methylated fatty acids were determined with a gas chromatograph (GC) coupled to a mass spectrometer (MS) (PerkinElmer Clarus 600 T GC-MS; Shelton, CT, USA) [32]. A 0.5 μL sample was injected into a 60 m × 0.25 mm i.d., 0.25 μm film thickness SUPELCOWAX 10 capillary column (Supelco Inc.). The operation conditions were as follows: injector temperature 210 °C; helium carrier gas flow rate 0.8 mL/min; split ratio 1:24; oven temperature 140 °C (hold 2 min) to 220 °C at 7 °C/min (hold 19 min); electron impact ionization voltage 70 eV; trap current 100 μA; ion source temperature 150 °C; mass range 22–395 m/z (0.14 scans/s with an intermediate time of 0.02 s between the scans).

The amount of each fatty acid was expressed as area percentages calculated from the total area of identified FAMEs.

2.3. Experimental Animal Protocol

Eighteen adult Wistar rats, male and female, weighing between 250 and 300 g were used in the experiment for five weeks. The ethics approval form number was 340/2022. During the experiment, all rats were kept in hygienic conditions: constant temperature and humidity, a light/darkness rhythm of 12/12 h and free access to water and food. The animals were handled gently, without causing them stress or pain.

In the initial phase, all rats (18 individuals) were kept (with food and water provided ad libitum) until diabetes was induced (2 weeks). They were divided from the beginning into three groups. Both weight and blood glucose were monitored 5 times for two weeks until blood glucose stabilized at values that allowed us to declare induced diabetes. It is important to mention that streptozotocin is a molecule from the nitroso-urea family, used in the treatment of neoplasia or as a natural antibiotic [33]. In 1960, the scientific community found that streptozotocin has selective toxicity on β-cells in the islets of Langerhans of the pancreas. This discovery raised interest in the use of streptozotocin to induce diabetes in experimental animals, especially rats, and mice, but also in rabbits and fish [34,35]. Repeated injections of 20 to 40 mg/kg of streptozotocin in rats and mice decrease insulin secretion without inducing cell death, which mimics insulitis seen in type 1 diabetes. Administered as a single dose of 100 to 200 mg/kg, streptozotocin causes rapid destruction of β-cells and causes diabetes [35].

Diabetes mellitus was induced in rats using the Streptozotocin (STZ) protocol previously described [34,35]. Firstly, all the animals were weighted (±1 g) and fasted for 6 h before the start of the experiment. The fasting blood glucose level of each Wistar rat was determined just before the experiment (day 0), using a commercial blood glucose meter (Accu-Chek® instant, Roche, Germany). The rats received a double intraperitoneal injection with STZ (65 mg/kg b.w., dissolved in 0.1 mM citrate buffer, pH 4.5). The rats were returned to their cages and provided with normal food and water. The diabetic state of the rats was evaluated 14 days after the STZ-NA administration by measuring fasting blood glucose levels from blood samples collected from the tail vein. The animals with blood glucose levels of more than 150 mg/dL were considered as diabetic and used in the present study.

Subsequently, they entered the pilot experiment for 5 weeks in which they received normal and specific food. Group one, the control group (CG), individuals without diabetes, and group two, the group with diabetic individuals (DNFG), received water and normal food ad libitum, while the third group, the group with diabetic individuals (DHFG), received specific food-hemp seeds and water ad libitum.

The experiment was carried out in accordance with the "Guide for the Care and Use of Laboratory Animals" (Department of Health and Welfare Education, National Institute of Health), 1996, and closely followed the directive of the European Council (86/609/1986) and the Order No. 37 of the Romanian Government of 2 February 2002 [36,37].

At the end of the experiment, blood samples from all rats were collected for biochemical analyses (0.5 mL). These were performed with microhematocrit tubes. The blood was taken from the orbital sinus into EDTA vacutainers and Eppendorf tubes to obtain plasma and blood serum. According to the literature, in the case of rats, blood glucose values that range between 85 and 132 mg/dL are considered normal [38]. At the beginning of the experiments, all rats had normal glycemia. In the second and third groups, rats received a single dose of the cytotoxic glucose analog intraperitoneally, more precisely STZ (60 mg/kg), freshly dissolved in 10 mM sodium citrate solution (pH = 4.5). The rats with constant blood glucose values of more than 300 mg/dL were considered diabetic [39]. Blood samples were collected from diabetic rats, using the same collection conditions as in all normo-glycaemic rats.

Blood Biochemical Parameters

The following parameters were considered: triglycerides (TGs), cholesterol, glucose and fructosamine concentrations using the methods from [40]. The emission spectra were recorded using the Jasco FP-8200 spectrofluorimeter. The method used for the determination of triglycerides was Enzymatic Glycerol Phosphate Oxidase/Peroxidase [41], whereas for determining cholesterolemia, the Enzymatic Cholesterol Oxidase/Peroxidase method was used [42]. As for the blood glucose concentration, the Enzymatic Glucose Oxidase/Peroxidase method was used [43] and for blood fructosamine the Enzymatic glycated-serum-protein assay was used [44].

2.4. Statistical Analysis

GraphPad Prism software was used for statistical analysis. To test the normality data, the Shapiro–Wilk test was used. Parametric (ANOVA) or non-parametric (Kruskal–Wallis) tests were performed to compare the groups. When F was statistically significant ($p \leq 0.05$), Tukey's Honest Significant Difference test was applied. Means ± SD were used to express the results.

3. Results

3.1. The Composition of *Cannabis sativa* 'Zenit', Antioxidant Activity, Identification and Quantification of Phenols and Fatty Acids from Hemp Oil

3.1.1. The Composition of *Cannabis sativa* 'Zenit', Antioxidant Activity, Identification and Quantification of Phenols and Fatty Acids from Hemp Oil

Strong causality connections can be found between all the results. Dry matter, lipids, proteins and ash content of hemp seeds are presented in Table 1. The average protein content, 20.95%, and the average lipid content, 29.17%, are the important ones because all the active compounds are proteins and lipids.

Table 1. *Cannabis sativa* 'Zenit' composition.

Analyses	Dry Matter (%)	Lipid Content (%)	Protein Content (%)	Ash Content (%)
Mean ± SD	94.57 ± 0.16	29.17 ± 0.46	20.95 ± 0.21	4.61 ± 0.33

The results obtained were checked against the dry substance; all the results are expressed as means ± standard deviation (M ± SD); SD = standard deviation.

3.1.2. The Composition of Total Phenols and Total Flavonoids

Analyses of phenols and flavonoids are essential because these substances act as antioxidants, which results in their anti-inflammatory role in various diseases [45]. The results are presented in Table 2 as means ± standard deviation and are expressed in mg GAE/g for phenols (2.75 ± 0.12) and ug CE/g for flavonoids (106.21 ± 5.92).

Table 2. The composition of total phenols and total flavonoids.

Type of Analyses	Total Phenols	Total Flavonoids
UM	2.75 ± 0.12 mg GAE/g	106.21 ± 5.92 µg CE/g

UM—unit of measurement; the results are presented as means ± standard deviation (M ± SD). GAE, milligrams of gallic acid equivalents per gram; CE, micrograms catechin equivalents per gram.

3.1.3. The Antioxidant Activity

As is seen in Table 3, the antioxidant activity is increased in methanol extract (112.28 µmol Trolox/100 g sample for methanolic extract and 97.09 µmol Trolox/100 g sample for ethanolic extract). Both methods (methanolic and ethanolic extraction) were used to identify the most sensitive for antioxidant identification and quantification. It is shown that the methanolic method is more specific for hemp seeds.

Table 3. Antioxidant activity of hemp extracts (methanolic and ethanolic extracts).

Samples	uM Trolox/100 g (Mean ± SD)
Methanol extract	112.28 ± 5.61 µmol Trolox/100 g sample
Ethanolic extract	97.09 ± 4.85 µmol Trolox/100 g sample

The values are expressed as micromol Trolox equivalents (µmol TE)/100 g sample. The results are presented as means ± standard deviation (M ± SD).

3.1.4. Phenolic Compounds

As presented in Table 4, 14 phenol compounds were identified and quantified. Lignanamides were found in the largest amount (total lignanamides 2493.17 µg/g hemp seeds for methanol and total 1500.48 µg/g hemp seeds for ethanol) compared to total hydroxycinnamic acid amides (814.78 µg/g hemp seeds for methanol and 444.62 µg/g hemp seeds for ethanol), total hydroxybenzoic acid (504.29 µg/g hemp seeds for methanol and 423.70 µg/g hemp seeds for ethanol) and benzoic acid (93.78 µg/g hemp seeds for methanol and 116.08 µg/g hemp seeds for ethanol). From total lignanamides, we had cannabisins A, B, C, D, F, Q, cannabisin B isomer and grossamide.

Table 4. Identification and quantification of phenolic compounds in the hemp seed sample (μg/g hemp seeds) via HPLC-DAD-MS-ESI+ analysis.

Peak	R_t (min)	UV λ_{max} (nm)	$[M + H]^+$ (m/z)	Phenolic Compound	Subclass	Hemp Seeds Methanol (Mean ± SD)	Hemp Seeds Ethanol (Mean ± SD)
1	2.78	270	123	Benzoic acid	Benzoic acid	93.78 ± 9.40	116.08 ± 11.20
2	2.97	270	139	2-Hydroxybenzoic acid	Hydroxybenzoic acid	210.06 ± 10.20	271.33 ± 13.57
3	9.47	280	155	Protocatechuic acid	Hydroxybenzoic acid	294.23 ± 14.56	152.37 ± 7.60
4	10.87	290	595	Cannabisin A	Lignanamide	82.68 ± 7.34	56.11 ± 4.80
5	11.48	290	597	Cannabisin B	Lignanamide	130.05 ± 10.07	149.54 ± 12.46
6	11.78	320,290	284	*N-trans*-Coumaroyltyramine	Hydroxycinnamic acid amide	196.91 ± 9.85	172.56 ± 8.63
7	14.01	290	611	Cannabisin C	Lignanamide	179.65 ± 8.98	203.55 ± 10.15
8	14.62	290	597	Cannabisin B isomer	Lignanamide	153.96 ± 7.50	146.88 ± 7.23
9	18.20	322,290	301	*N-trans*-Caffeoyltyramine	Hydroxycinnamic acid amide	339.04 ± 16.95	181.42 ± 9.07
10	19.29	340,290	597	Cannabisin Q	Lignanamide	201.34 ± 10.03	171.67 ± 8.85
11	20.40	330,290	314	*N*-Feruloyltyramine	Hydroxycinnamic acid amide	278.83 ± 11.15	90.65 ± 8.53
12	22.09	290	625	Cannabisin D	Lignanamide	444.42 ± 17.71	119.87 ± 10.20
13	22.79	290	625	Cannabisin F	Lignanamide	470.99 ± 23.50	231.01 ± 11.55
14	23.39	320,290	625	Grossamide	Lignanamide	830.08 ± 41.50	421.84 ± 21.09
				Total phenolics		3906.02	2484.88

3.1.5. Fatty Acid Composition from Seed Oil

The results are shown in percentages from areas and presented in Tables 5 and 6, and Figure 2. As can be observed in Table 6, the highest percentage of fatty acids is represented by polyunsaturated fatty acids, PUFAs, which are crucial, as they are considered indirect antioxidants [46]. The ratio between *n*-6 and *n*-3 (*n*-6/*n*-3), which is the most important for the prevention of various diseases [47] is also presented in Table 6. The *n*-6/*n*-3 ratio is 4.41 and appears to be optimal for preventing and inhibiting disease progression.

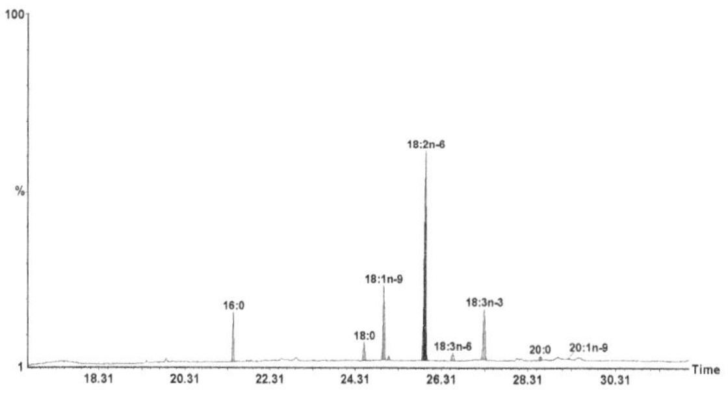

Figure 2. GC-MS chromatogram of FAMEs from hemp seeds analyzed with a Supelcowax 10 capillary column. Peak identification: 16:0, palmitic; 18:0, stearic; 18:1*n*-9, oleic; 18:1*n*-7, vaccenic; 18:2*n*-6, linoleic; 18:3*n*-6, γ-linolenic; 18:3*n*-3, α-linolenic; 20:0, arachidic; 20:1*n*-9, 11-eicosenoic acids.

Table 5. The specific fatty acids from *Cannabis sativa* seed oil.

Specific Fatty Acids	Concentration (% by Area)	Standard Deviation (SD)
16:0	7.84	0.27
18:0	3.68	0.15
18:1*n*-9	15.73	0.66
18:1*n*-7	0.71	0.05
18:2*n*-6	55.93	2.49
18:3*n*-6	1.73	0.08
18:3*n*-3	13.06	0.58
20:0	0.97	0.06
20:1*n*-9	0.35	0.03

16:0, palmitic; 18:0, stearic; 18:1*n*-9, oleic; 18:1*n*-7, vaccenic; 18:2*n*-6, linoleic; 18:3*n*-6, γ-linolenic; 18:3*n*-3, α-linolenic; 20:0, arachidic; 20:1*n*-9, 11-eicosenoic acids. The results are presented in percentages. The values represent the means of three samples, analyzed individually in triplicate ($n = 3 \times 3$). SD—standard deviation.

Table 6. Fatty acid type from *Cannabis sativa* seed oil.

Fatty Acid Type	Concentration (% by Area)	Standard Deviation (SD)
SFA	12.49	0.61
MUFA	16.78	0.69
PUFA	70.72	2.97
n-3 PUFA	13.06	0.59
n-6 PUFA	57.66	2.57
n-6/*n*-3	4.41	
PUFAs/SFAs	5.66	

SFAs, saturated fatty acids; MUFAs, monounsaturated fatty acids; PUFAs, polyunsaturated fatty acids. The values represent the means of three samples, analyzed individually in triplicate ($n = 3 \times 3$). SD—standard deviation.

3.2. *Glycemia and Weight during Diabetes Induction in Rats*

Glycaemia and weight presented a different trend (Table 7). Both weight and blood glucose were monitored 5 times every two weeks before the stabilization of diabetes. Until the diabetes was stabilized, the blood glucose values varied as seen in Table 7 [48]. Regarding the variation in glucose concentration during diabetes induction, it was shown that the concentrations were higher in both diabetic groups compared to the control group. In contrast, the weight trend was different between means, but the confidence intervals overlap after diabetes induction, namely 246.8 ± 9.2 g in DNFG, and 249.3 ± 17.8 g in DHFG, compared to CG, which had 254.9 ± 17.9.

Table 7. Glucose and weight prior to the experiment.

Parameters	CG	DNFG	DHFG
Glycemia (mg/dL)	112 ± 11.9	422.3 ± 120.2	398.8 ± 98.9
Weight (g)	254.9 ± 17.9	246.8 ± 9.2	249.3 ± 17.8

CG, the Control Group, received water and normal food; in the DNFG, the Diabetic Normal Feed Group, the rats were fed with normal feed; in the DHFG, the Diabetic Hemp Seeds Feed Group, the animals received only hemp seeds as feed; the values are presented as mean ± standard deviation (M ± SD), $n = 6$.

3.3. Blood Biochemical Changes after Hemp Seed Treatment in Diabetic Rats Compared to CG

There were significant differences in glucose and cholesterol concentrations (between the DNFG and DHFG). Regarding triglyceride blood concentration, there were significant changes $p \leq 0.05$ (*) between the CG and DNFG groups. The cholesterol concentration increased significantly in the diabetic groups compared to the control, with a specification that the hemp seeds decrease the effects of diabetes, $p \leq 0.001$ (***). As expected, glucose and fructosamine followed the same trend; more significantly, fructosamine had $p \leq 0.001$, (***) compared to glucose, which had $p \leq 0.01$ (**), regarding CG and DNFG.

4. Discussion

4.1. The Composition of Cannabis sativa 'Zenit', Antioxidant Activity, Identification and Quantification of Phenols and Fatty Acids from Hemp Oil

As diabetes mellitus represents a major problem for public health, beneficial alternatives are sought in order to prevent the occurrence of this disease [49]. One of them is linked to embracing a healthy lifestyle. In this respect, diet plays a crucial role. The nutritional value of diets is of the utmost importance as daily menus nowadays are poor in nutrients and very rich in calories. In addition, the majority of foods are improved with salts, carbohydrates and fats [50]. A good alternative is to consume more vegetables, fruits, nuts (like hemp seeds), healthy meat, milk and eggs. Therefore, hemp seeds were selected to use in this respect. From the composition of hemp seeds, flavonoids, among others compounds such as lignamides, PUFAs and proteins are recognized for their antidiabetic effects [19]. For instance, flavonoids are involved in the regulation of the insulin-signaling pathway, glucose uptake, as well as adipogenesis [51]. Flavonoids are important antioxidants with antimicrobial, anti-allergic and anti-inflammatory action in both plants and animals. This action is due to their ability to reduce oxidative stress. Moreover, their capacity is doubled by the fact that, once ingested, flavonoids are degraded to phenolic acids, which potentiate this capacity to capture free radicals.

As such, through their antioxidant role in vivo, they determine the decrease of low-density lipoproteins and restore the homeostasis of polyunsaturated fatty acids in plasma membranes [45]. Moreover, they can have a good influence on β-cell proliferation. In this way, insulin production increases at the same time as hyperglycemia decreases [52]. Phenol compounds such as lignan amides from seeds are capable of anti-inflammatory effects [53]. Since oxidative stress is linked to various pathological conditions, the administration of products with antioxidant activity has been pursued. In this way, as is seen in Table 3, hemp seeds have increased antioxidant capacity [54]. These results may explain health trends from biochemical analyses. Moreover, it is very important to specifically observe which antioxidant compound was responsible for these changes. In this way, the consumption of hydroxybenzoic acid was correlated with hypoglycemic activity [55].

Generally, in plants, lipid composition, and especially fatty acids, differ between taxa. This is due to several factors such as geographical area, soil composition and good environmental conditions [56]. As previously observed, PUFA levels from hemp 'Zenit' are higher (Table 6). They are involved in insulin action, membrane cell integrity, cellular-signaling pathway and immune function [57]. According to the literature, there are two groups of long-chain PUFAs: ω-3 and ω-6. The most representative PUFAs are the following: α-linolenic, γ-linolenic, eicosenoic acids, which belong to ω-3, while the ω-6 group includes arachidonic and linoleic acids [58]. As far as PUFAs are concerned, literature data focus on in vitro as well as in vivo studies, which demonstrates that omega-3 acids in particular act as indirect antioxidants in vascular endothelial cells. In this way, they are anti-inflammatory compounds, and therefore their administration reduces the risk of atherosclerosis [46]. As can be observed in Table 5, hemp seeds contain γ-linolenic acid (GLA), which is known for its beneficial effects on health. Firstly, the GLA from the diet is used as a prophylactic compound in different states of some chronic diseases. This is possible due to its ability to change the lipid content from cellular membranes such as prostaglandin E1 and 15-(S)-hydroxy-8,11,13-eicosatrienoic. This change is very healthful because of its antiproliferative and anti-inflammatory effects [59]. Moreover, GLA can be used as treatment for diabetic nephropathy [60]. Regarding the roles of PUFAs, the ratio between two categories of fatty acids needs to be taken into account. For example, in cardiovascular diseases, a 4/1 ratio is correlated with a significant change in mortality, to approximately 70%. At the same time regarding colorectal cancer, a 2.5/1 ratio was associated with a decrease in cell proliferation [47].

4.2. Glycemia and Weight during the Experiment

As can be seen in Table 7, there were differences between the weight means of groups, but the confidence intervals overlap (Table 7). This may be due to polyuria determined by diabetes. Additionally, STZ is a toxic substance, and the rats' appetite might have been lower, especially during diabetes induction.

A different trend was observed regarding glycemia variation. The increased glucose concentration in DNFG rats compared to CG is due to STZ diabetic effects [61]. As already shown, STZ through alkylation causes the death of β pancreatic cells via glucose transporters 2—GLUT2 [61]. GLUT2s act as sensors and they are responsible for the control of glucose intake by different tissues [62]. Thus, the glucose concentration was different.

4.3. Blood Biochemical Changes after the Treatment with Hemp Seeds in Diabetic Rats

Biochemical changes are essential because they closely reflect the health of the body. Certain parameters were chosen due to their evolution and usefulness in diabetes assessment (Figure 3). As shown in Figure 3a, a significant decrease in triglyceride concentration was observed after the administration of STZ. On the one hand, STZ administration in rats might have produced a decrease in their appetite and in the case of malnutrition, triglycerides were lowered [63]. This aspect may be supported by the same trend in weight with the same modification observed in children who suffer from protein energy malnutrition. On the other hand, there were studies that reported a high level of triglycerides after diabetes induction of STZ, but following a long period of treatment, from 6 to 12 weeks [64,65]. The rats in this study were treated for five weeks. Hemp seed administration may have increased TG concentration compared to DNFG. The only explanation may be linked to the fact that some compounds from hemp seeds are used as therapeutic agents such as phenolic ones and PUFAs in different diseases and are capable of restoring membrane, structural and biochemical homeostasis [18,66,67]. As expected, cholesterol concentration (Figure 3b) was significantly increased after diabetes induction. Previous studies have shown that this change is normal due to STZ [68]. However, these cholesterol abnormalities can be attributed to insulin deficiency caused by the death of pancreatic β cells. Due to these modifications, cholesterol synthesis is increased. Moreover, the higher rate of LDL glycation is responsible for suppressing cholesterol degradation, which also explains the increase in glycemia after diabetes induction [69].

Considering that the total cholesterol concentration was determined, this variation can be somewhat correlated with the low concentration of triglycerides in the blood. This can be explained by the fact that in the liver, they can bind to VLDL and form VLDL-TG complexes [70]. Interestingly, after the consumption of hemp seeds, cholesterolemia significantly decreased. This significant finding confirms previous studies which showed the same trend in the case of omega-3 fatty acid treatment [71,72]. The mechanism is linked to the capacity of omega-3 fatty acids to stimulate the degradation of apolipoprotein B-100-containing lipoproteins. Finally, cholesterol synthesis is reduced [73]. This is possible due to the mediated capacity of PUFAs, especially omega-3 acids, to suppress the signaling pathway of sterol regulatory element-binding protein-1 (SREBP-1) [74] (Figure 4).

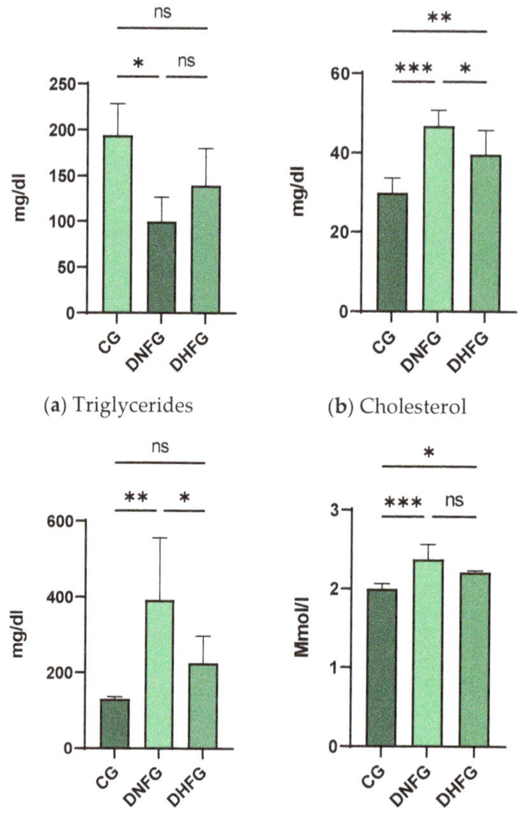

(**a**) Triglycerides (**b**) Cholesterol

(**c**) Glucose (**d**) Fructosamine

Figure 3. Blood biochemical changes for (**a**) triglyceride, (**b**) cholesterol, (**c**) glucose and (**d**) fructosamine after 5 weeks of treatment in CG, Control Group, which received water and normal feed; DNFG, Diabetic Normal Feed Group—the rats were fed with normal feed; DHFG, Diabetic Hemp Seeds Feed Group—the animals received only hemp seeds as feed. Values are means ± standard deviation (M ± SD), $n = 6$, $p \leq 0.05$ (*), $p \leq 0.01$ (**) and $p \leq 0.001$ (***) (Kruskal–Wallis test for triglycerides and one-way ANOVA for the other parameters).

As previously reported, STZ administration led to diabetes and in this respect, glycemia was significantly increased in DNFG (Figure 3c) compared to CG [78]. The administration of hemp seeds determined a decrease in glycemia most probably due to their antioxidant content such as phenol compounds, flavonoids and PUFAs, which are recognized for their hypoglycemic effects [18,79] (Figure 5). It is well established that flavonoids are capable of repairing DNA damage. As such, the alkylation induced via STZ may have been suppressed by hydroxycinnamic acid and flavonoid action from hemp seeds [80]. Moreover, Hydroxybenzoic acid has protective effects on pancreatic β cells [81]. In addition, the high level of protein from hemp seeds may be responsible for the decrease in glucose concentration via a mechanism that involves two peptides (Pro-Leu-Met-Leu-Pro, Leu-Arg) which can cause the inhibition of α-glucosidase activity [26]. To some extent, this would still be possible in vivo, only if stem cells were transplanted, and were transformed into pancreatic β cells based on the signaling pathways [82]. As such, glycemia decreased.

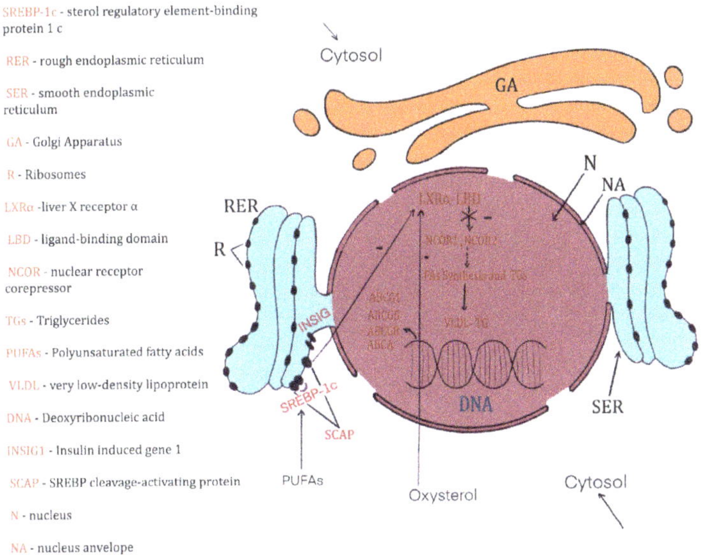

Figure 4. The mechanism by which PUFAs suppress cholesterol synthesis via inhibition of the signaling pathway of sterol regulatory element-binding protein-1 (SREBP-1) [74]. PUFAs through their action on the SCAP complex inhibit LXRα-LBD, which further generates suppression of nuclear corepressor receptors NCOR1 and NCOR2 [75]. All these inhibit the synthesis of fatty acids (FAs) as well as triglycerides (TGs). In this way, VLDL-TG is no longer synthesized. Similar to PUFAs, the high concentration of oxysterol has the same effect; it inhibits LXRα-LBD [76,77].

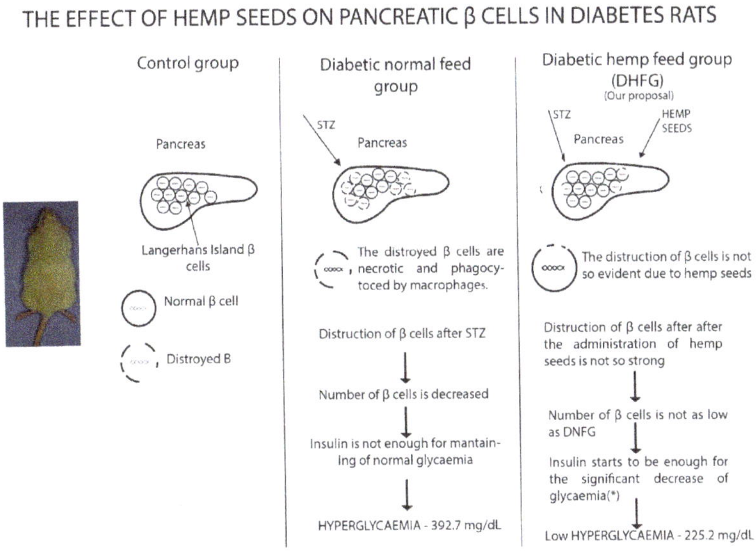

Figure 5. The effect of hemp seeds on pancreatic β cells in diabetes rats. * significant decrease.

The CG, the Control Group, received water and normal food; in the DNFG, the Diabetic Normal Feed Group, the rats were fed with normal feed; in the DHFG, the Diabetic Hemp Seeds Feed Group, the animals received only hemp seeds as feed. Diabetes was induced via streptozotocin (STZ) in Wistar rats. Pancreatic β cells were destroyed. As a result, the number of pancreatic β cells decreased, and the secreted insulin was not sufficient to maintain a normal blood glycemia level. Therefore, all rats had hyperglycemia (392.7 mg/dL). In the rats that ate only hemp seeds, blood glycemia decreased significantly compared to the previous ones (225.2 mg/dL). It is thus assumed that the effects of streptozotocin were inhibited by compounds found in hemp seeds.

Fructosamine (Figure 3d) was used as it is a useful test to verify glucose homeostasis [83,84]. Generally, it is frequently used for the assessment of diabetes mellitus [85]. The term fructosamine is specific for all glycated proteins from serum. The formation of fructosamine compounds is based on a non-enzymatic reaction [86]. As we see in Figure 3d, fructosamine increased after STZ. This change is normal in diabetes and follows the same trend as glycemia [83]. In the DHFG, the consumption of hemp seeds insignificantly decreased fructosamine concentration. Nevertheless, in this case the trend was the same as that of glucose. In this respect, a possible explanation is PUFA hemp seeds. Yoshimura and his coworkers have demonstrated that PUFAs decreased blood glucose concentration as well as fructosamine [87]. Moreover, this modification can be produced via GLA, because it is responsible for insulin sensitivity in rat muscle tissue and ultimately for diabetes suppression [88–90].

5. Conclusions

In conclusion, our results show that the administration of hemp seeds can modify different biochemical parameters in rats. The significant decreases in glucose and cholesterol concentrations after hemp seed consumption are the most relevant changes in this sense. Once they occur, we can hypothesize that different active compounds, which are found in higher quantity in hemp seeds and are recognized for their anti-diabetic effects, can inhibit pancreatic β-cell destruction, even if the mechanism remains unknown. The high content of antioxidants such as phenol compounds, PUFAs, flavonoids and proteins may be responsible for these health effects. Considering that the present research is a pilot study, it would be necessary to investigate long-term effects with respect to flavonoids, specific phenolic compounds, GLA and PUFAs to find which of them are responsible for these health changes and the mechanisms underlying them. However, hemp seeds may represent a healthy alternative for type 2 diabetes. Extrapolating the results to other animals and humans will represent a necessary requirement.

Author Contributions: Conceptualization, C.M. and I.P.; methodology, S.M.M., C.U., F.R. and F.D.; software, A.O.; validation, L.M., D.P. and M.M.; formal analysis, C.M.; investigation, M.M., F.D. and S.M.M.; resources, L.M., D.P. and S.M.M.; data curation, C.M., F.R., T.M. and C.U.; writing—original draft preparation, C.M. and A.O.; writing—review and editing, C.M., A.O. and I.P.; visualization, T.M. and L.M.; supervision, D.P. and M.M.; project administration, C.M. and I.P.; funding acquisition, F.D. All authors have read and agreed to the published version of the manuscript.

Funding: This work was supported by a grant of the Ministry of Research. Innovation and Digitization, CNCS—UEFISCDI, project number PN-III-P4-PCE-2021-0750, within PNCDI III.

Institutional Review Board Statement: The animal study protocol was approved by the Bioethics Committee of UASVM Cluj-Napoca (340/2022).

Informed Consent Statement: Not applicable.

Data Availability Statement: Data is contained within the article.

Conflicts of Interest: The authors declare no conflict of interest.

References

1. Kleinert, M.; Clemmensen, C.; Hofmann, S.M.; Moore, M.C.; Renner, S.; Woods, S.C.; Huypens, P.; Beckers, J.; de Angelis, M.H.; Schürmann, A.; et al. Animal models of obesity and diabetes mellitus. *Nat. Rev. Endocrinol.* **2018**, *14*, 140–162. [CrossRef] [PubMed]
2. Kahn, S.E.; Hull, R.L.; Utzschneider, K.M. Mechanisms linking obesity to insulin resistance and type 2 diabetes. *Nature* **2006**, *444*, 840–846. [CrossRef] [PubMed]
3. Majewski, M.K.; Ognik, K.; Juśkiewicz, J. The interaction between resveratrol and two forms of copper as carbonate and nanoparticles on antioxidant mechanisms and vascular function in *Wistar rats*. *Pharmacol. Rep.* **2019**, *71*, 862–869. [CrossRef]
4. Majewski, M.K.; Ognik, K.; Juśkiewicz, J. The antioxidant status, lipid profile, and modulation of vascular function by fish oil supplementation in nano-copper and copper carbonate fed *Wistar rats*. *J. Funct. Foods* **2020**, *64*, 103595. [CrossRef]
5. Kaushal, N.; Dhadwal, S.; Kaur, P. Ameliorative effects of hempseed (*Cannabis sativa*) against hypercholesterolemia associated cardiovascular changes. *Nutr. Metab. Cardiovasc. Dis.* **2020**, *30*, 330–338. [CrossRef]
6. IT IS. *Cannabis sativa* TSN 19109. 2023. Available online: https://explorer.natureserve.org/Taxon/ELEMENT_GLOBAL.2.131607/Cannabis_sativa (accessed on 23 February 2023).
7. Farinon, B.; Molinari, R.; Constantini, L.; Merendino, N. The seed of industrial hemp (*Cannabis sativa* L.): Nutritional Quality and Potential Functionality for Human Health and Nutrition. *Nutrition* **2020**, *12*, 1935. [CrossRef]
8. Placido, D.F.; Lee, C.C. Potential of industrial hemp for phytoremediation of heavy metals. *Plants* **2022**, *11*, 595. [CrossRef]
9. Salentijn, E.M.J.; Zhang, Q.; Amaducci, S.; Yang, M.; Trindade, L.M. New developments in fiber hemp (*Cannabis sativa* L.) breeding. *Ind. Crops Prod.* **2015**, *68*, 32–41. [CrossRef]
10. Zhou, Q.; Huang, F.; Zheng, C.; Guo, P.; Li, W.; Liu, C.; Wan, C. Physicochemical properties and volatile terpene of hempseed oils in Bama region. *Oil Crop Sci.* **2017**, *1*, 13–22.
11. Callaway, J.C. Hempseed as a nutritional resource: An overview. *Euphytica* **2004**, *140*, 65–72. [CrossRef]
12. Szlas, A.; Kurek, J.M.; Krejpcio, Z. The Potential of L-Arginine in Prevention and Treatment of Disturbed Carbohydrate and Lipid Metabolism—A Review. *Nutrients* **2022**, *14*, 961. [CrossRef]
13. Rodriguez-Leyva, D.; Pierce, G.N. The cardiac and haemostatic effects of dietary hempseed. *Nutr. Metab.* **2010**, *7*, 32. [CrossRef]
14. Sokoła-Wysoczańska, E.; Wysoczanski, T.; Wagner, J.; Czyz, K.; Bodkowski, R.; Lochynski, S.; Patkowska-Sokola, B. Polyunsaturated fatty acids and their potential therapeutic role in cardiovascular system disorders—A review. *Nutrients* **2018**, *10*, 1561. [CrossRef]
15. Holub, B.J. Clinical nutrition: 4. Omega-3 fatty acids in cardiovascular care. *Cmaj* **2002**, *166*, 608–615.
16. Lunn, J.; Theobald, H.E. The health effects of dietary unsaturated fatty acids. *Nutr. Bull.* **2006**, *31*, 178–224. [CrossRef]
17. Opyd, P.M.; Jurgonski, A.; Fotschki, B.; Juskiewicz, J. Dietary hemp seeds more effectively attenuate disorders in genetically obese rats than their lipid fraction. *J. Nutr.* **2020**, *150*, 1425–1433. [CrossRef] [PubMed]
18. Singh, N.J.; Sood, S.V. Effect of gamma linolenic acid pretreatment on diabetic neuropathy in rats. *Int. J. Basic Clin. Pharmacol.* **2013**, *2*, 320–325. [CrossRef]
19. Pollastro, F.; Minassi, A.; Fresu, L.G. Cannabis phenolics and their bioactivities. *Curr. Med. Chem.* **2018**, *25*, 1160–1185. [CrossRef] [PubMed]
20. Radwan, M.M.; ElSohly, M.A.; Slade, D.; Ahmed, S.A.; Wilson, L.; El-Alfy, A.T.; Khan, I.A.; Ross, S.A. Non-cannabinoid constituents from a high potency *Cannabis sativa* variety. *Phytochemistry* **2008**, *69*, 2627–2633. [CrossRef]
21. Mkpenie, V.N.; Essien, E.E.; Udoh, I.I. Effect of extraction conditions on total polyphenol contents, antioxidant and antimicrobial activities of *Cannabis sativa* L. *Electron. J. Environ. Agric. Food Chem.* **2011**, *11*, 300–307.
22. Mandal, S.M.; Chakraborty, D.; Dey, S. Phenolic acids act as signaling molecules in plant-microbe symbioses. *Plant Signal. Behav.* **2010**, *5*, 359–368. [CrossRef] [PubMed]
23. Pan, A.; Sun, J.; Chen, Y.; Ye, X.; Li, H.; Yu, Z.; Wang, Y.; Gu, W.; Zhang, X.; Chen, X.; et al. Effects of a flaxseed-derived lignan supplement in type 2 diabetic patients: A randomized, double-blind, cross-over trial. *PLoS ONE* **2007**, *2*, e1148. [CrossRef] [PubMed]
24. Kasetti, R.B.; Nabi, S.A.; Swapna, S.; Apparao, C. Cinnamic acid as one of the antidiabetic active principle(s) from the seeds of Syzygium alternifolium. *Food Chem. Toxicol.* **2012**, *50*, 1425–1431. [CrossRef] [PubMed]
25. Sales, P.M.; Souza, P.M.; Simeoni, L.A.; Silveira, D. α-Amylase inhibitors: A review of raw material and isolated compounds from plant source. *J. Pharm. Pharm. Sci.* **2012**, *15*, 141–183. [CrossRef] [PubMed]
26. Ren, Y.; Liang, K.; Zhang, M.; Chen, Y.; Wu, H.; Lai, F. Identification and characterization of two novel α-glucosidase inhibitory oligopeptides from hemp (*Cannabis sativa* L.) seed protein. *J. Funct. Foods* **2016**, *26*, 439–450. [CrossRef]
27. George, W.; Latimer, J.R. *Official Methods of Analysis of AOAC*, 21st ed.; AOAC International: Gaithersburg, MD, USA, 2019.
28. Brand-Williams, W.; Cuvelier, M.E.; Berset, C.L.W.T. Use of a free radical method to evaluate antioxidant activity. *LWT-Food Sci. Technol.* **1995**, *28*, 25–30. [CrossRef]
29. Benkirane, C.; Moumen, A.B.; Fauconnier, M.L.; Belhaj, K.; Abid, M.; Caid, H.S.; Elamrani, A.; Mansouri, F. Bioactive compounds from hemp (*Cannabis sativa* L.) seeds: Optimization of phenolic antioxidant extraction using simplex lattice mixture design and HPLC-DAD/ESI-MS 2 analysis. *RSC Adv.* **2022**, *12*, 25764–25777. [CrossRef]
30. Christie, W.W.; Brechany, E.Y.; Shukla, V.K.S. Analysis of seed oils containing cyclopentenyl fatty acids by combined chromatographic procedures. *Lipids* **1989**, *24*, 116–120. [CrossRef]

31. Dulf, F.V.; Vodnar, D.C.; Tosa, M.I.; Dulf, E.H. Simultaneous enrichment of grape pomace with γ-linolenic acid and carotenoids by solid-state fermentation with Zygomycetes fungi and antioxidant potential of the bioprocessed substrates. *Food Chem.* **2020**, *310*, 125927. [CrossRef]
32. Bindea, M.; Rusu, B.; Rusu, A.; Trif, M.; Leopold, L.F.; Dulf, F.; Vodnar, D.C. Valorification of crude glycerol for pure fractions of docosahexaenoic acid and β-carotene production by using Schizochytrium limacinum and Blakeslea trispora. *Microb. Cell Factories* **2018**, *17*, 1–13. [CrossRef]
33. Lewis, C.; Barbiers, A.R. Streptozotocin, a new antibiotic. In vitro and in vivo evaluation. *Antibiot. Annu.* **1959**, *7*, 247–254. [PubMed]
34. Rerup, C.C. Drugs producing diabetes through damage of the insulin secreting cells. *Pharmacol. Rev.* **1970**, *22*, 485–518.
35. Furman, B.L. Streptozotocin-induced diabetic models in mice and rats. *Curr. Protoc. Pharmacol.* **2015**, *70*, 5–47. [CrossRef] [PubMed]
36. National Research Coucil. *Guide for the Care and Use of Laboratory Animals*; The National Academies Press: Washington, DC, USA, 2010.
37. Council Directive 86/609/EEC of 24 November 1986 on the approximation of laws, regulations and administrative provisions of the Member States regarding the protection of animals used for experimental and other scientific purposes. *Off. J. Eur. Commun.* **1986**, *29*, L358.
38. Kohn, D.F.; Clifford, C.B. Biology and diseases of rats. In *Laboratory Animal Medicine*; Academic Press: Cambridge, MA, USA, 2002; pp. 121–165. [CrossRef]
39. Kim, Y.S.; Kim, N.H.; Lee, S.W.; Lee, Y.M.; Jang, D.S.; Kim, S.H. Effect of protocatechualdehyde on receptor for advanced glycation end products and TGF-β1 expression in human lens epithelial cells cultured under diabetic conditions and on lens opacity in streptozotocin-diabetic rats. *Eur. J. Pharmacol.* **2007**, *569*, 171–179. [CrossRef]
40. Johnson-Delaney, C.A. *Exotic Companion Medicine Handbook for Veterinarians*; Wingers Publishing Incorporated: Lake Worth, FL, USA, 1996.
41. Baggio, G.; Donazzan, S.; Monti, D.; Mari, D.; Martini, S.; Gabelli, C.; Dalla Vestra, M.; Previato, L.; Guido, M.; Pigozzo, S.; et al. Lipoprotein(a) and lipoprotein profile in healthy centenarians: A reappraisal of vascular risk factors. *FASEB J.* **1998**, *12*, 433–437. [CrossRef] [PubMed]
42. Allain, C.C.; Poon, L.S.; Chan, C.S.; Richmond, W.; Fu, P.C. Enzymatic determination of total serum cholesterol. *Clin. Chem.* **1974**, *20*, 470–475. [CrossRef] [PubMed]
43. Tietz, N.W. *Clinical Guide to Laboratory Tests*; W.B. Saunders, Co.: Philadelphia, PA, USA, 1995; p. 1096.
44. O'Brien, J.E.; Brookes, M. Determination of reference values for a novel ketoamine-specific fructosamine assay for assessment of diabetic glycemic control. *Diabetes Technol. Ther.* **1999**, *1*, 447–455. [CrossRef]
45. Pietta, P.G. Flavonoids as antioxidants. *J. Nat. Prod.* **2000**, *63*, 1035–1042. [CrossRef]
46. Richard, D.; Kefi, K.; Barbe, U.; Bausero, P.; Visioli, F. *Polyun saturated* fatty acids as antioxidants. *Pharmacol. Res.* **2008**, *57*, 451–455. [CrossRef]
47. Simopoulos, A.P. The importance of the ratio of omega-6/omega-3 essential fatty acids. *Biomed. Pharm.* **2002**, *56*, 365–379. [CrossRef] [PubMed]
48. Rossini, A.A.; Like, A.A.; Chick, W.L.; Appel, M.C.; Cahill, G.F., Jr. Studies of streptozotocin-induced insulitis and diabetes. *Proc. Natl. Acad. Sci. USA* **1977**, *74*, 2485–2489. [CrossRef] [PubMed]
49. Chen, L.; Magliano, D.J.; Zimmet, P.Z. The worldwide epidemiology of type 2 diabetes mellitus—Present and future perspectives. *Nat. Rev. Endocrinol.* **2012**, *8*, 228–236. [CrossRef] [PubMed]
50. Willett, W.C.; Koplanet, J.P.; Nugent, R.; Dusenbury, C.; Puska, P.; Gaziano, T.A. Prevention of Chronic Disease By means of Diet and Lifestyle Changes. In *Disease Control Priorities in Developing Countries*, 2nd ed.; World Bank: Washington, DC, USA, 2006.
51. Vinayagam, R.; Xu, B. Antidiabetic properties of dietary flavonoids: A cellular mechanism review. *Nutr. Metab.* **2015**, *12*, 60. [CrossRef]
52. Graf, B.A.; Milbury, P.E.; Blumberg, J.B. Flavonols, flavones, flavanones, and human health: Epidemiological evidence. *J. Med. Food* **2005**, *8*, 281–290. [CrossRef]
53. Flores-Sanchez, I.J.; Verpoorte, R. Secondary metabolism in cannabis. *Phytochem. Rev.* **2008**, *7*, 615–639. [CrossRef]
54. Girgih, A.T.; Alashi, A.M.; He, R.; Malomo, S.A.; Raj, P.; Netticadan, T.; Aluko, R.E. A novel hemp seed meal protein hydrolysate reduces oxidative stress factors in spontaneously hypertensive rats. *Nutrients* **2014**, *6*, 5652–5666. [CrossRef]
55. Peungvicha, P.; Temsiririrkkul, R.; Prasain, J.K.; Tezuka, Y.; Kadota, S.; Thirawarapan, S.S.; Watanabe, H. 4-Hydroxybenzoic acid: A hypoglycemic constituent of aqueous extract of *Pandanus odorus* root. *J. Ethnopharmacol.* **1998**, *62*, 79–84. [CrossRef]
56. Giovino, A.; Marino, P.; Domina, G.; Rapisarda, P.; Rizza, G.; Saia, S. Fatty acid composition of the seed lipids of Chamaerops humilis L. natural populations and its relation with the environment. *Plant Biosyst.* **2015**, *149*, 767–776. [CrossRef]
57. Benatti, P.; Peluso, G.; Nicolai, R.; Calvani, M. Polyunsaturated fatty acids: Biochemical, nutritional and epigenetic properties. *J. Am. Coll. Nutr.* **2004**, *23*, 281–302. [CrossRef]
58. Michalak, A.; Mosińska, P.; Fichna, J. Polyunsaturated fatty acids and their derivatives: Therapeutic value for inflammatory, functional gastrointestinal disorders, and colorectal cancer. *Front. Pharmacol.* **2016**, *7*, 459. [CrossRef]
59. Schroeder, H.W., Jr.; Cavacini, L. Structure and function of immunoglobulins. *J. Allergy Clin. Immunol.* **2010**, *125*, S41–S52. [CrossRef] [PubMed]

60. Kim, D.H.; Yoo, T.H.; Lee, S.H.; Kang, H.Y.; Nam, B.Y.; Kwak, S.J.; Kim, J.K.; Park, J.T.; Han, S.J.; Kang, S.W. Gamma linolenic acid exerts anti-inflammatory and anti-fibrotic effects in diabetic nephropathy. *Yonsei Med. J.* **2012**, *536*, 1165–1175. [CrossRef] [PubMed]
61. Lenzen, S. The mechanisms of alloxan-and streptozotocin-induced diabetes. *Diabetologia* **2008**, *51*, 216–226. [CrossRef]
62. Navale, A.M.; Paranjape, A.N. Glucose transporters: Physiological and pathological roles. *Biophys. Rev.* **2016**, *8*, 5–9. [CrossRef]
63. Verma, G.K.; Yadav, Y.S.; Yadav, R.K.; Sharma, I.K.; Bharat, K.; Yadav, K.K. Study of lipid profile levels in malnourished and healthy children: A case control study. *Pediatr. Rev. Int. J. Pediatr. Res.* **2018**, *5*, 156–161. [CrossRef]
64. Aloud, A.A.; Chinnadurai, V.; Govindasamy, C.; Alsaif, M.A.; Al-Numair, K.S. Galangin, a dietary flavonoid, amelio-rates hyperglycaemia and lipid abnormalities in rats with streptozotocin-induced hyperglycaemia. *Pharm. Biol.* **2018**, *56*, 302–308. [CrossRef]
65. Khadke, S.; Mandave, P.; Kuvalekar, A.; Pandit, V.; Karandikar, M.; Mantri, N. Synergistic effect of omega-3 fatty acids and oral-hypoglycemic drug on lipid normalization through modulation of hepatic gene expression in high fat diet with low streptozotocin-induced diabetic rats. *Nutrients* **2020**, *12*, 3652. [CrossRef]
66. Kaushal, N.; Gupta, M.; Kulshreshtha, E. Hempseed (*Cannabis sativa*) lipid fractions alleviate high-fat diet-induced fatty liver disease through regulation of inflammation and oxidative stress. *Heliyon* **2020**, *6*, e04422. [CrossRef]
67. Testa, R.; Bonfigli, A.R.; Genovese, S.; De Nigris, V.; Ceriello, A. The possible role of flavonoids in the prevention of diabetic complications. *Nutrients* **2016**, *8*, 310. [CrossRef]
68. Andallu, B.; Kumar, A.V.; Varadacharyulu, N.C. Lipid abnormalities in streptozotocin-diabetes: Amelioration by *Morus indica* L. cv *Suguna leaves*. *Int. J. Diabetes Dev. Ctries.* **2009**, *29*, 123. [CrossRef] [PubMed]
69. Monnier, L.; Colette, C.; Percheron, C.; Descomps, B. Insulin, diabetes and cholesterol metabolism. *C R Seances Soc. Biol. Fil.* **1995**, *189*, 919–931. [PubMed]
70. Alves-Bezerra, M.; Cohen, D.E. Triglyceride metabolism in the liver. *Comprehensive Physiology* **2017**, *8*, 1. [CrossRef] [PubMed]
71. Adeyemi, W.J.; Olayaki, L.A. Calcitonin and omega–3 fatty acids exhibit antagonistic and non-additive effects in ex-perimental diabetes. *Pathophysiology* **2018**, *25*, 117–123. [CrossRef]
72. Safhi, M.M.; Anwer, T.; Khan, G.; Siddiqui, R.; Sivakumar, S.M.; Alam, M.F. The combination of canagliflozin and omega-3 fatty acid ameliorates insulin resistance and cardiac biomarkers via modulation of inflammatory cytokines in type 2 diabetic rats. *Korean, J. Physiol. Pharmacol.* **2018**, *22*, 493–501. [CrossRef]
73. Pan, M.; Cederbaum, A.I.; Zhang, Y.L.; Ginsberg, H.N.; Williams, K.J.; Fisher, E.A. Lipid peroxidation and oxidant stress regulate hepatic apolipoprotein B degradation and VLDL production. *J. Clin. Investig.* **2004**, *113*, 1277–1287. [CrossRef]
74. Zuliani, G.; Galvani, M.; Leitersdorf, E.; Volpato, S.; Cavalieri, M.; Fellin, R. The role of polyunsaturated fatty acids (PUFA) in the treatment of dyslipidemias. *Curr. Pharm. Des.* **2009**, *15*, 4087–4093. [CrossRef]
75. Yoshikawa, T.; Shimano, H.; Yahagi, N.; Ide, T.; Amemiya-Kudo, M.; Matsuzaka, T.; Nakakuki, M.; Tomita, S.; Okazaki, H.; Tamura, Y.; et al. Poly-unsaturated fatty acids suppress sterol regulatory element-binding protein 1c promoter activity by inhibition of liver X receptor (LXR) binding to LXR response elements. *J. Biol. Chem.* **2002**, *277*, 1705–1711. [CrossRef]
76. Chen, W.; Chen, G.; Head, D.L.; Mangelsdorf, D.J.; Russell, D.W. Enzymatic reduction of oxysterols impairs LXR sig-naling in cultured cells and the livers of mice. *Cell Metab.* **2007**, *5*, 73–79. [CrossRef]
77. Janowski, B.A.; Willy, P.J.; Devi, T.R.; Falck, J.R.; Mangelsdorf, D.J. An oxysterol signalling pathway mediated by the nuclear receptor LXRα. *Nature* **1996**, *383*, 728–731. [CrossRef]
78. Akbarzadeh, A.; Norouzian, D.; Mehrabi, M.R.; Jamshidi, S.H.; Farhangi, A.; Verdi, A.A.; Mofidian, S.M.A.; Rad, B.L. Induction of diabetes by streptozotocin in rats. *Indian J. Clin. Biochem.* **2007**, *22*, 60–64. [CrossRef] [PubMed]
79. Al-Ishaq, R.K.; Abotaleb, M.; Kubatka, P.; Kajo, K.; Büsselberg, D. Flavonoids and their anti-diabetic effects: Cellular mechanisms and effects to improve blood sugar levels. *Biomolecules* **2019**, *9*, 430. [CrossRef] [PubMed]
80. Charles, C.; Chemais, M.; Stévigny, C.; Dubois, J.; Nachergael, A.; Duez, P. Measurement of the influence of flavonoids on DNA repair kinetics using the comet assay. *Food Chem.* **2012**, *135*, 2974–2981. [CrossRef] [PubMed]
81. Costabile, A.; Corona, G.; Sarnsamak, K.; Atar-Zwillenberg, D.; Yit, C.; King, A.J.; Vauzour, D.; Barone, M.; Turroni, S.; Brigidi, P.; et al. Wholegrain fermentation affects gut microbiota composition, phenolic acid metabolism and pancreatic beta cell function in a rodent model of type 2 diabetes. *Front. Microbiol.* **2022**, *13*, 1004679. [CrossRef]
82. Goodrich, A.D.; Ersek, A.; Varain, N.M.; Groza, D.; Cenariu, M.; Thain, D.S.; Almeida-Porada, G.; Porada, C.D.; Zanjani, E.D. In vivo generation of beta-cell-like cells from CD34(+) cells differentiated from human embryonic stem cells. *Exp. Hematol.* **2010**, *38*, 516–525.e4. [CrossRef]
83. Neelofar, K.; Ahmad, J. A comparative analysis of fructosamine with other risk factors for kidney dysfunction in dia-betic patients with or without chronic kidney disease. *Diabetes Metab. Syndr.* **2019**, *13*, 240–244. [CrossRef]
84. Pedrosa, W.; Diniz, M.D.F.H.S.; Barreto, S.M.; Vidigal, P.G. Establishing a blood fructosamine reference range for the Brazilian population based on data from ELSA–Brasil. *Pract. Lab. Med.* **2019**, *13*, e00111. [CrossRef]
85. Baker, J.R.; Johnson, R.N.; Scott, D.J. Serum fructosamine concentrations in patients with type II (non-insulin-dependent) diabetes mellitus during changes in management. *Br. Med. J.* **1984**, *288*, 1484–1486. [CrossRef]
86. Venos, E.; de Koning, L. Endocrine markers of diabetes and cardiovascular disease risk. In *Endocrine Biomarkers*; Hossein, S., Gregory, K., Eds.; Elsevier: Amsterdam, The Netherlands, 2017; pp. 251–299. [CrossRef]

87. Yoshimura, E.H.; Santos, N.W.; Machado, E.; Agustinho, B.C.; Pereira, L.M.; de Aguiar, S.C.; Sá-Nakanishi, A.B.; Mareze-da-Costa, C.E.; Zeoula, L.M. Functionality of cow milk naturally enriched with polyunsaturated fatty acids and polyphenols in diets for diabetic rats. *PLoS ONE* **2018**, *13*, e0195839. [CrossRef]
88. Iggman, D.; Ärnlöv, J.; Vessby, B.; Cederholm, T.; Sjögren, P.; Risérus, U. Adipose tissue fatty acids and insulin sensitivity in elderly men. *Diabetologia* **2010**, *53*, 850–857. [CrossRef]
89. Khamaisi, M.; Rudich, A.; Beeri, I.; Pessler, D.; Friger, M.; Gavrilov, V.; Tritschler, H.; Bashan, N. Metabolic Effects of γ-Linolenic Acid–α-Lipoic Acid Conjugate in Streptozotocin Diabetic Rats. *Antioxid. Redox Signal.* **1999**, *1*, 523–535. [CrossRef] [PubMed]
90. Krachler, B.; Norberg, M.; Eriksson, J.W.; Hallmans, G.; Johansson, I.; Vessby, B.; Weinehall, L.; Lindahl, B. Fatty acid profile of the erythrocyte membrane preceding development of Type 2 diabetes mellitus. *Nutr. Metab. Car-Diovascular Dis.* **2008**, *18*, 503–510. [CrossRef] [PubMed]

nutrients

MDPI

Article

Rare Sugar Metabolism and Impact on Insulin Sensitivity along the Gut–Liver–Muscle Axis In Vitro

Amar van Laar [1], Charlotte Grootaert [1], Andreja Rajkovic [1,2], Tom Desmet [3], Koen Beerens [3] and John Van Camp [1,*]

[1] NutriFOODChem, Department of Food Technology, Safety and Health, Faculty of Bioscience Engineering, Ghent University, 9000 Ghent, Belgium
[2] Food Microbiology and Food Preservation, Department of Food Technology, Safety and Health, Faculty of Bioscience Engineering, Ghent University, 9000 Ghent, Belgium
[3] Centre for Synthetic Biology, Department of Biotechnology, Faculty of Bioscience Engineering, Ghent University, 9000 Ghent, Belgium
* Correspondence: john.vancamp@ugent.be

Abstract: Rare sugars have recently attracted attention as potential sugar replacers. Understanding the biochemical and biological behavior of these sugars is of importance in (novel) food formulations and prevention of type 2 diabetes. In this study, we investigated whether rare sugars may positively affect intestinal and liver metabolism, as well as muscle insulin sensitivity, compared to conventional sugars. Rare disaccharide digestibility, hepatic metabolism of monosaccharides (respirometry) and the effects of sugars on skeletal muscle insulin sensitivity (impaired glucose uptake) were investigated in, respectively, Caco-2, HepG2 and L6 cells or a triple coculture model with these cells. Glucose and fructose, but not L-arabinose, acutely increased extracellular acidification rate (ECAR) responses in HepG2 cells and impaired glucose uptake in L6 cells following a 24 h exposure at 28 mM. Cellular bioenergetics and digestion experiments with Caco-2 cells indicate that especially trehalose (α1-1α), D-Glc-α1,2-D-Gal, D-Glc-α1,2-D-Rib and D-Glc-α1,3-L-Ara experience delayed digestion and reduced cellular impact compared to maltose (α1-4), without differences on insulin-stimulated glucose uptake in a short-term setup with a Caco-2/HepG2/L6 triple coculture. These results suggest a potential for L-arabinose and specific rare disaccharides to improve metabolic health; however, additional in vivo research with longer sugar exposures should confirm their beneficial impact on insulin sensitivity in humans.

Keywords: rare sugars; diabetes; skeletal muscle; bioenergetics; cell research

1. Introduction

Diabetes is a metabolic disease that was estimated to affect more than 500 million in 2021, with a further increase projected for the future [1]. In type 2 diabetes, as the most common and lifestyle-dependent variant, the cellular actions of insulin including its stimulatory effect on glucose uptake is impaired, resulting in increased blood glucose levels (hyperglycemia) [2–4]. Metabolic abnormalities in diabetes may induce oxidative stress, inflammation and production of advanced glycation end products (AGEs), which all contribute to vascular damage and thereby increase the risk for cardiovascular diseases, kidney failure and blindness [5,6]. Before diabetes is diagnosed, peripheral insulin sensitivity is often already impacted, and lifestyle-related prevention is especially effective in this prediabetic state [7]. Furthermore, prediabetes already increases the risk for vascular complications [8]. In this context, sugar replacement is of interest, as (I) excessive sugar intake may contribute to obesity as an important risk factor for type 2 diabetes, (II) low-glycemic index diets may decrease the risk for type 2 diabetes development, (III) a lower intake of glycemic sugars could limit the high blood glucose people in people with impaired insulin sensitivity and (IV) a mechanistic basis exists by which high amounts of fructose impact hepatic de novo lipogenesis and the development of hepatic insulin resistance [9–12]. Relatively new candidates to substitute conventional sugars can be found among rare sugars, which have been defined by the International Society of Rare Sugars as 'monosaccharides and their derivatives that are present in limited quantities in nature' [13]. Several studies have reported benefits of rare compared to conventional sugars such as a slower digestion and reduced impact of isomaltulose on blood glucose levels compared to sucrose [14].

The regulation of metabolic health to prevent metabolic abnormalities involves an intensive interaction between the liver and skeletal muscles, which depends on available metabolic substrates and inter-organ crosstalk [15,16]. Liver health impacts the ability of skeletal muscles to maintain glucose homeostasis, which is crucial as (resting) skeletal muscles are responsible for 75% of insulin-mediated whole-body glucose disposal and regulate many metabolic health factors via secreted myokines [17,18]. Skeletal muscle insulin resistance impairs this insulin-mediated glucose uptake and thereby increases the glycemic effect of foods which for sugars depends partially on intestinal digestion speed and resulting monosaccharide uptake, and partially on how fast glucose is cleared from the circulation [19]. Additionally, glucose clearance in skeletal muscles is impacted by exercise, which promotes insulin-independent uptake of glucose and temporarily improves insulin sensitivity [20].

Insulin sensitivity is closely related to mitochondrial respiration. Mitochondrial dysfunction may contribute to insulin resistance due to inflammation, ROS production and β-oxidation impairment, whereas treatments to improve mitochondrial function also alleviate insulin resistance [21]. Furthermore, insulin as such improves the mitochondrial function, which is prevented by palmitate as a known inducer of insulin resistance [22]. Insulin resistance is often characterized by a reduced metabolic flexibility as well, which is the capacity to adapt fuel oxidation depending on the available substrates, such as glucose and fatty acids; therefore, decreased metabolic flexibility is recognized as a predictor for metabolic disease development [23]. Therefore, cellular bioenergetics, which is the study of pathways via which cells generate and invest energy [24], is an interesting approach to investigate insulin sensitivity, along with insulin-mediated glucose uptake. Cellular bioenergetics can be studied with respirometry, using probes that measure oxygen and protons [24], which result in an extracellular acidification rate (ECAR) as an indirect measure of anaerobic lactate production and oxygen consumption rate (OCR) as an indirect measure of mitochondrial respiration. Metabolic stressors are often used to obtain further insight in metabolic pathways by eliminating the metabolism of specific macronutrients or by inhibiting the electron transport chain, as visualized in Figure 1. Hereby, respirometry has revealed remarkable metabolic adaptations upon the replacement of glucose as the dominant sugar in cell culture. For instance, chronic replacement of glucose by galactose skewed muscle cells towards a more aerobic metabolism, a transition that was exclusively observed in metabolically healthy myotubes and absent in post-diabetic myotubes [25]. However, knowledge on cellular adaptations upon chronic replacement of glucose by other sugars is scarce.

In this study, we aim to investigate the impact of rare and conventional sugars on metabolic health by studying (I) skeletal muscle insulin sensitivity, (II) acute sugar-induced changes in cellular bioenergetics and (III) metabolic adaptation upon chronic exposure of liver cells to particular monosaccharides.

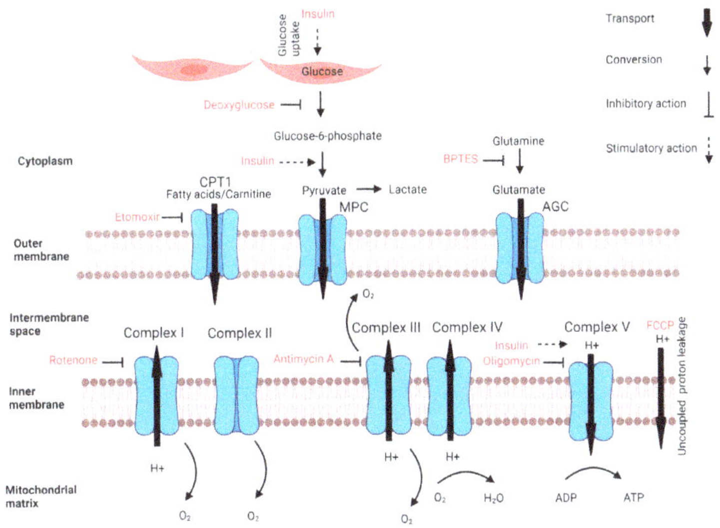

Figure 1. Effects of insulin and metabolic stressors on mitochondrial and macronutrient metabolism (Biorender).

2. Materials and Methods

2.1. Materials

Caco-2, HepG2 and L6 cell lines were obtained from ATCC (Manassas, VA, USA). Dulbecco's Modified Eagle Medium (DMEM), Phosphate Buffered Saline (PBS), non-essential amino acids, trypsin-EDTA and Penicillin/Streptomycin were purchased from Gibco (Paisley, UK). XF base medium (with and without (4-(2-hydroxyethyl)-1-piperazineethanesulfonic acid (HEPES)), Seahorse plates (24-well), Seahorse cartridges (24-well), Seahorse calibrant and mito-stress assay kits were all obtained from Agilent (Machelen, Belgium). Fetal Bovine Serum (FBS) and Trypan Blue were obtained from VWR (Leuven, Belgium). Cell culture flasks (25 and 75 cm^2) and some cell culture plates (96-well, 6-well and 6-well coculture ThinCerts) were obtained from Greiner (Vilvoorde, Belgium). Additionally, 24-well plates and Transwell plates with 24-well insert were obtained from Novolab (Diebeke, Belgium), whereas 12-well coculture inserts were purchased from Sigma-Aldrich (Overijse, Belgium). Human recombinant insulin, 2-deoxyglucose, glycerol, 2-mercaptoethanol, bovine serum albumin (BSA), glucose, fructose, maltose (D-Glcp-α1,4-D-Glcp, 99% pure), galactose, mannitol, resazurin, trishydroxymethylaminomethane (Tris), sulforhodamine B (SRB), horseradish peroxidase and glucose oxidase from Aspergillus niger were also purchased from Sigma-Aldrich. Rare disaccharides nigerose (D-Glcp-α1,3-D-Glcp, 88% pure), kojibiose (D-Glcp-α1,2-D-Glcp, 99% pure), D-Glcp-α1,2-D-Galp (96% pure), D-Glcp-α1,2-D-Ribp (95% pure) and D-Glcp-α1,3-L-Arap (95% pure) were synthesized as described previously [26,27], whereas trehalose (D-Glcp-α1,1α-D-Glcp, 99% pure) was kindly provided by Cargill (Mechelen, Belgium). L-arabinose was obtained from Merck (Darmstadt, Germany). NaHCO3 and glacial acetic acid and sodium acetate were obtained from Chem-Lab (Zedelgem, Belgium). Dodecyl sulfate (SDS), 2,2′-azino-bis(3-ethylbenzothiazoline-6-sulfonic acid (ABTS) and OnGuard II Ag Crtgs were purchased from ThermoFisher (Merelbeke, Belgium). Trichloroacetic acid (TCA) was purchased from Acros Organics (Geel, Belgium). DC protein assay kit was purchased from Bio-Rad (Temse, Belgium).

2.2. Cell Culture and Exposure

Caco-2, HepG2 and L6 cells were cultured in DMEM with 25 mM (Caco-2 cells) or 5.5 mM glucose (HepG2 and L6 cells), supplemented with 10% heat-inactivated and sterile-filtered FBS, 1% non-essential amino acids and 1% Penicillin/Streptomycin. Furthermore, alternative culture conditions were tested for HepG2 cells (5.5 mM galactose, 5.5 mM fructose, 5.5 mM L-arabinose or 25 mM glucose as sole carbohydrate source), which were applied for 2–3 weeks prior to cell seeding. Cells were incubated in a CO_2 incubator (Memmert; VWR, Belgium) at 37 °C and 10% CO_2, and the cell medium was refreshed every two or three days. At 80% confluency, cells were trypsinized and split at a ratio of 1:3 (Caco-2), 1:5 (HepG2) or 1:10 (L6 cells). The cell suspension was mixed 1:1 with Trypan Blue and cells were counted using a Bürker counting chamber (VWR; Leuven, Belgium).

Monocultures of HepG2, L6 and Caco-2 cells were seeded at a density of 3×10^4 cells per well in 24-well Seahorse plates or 2×10^4 cells per well in 96-well plates. HepG2 and L6 cells were cocultured in a 24-well coculture setup, in which HepG2 cells were seeded on the transwell insert and L6 cells in the basolateral compartment, both at a density of 5×10^4/well. Finally, triple cocultures with Caco-2, HepG2 and L6 cells were established (Figure 2). In this setup, Caco-2 cells were seeded in the upper 12-well inserts at 1.5×10^5 cells per well, HepG2 cells were seeded in the middle 6-well inserts at 3×10^5 cells per well, and L6 cells on the bottom of the setup were seeded at 3×10^5 cells per well in normal 6-well cell culture plates. Cell layers were divided by two metal supports holding, respectively, the Caco-2 and HepG2 inserts. HepG2 and L6 cells were used upon confluency, whereas Caco-2 cells were used after two weeks of differentiation. The cells were exposed in sugar-free XF base medium with 3.7 g/L NaHCO3 and sugars of interest at concentrations up to 28 mM for monosaccharides or 14 mM as the disaccharide equivalent, as the only supplements.

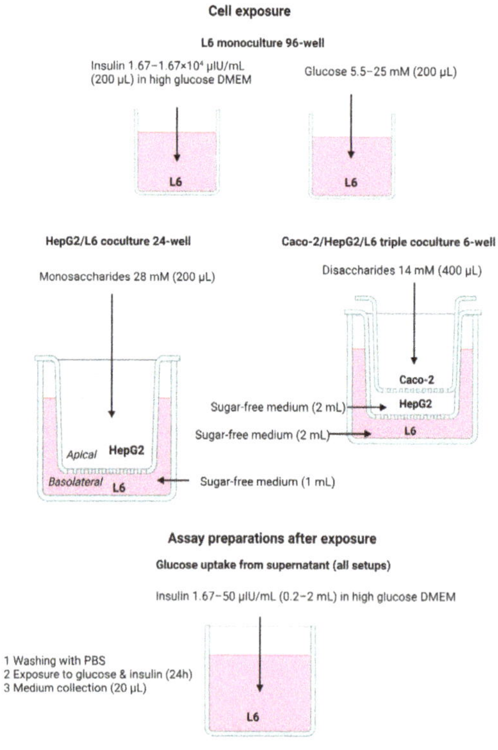

Figure 2. Schematic overview showing the different exposure conditions in the monoculture, coculture and triple coculture setup, as well as the preparation steps taken for the different methods to determine insulin sensitivity.

2.3. Development of Insulin Resistance Models

Cellular insulin sensitivity was investigated in multiple setups using various methods, as visualized in Figure 2.

As a first experiment, the effect of insulin on glucose uptake was tested for a range of insulin concentrations ($1.67–1.67 \times 10^4$ μIU/mL). In a next step, the effect of 24 h exposure to physiological (5.5 mM) versus high glucose (25 mM) was tested in a L6 96-well monoculture model. These are the glucose concentrations that are typically found in commercial DMEM cell culture media. This monoculture model was expanded by adding HepG2 liver cells in a 24-well coculture setup. HepG2 cells were exposed for 24 h to monosaccharides at 28 mM, allowing an indirect exposure of metabolized monosaccharides on the L6 muscle cells. Finally, Caco-2 cells were added to the model, resulting in a 6-well triple-coculture setup with Caco-2, HepG2 and L6 cells. In this model, Caco-2 cells were exposed for 24 h to different disaccharides at 14 mM, and their impact on insulin-stimulated glucose uptake and insulin pathway signaling was investigated in L6 muscle cells. Following exposures in different models, an additional 24 h exposure to 25 mM glucose in combination with 1.67 or 50 μIU/mL insulin was performed to determine insulin-mediated glucose uptake from measurements of glucose concentration in the cell-treated medium, as a measure for insulin sensitivity in L6 cells. In addition, alterations of insulin-mediated aerobic and anaerobic responses following 24 h sugar exposure were investigated using respirometry with a Seahorse XFe24 analyzer.

2.4. HPAEC-PAD Sugar Analysis to Study Digestibility of Rare Disaccharides

Medium samples of 20 μL were taken from the Caco-2 compartment of the triple coculture model after 0, 2, 6 and 24 h of 14 mM disaccharide exposure. These samples were diluted 1000 times in distilled water and 1 mL of diluted sample was filtered through OnGuard II Ag Crtgs to remove negatively charged ions. Filtered samples were used to determine glucose concentrations in the samples using high-pressure anion-exchange chromatography with pulsed amperometric detection (HPAEC-PAD) [28]. Sample analysis was performed using a HPAEC-PAD system (Dionex ICS-3000, Thermo Scientific, (Merelbeke, Belgium)) with a CarboPac PA20 pH-stable anion exchange column for carbohydrate separation at a flow rate of 0.5 $\text{mL}\cdot\text{min}^{-1}$. An isocratic elution with 99% eluent B (100 mM NaOH) and 1% eluent C (1 M NaOAc and 100 mM NaOH) was used for the first 9 min, followed by 70% eluent B and 30% eluent C for 3 min. The initial eluent composition was then restored to run samples for 1.5 min. Glucose concentrations were quantified using a standard curve (1–30 μM) with an LOD of 0.14 μM and LOQ of 0.43 μM.

2.5. Aerobic and Anaerobic Metabolism

Respirometry experiments were performed with different setups and stressors as visualized in Figure 3a. As visualized, respirometry was used for three different end points: to characterize (I) the speed and potency of the effects of different (rare) disaccharides on anaerobic responses in intestinal Caco-2 cells, as an indirect measure of brush border digestion; (II) origin of ATP production (aerobic and/or anaerobic) upon short-term incubation with different monosaccharides in HepG2 liver cells; and (III) insulin-stimulated anaerobic glycolysis and maximal respiration in L6 cells within the context of skeletal muscle insulin resistance. Examples of ECAR and OCR profiles with these setups, based on our experimental data under standard conditions, are shown in Figure 3b.

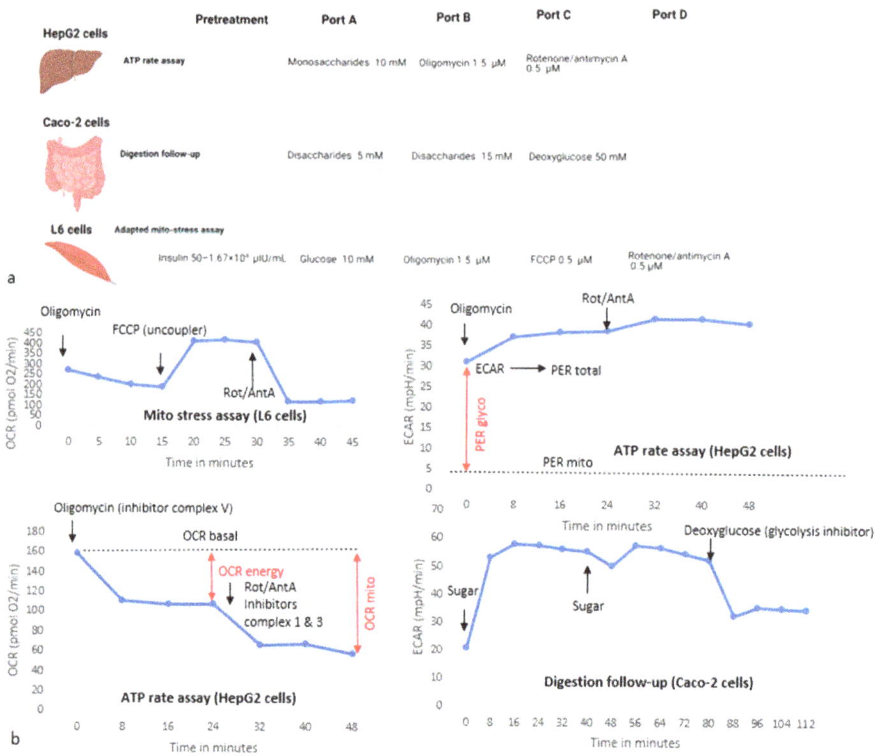

Figure 3. Overview figure for exposures and setups in cellular bioenergetics experiments. Subfigures show the stimuli and stressors that were used in different setups (**a**) as well examples of general responses to different stressors (**b**), based on our experimental data. Measures for calculation of ATP rates are indicated in the graphs for the ATP rate assay.

2.5.1. General Seahorse Setup

In all setups, Seahorse cartridges were hydrated with 1 mL/well calibrant solution and placed in a closed plastic bag with wet paper to prevent dehydration during a 24 h incubation in an incubator at 37 °C. On the day of the assay, calibrant solution was refreshed 1 h before loading cartridges, and the injection ports of the cartridges were loaded with treatment solutions at ten times the final concentration, with increasing volumes for each port (A 20 µL, B 22 µL, C 25 µL and D 28 µL). Cells were placed on nutrient-free (no glucose, glutamine or pyruvate) base XF medium with HEPES pH7.4 for one hour under CO_2-free conditions at 37 °C. Immediately before the assay, cells were washed again and received 180 µL sugar-free XF medium with HEPES. The Seahorse procedure was run with fixed durations for all assays: three to five loops (fiver loops were only used for one of the disaccharide digestion experiments) of mixing (±1 min), waiting (±2 min) and measuring (±5 min) after each injection.

2.5.2. Origin of ATP Production: Setup and Calculations

The ATP rate assay (Agilent) was used to determine the amount and origin of ATP production during exposures with glucose, fructose, galactose, L-arabinose and mannitol in HepG2 cells, according to the manufacturer's instructions. To investigate the impact of monosaccharides on the ATP rate in standard HepG2 cells and cells that were chronically exposed to alternative monosaccharides, they were injected at 10 mM. Next, the response on oligomycin (1.5 µM) and rotenone/antimycin A (0.5 µM) was measured. ATP rates were calculated according to the assay manual [29] and Seahorse white papers [30], and are based on changes within three readings (from the moment of injection until the final measurement before the next injection is applied) after injection of sugars or stressors. The mitochondrial energy production rate (ATPmito) is calculated as the component of oxidative phosphorylation that is inhibited by introduction of oligomycin (OCR energy, Figure 3 and Equation (1)), multiplied by the number of oxygen molecules (2) and the phosphate per oxygen ratio (P/O is 2.75 on average, Equation (2)). Calculation of the glycolytic ATP production rate requires the total proton efflux rate (PERtotal), which is obtained by multiplying ECAR with the volume of the measurement chamber (5.65 μL in 24-well format), volume scaling factor (1.19 μL in 24-well format) and the buffer factor (depends on medium composition and sensor, can be determined manually, and should be between 2.6–4, Equation (3)). OCRmito is then calculated as the part of OCR that is inhibited after the introduction of both oligomycin and rotenone/antimycin A (Figure 3 and Equation (4)). PERmito is the CO_2-dependent non-glycolytic proton efflux and can be calculated by multiplying the mitochondrial OCR (OCRmito) with the CO_2 conversion factor, which Agilent determined to be 0.6 for the 24-well system (Equation (5)). The glycolytic proton efflux rate (PERglyco) is calculated by subtracting the mitochondrial proton efflux (PERmito) from the total proton efflux rate (PERtotal) (Figure 3 and Equation (6)), and is the equivalent to the glycolytic proton efflux rate (PERglyco) (Equation (7)), in case the medium and stressors indicated in the ATP rate manual protocol are used. Finally, the ATP rate (ATPtotal) was calculated as the sum of the mitochondrial (ATPmito) and glycolytic ATP production rate (ATPglyco) (Equation (8)).

$$\text{OCRenergy} = \text{OCRbasal} - \text{OCRoligo} \tag{1}$$

$$\text{ATPmito} = \text{OCRenergy} \times 2 \times 2.75\ (5.5 \times \text{OCRenergy}) \tag{2}$$

$$\text{PERtotal} = \text{ECAR} \times 1.19 \times 5.65 \times 2.8\ (18.8 \times \text{ECAR}) \tag{3}$$

$$\text{OCRmito} = \text{OCRbasal} - \text{OCRrot/antA} \tag{4}$$

$$\text{PERmito} = \text{OCRmito} \times 0.6 \tag{5}$$

$$\text{PERglyco} = \text{PERtotal} - \text{PERmito} \tag{6}$$

$$\text{ATPglyco} = \text{PERglyco} \tag{7}$$

$$\text{ATPtotal} = \text{ATPmito} + \text{ATPglyco} \tag{8}$$

2.5.3. Disaccharide Digestion

Caco-2 cells were exposed to disaccharides twice (15 or 25 min exposure each) within the Seahorse running time, after which the anaerobic responses were stopped with 50 mM 2-deoxyglucose (Figure 3a). Cellular respirometry is usually performed with semi-confluent cells, but we determined ECAR responses in a confluent layer of differentiated Caco-2 cells to induce additional α-glucosidase expression.

2.5.4. Insulin-Mediated Responses and Insulin Sensitivity

L6 cells were exposed to insulin for 20 min at 50–1.67 × 10^4 μIU/mL. Then, a variant of the mito-stress assay was performed with glucose as first injection, followed by oligomycin (1.5 μM), FCCP (0.5 μM) and rotenone/antimycinA (0.5 μM) (Figure 3a). In experiments to test the effect of high glucose pre-treatment on insulin-mediated responses, an additional 24-h glucose exposure at 28 mM was performed prior to the insulin exposure.

2.6. Insulin-Mediated Glucose Uptake Based on GOD-POD Measurements

A glucose oxidase-peroxidase (GOD-POD) mixture [31] was prepared by adding 50 mg ABTS, 45.23 mg glucose oxidase and 6.92 mg peroxidase in 100 mL 0.2 M acetic acid (pH 4.5), which was stored in aliquots at −20 °C. The 50 μL undiluted cell-treated medium from L6 cells was added to a clear plate, after which 200 μL GOD-POD reagent mixture was added. The plate was incubated for 5 min at 37 °C and absorbance was measured at 420 nm. A glucose standard curve (0–2400 μM in distilled water) was prepared to quantify the amount of glucose in cell-treated medium with a LOD and LOQ of 53 μM and 160 μM, respectively. Glucose uptake was calculated for each exposure condition by subtracting the calculated glucose content in cell-treated medium from the calculated glucose content in medium from wells without cells as a blank.

2.7. Resazurin Assay for Cellular Reductase Activity

Resazurin stock solution (1 mg/mL in distilled water) was added to the cell medium at 1:100 *v*/*v* [32]. The plate was incubated for two hours at 37 °C and fluorescence was measured (λexc/λem = 560/590 nm) in a black 96-well plate.

2.8. Protein Correction

Results were corrected for protein content by performing an SRB assay [33] or BioRad protein assay [34].

2.8.1. SRB Assay

After the assays, cells were fixated with 1:4 *v*/*v* 50% TCA in medium for at least 1 h at 4 °C. The cells were washed with tap water at least three times and SRB solution was added in excess. After 30 min, the plate was washed at least three times with 1% glacial acetic acid. Next, the protein-adhered SRB stain was dissolved by adding 200 μL 10 mM Tris and pipetted up and down to homogenize the stain. Absorbance was measured at 490 nm.

2.8.2. Lysate Preparation and Bio-Rad Protein Assay

After 24 h exposure, L6 cells were first washed with cold PBS and 600 μL Laemmli buffer (1.5×) was added per well. Then, the cell layer was disrupted with cell scraper and the lysate was transferred to an Eppendorf tube and centrifuged for 10 min at 14,000 rpm and 4 °C. The supernatant was transferred to a second Eppendorf tube on ice, and stored at −20 °C prior to analysis. Protein content of the lysates was determined with the Bio-Rad DC protein assay according to the manufacturer's instructions. Standard curves (0.2–3 mg/mL) were constructed from bovine serum albumin (BSA) standard solution, showing a linear relationship between absorbance and protein content with a LOD and LOQ of 0.09 and 0.27 mg/mL, respectively. Absorbance at 750 nm was measured after 15 min incubation of the lysate with the reaction mixture.

2.9. Statistics and Calculations

Statistical analyses were performed with SPSS 26 using a significance cut-off of $p < 0.05$. Levene's tests were performed to check for homogeneity of variance. Conditions were compared with one-way analysis of variance (ANOVA), using the Tukey correction for homogeneous data or Games–Howell correction for non-homogenous data. As an exception, the significance of changes in glucose concentrations (HPAEC-PAD data) was determined within exposure conditions using two-way ANOVA, with time and concentration as input parameters. Mitochondrial, glycolytic and total ATP rates were calculated as explained stepwise in the paragraph on 'aerobic and anaerobic metabolism'. Insulin concentrations were converted from pM to μIU/mL using a conversion factor of 6, based on the molecular weight of 5808 kDa.

3. Results

3.1. Disaccharide Digestion and Related ECAR Responses in Intestinal Caco-2 Cells

Respirometry was used to study how gradual glucose release from easily and slowly digestible disaccharides impacts ECAR responses. In differentiated Caco-2 cells, injection of 10 mM glucose and to a lesser degree 5 mM maltose resulted in an increase in ECAR, whereas no response was observed with 10 mM of the mannitol control or kojibiose and trehalose at 5 mM (Figure 4). A second injection at three times the initial concentration did not have clear additive effects. Sugar-induced ECAR responses were stopped upon injection of 50 mM deoxyglucose. The absence of an increase in ECAR with kojibiose and trehalose suggests that these rare sugars have a reduced metabolic impact compared to glucose, most likely related to delayed digestion.

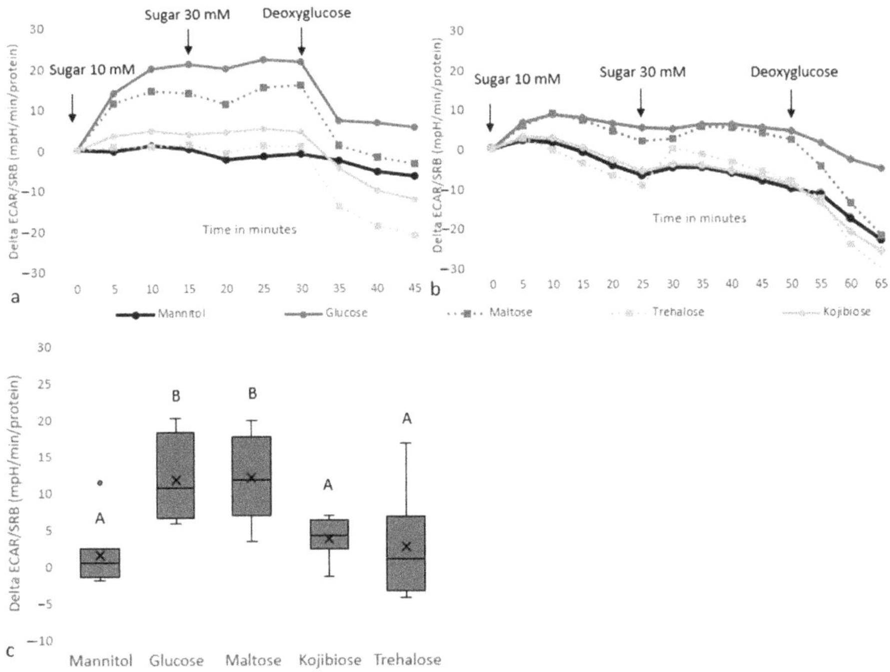

Figure 4. The effect of disaccharide injections on protein-corrected ECAR responses in differentiated Caco-2 cells. Subfigures (**a**,**b**) show the ECAR profiles of individual experiments with 3–5 replicates per condition, whereas (**c**) shows the boxplots for the 10 min ECAR increase upon injection of 10 mM sugar and consists of the pooled data from graphs (**a**,**b**). The assay medium at the start of the assay consisted of XF base medium, without sugars, L-glutamine or pyruvate. B indicates a statistically significant ($p < 0.05$) difference compared to the mannitol control, whereas A indicates that there is no significant difference compared to mannitol.

HPAEC-PAD glucose quantification showed differences in glucose release for different rare disaccharides. Maltose digestion resulted in a significant increase in glucose concentrations in the Caco-2 medium, which was not observed for the mannitol or rare disaccharides (Figure 5). This pronounced increase in medium glucose concentrations during maltose exposure suggests that the release of glucose exceeds cellular uptake, highlighting that this sugar is more easily digested than the rare disaccharides. Kojibiose and nigerose digestion resulted in stable glucose concentrations over time, whereas a decrease in glucose concentrations was observed during 24-h exposure to trehalose and the analogues of kojibiose (D-Glc-α1,2-D-Gal and D-Glc-α1,2-D-Rib) and nigerose (D-Glc-α1,3-L-Ara). The stable glucose concentrations during kojibiose and nigerose exposure suggests that these sugars are digested at an intermediate rate, whereas a decrease in glucose concentrations suggests slow digestion of a sugar, insufficient to supply the cells with basal levels of glucose.

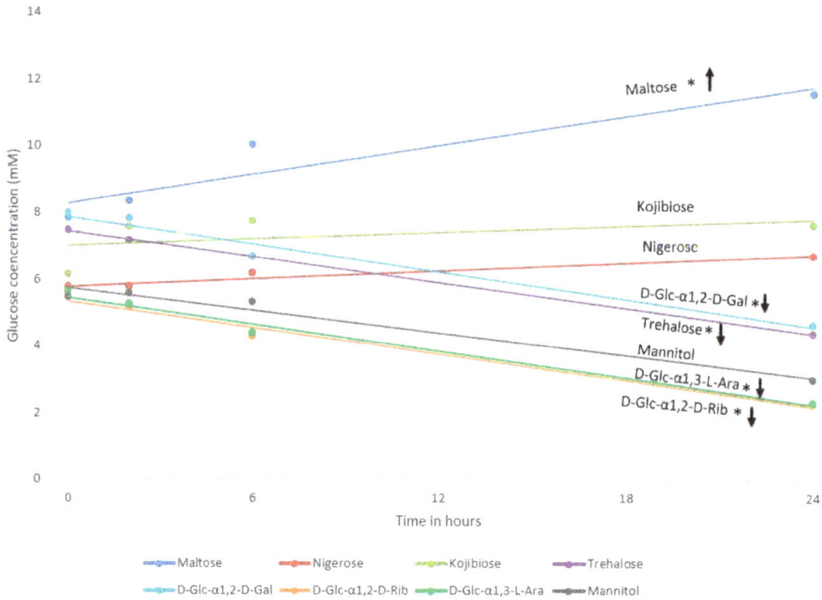

Figure 5. Changes in glucose concentrations within 24 h, following disaccharide exposures in a Caco-2/HepG2/L6 triple coculture model. This figure visualizes the glucose concentration in the Caco-2 medium, sampled at 0, 2, 4 and 24 h exposure. Data were generated from four different wells (each from a separate plate) per condition with * indicating a statistically significant ($p < 0.05$) change in glucose concentration within 24 h. Arrows indicate the direction (increase versus decrease) of the significant changes in glucose concentration.

3.2. Effects of Chronic Monosaccharide Exposure on Energy Metabolism in HepG2 Liver Cells

3.2.1. Cell Growth and Morphology

To study which aspects of the cellular metabolism are impacted by chronic exposure to only glucose, galactose, fructose and L-arabinose, chronically exposed HepG2 cells were characterized in terms of their basal state (in the absence of nutrients) and after additional exposure to the different monosaccharides (pre-treatment + extra 24-h exposure to different monosaccharides) using resazurin conversion and respirometry. Replacement of glucose by other monosaccharides reduced the growth rate of HepG2 cells, with the largest decrease following chronic L-arabinose exposure (at least four-fold slower, based on cell count and time till ±80% confluency), and an intermediate cell growth (two-to-three-fold slower) following chronic exposure to galactose and fructose. Two days after splitting, cells that were chronically exposed to fructose, galactose or L-arabinose were in a different stage of growth with a different morphology (Figure 6), although they eventually obtained 'normal' HepG2 morphology as confluency increased. These findings suggest that cells grow most efficiently in a glucose-containing medium, but are able to adapt to the presence of other sugars.

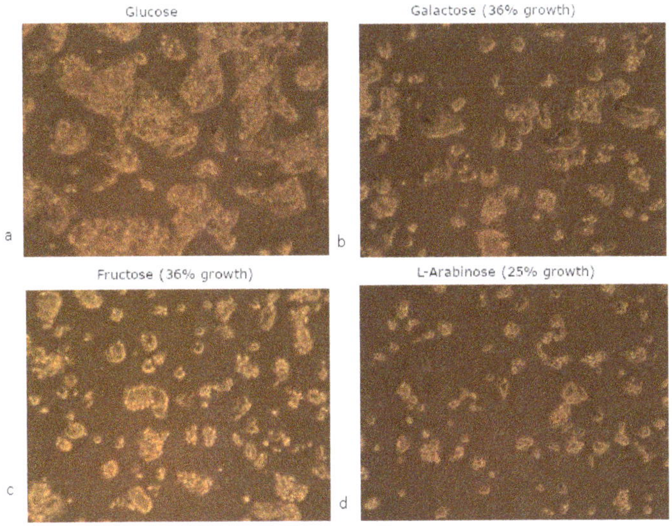

Figure 6. Morphology of HepG2 cells following chronic exposure to 5.5 mM glucose (**a**), galactose (**b**), fructose (**c**) or L-arabinose (**d**). Microscopic pictures (10× magnification) were taken after 2 weeks of culture in the different media and 2 days after splitting. Cell growth on alternative media is indicated as % compared to glucose, based on cell count and time until ±80% confluency. Chronic exposures were performed in sugar-free medium with FBS and L-glutamine.

3.2.2. Resazurin Conversion

Chronic exposure to alternative monosaccharides at 5.5 mM altered protein-corrected resazurin conversion in HepG2 cells as well, although the same monosaccharides at 28 mM (glucose and fructose) provided stimulatory 24 h energy effects (compared to mannitol) in all chronically pre-treated HepG2 cells (Table 1). Galactose approached the fructose and glucose response specifically in cells that were chronically pre-treated with galactose, suggesting an adaptation to the presence of galactose. Galactose pre-treated cells also produced more energy than the traditionally cultured HepG2 cells (with 5.5 mM glucose) during all of the 24 h exposures (including mannitol) (Table 1), indicating an alteration of the basal cellular metabolism. These increased responses were also observed in L-arabinose pre-treated cells, except during exposure to L-arabinose. Fructose pre-treated HepG2 cells responded more strongly to specifically a glucose exposure compared to traditionally cultured cells.

Table 1. Protein-corrected resazurin conversion in HepG2 cells cultured in media with 5.5 mM monosaccharides (glucose, fructose, galactose or L-arabinose) following 24 h exposure to these different sugars, presented as % compared to mannitol exposure in glucose pre-treated HepG2 cells. Data were generated from 3 experiments with a total of 18 replicates and are presented as mean ± standard deviation with * indicating a statistically significant ($p < 0.05$) effect of the monosaccharide exposure compared to mannitol and # indicating an effect of the pre-treatment compared to glucose.

Chronic Pre-Treatment (5.5 mM)				
	Glucose	**Fructose**	**Galactose**	**L-arabinose**
24 h Exposures	**RESAZURIN/SRB RATIO %**			
Mannitol	100 ± 15	118 ± 32	146 ± 28 #	137 ± 42 #
Glucose	158 ± 24 *	204 ± 39 * #	204 ± 25 * #	195 ± 40 * #
Fructose	150 ± 39 *	171 ± 32 *	184 ± 43 * #	207 ± 45 * #
Galactose	105 ± 16	128 ± 34	180 ± 43 #	150 ± 26 #
L-arabinose	101 ± 19	117 ± 26	124 ± 27 #	113 ± 26

3.2.3. Acute OCR and ECAR Responses to Sugar Injections in Galactose Versus Glucose Pre-Treated Cells

To study differences in cellular metabolism and contribution of glycolysis and/or mitochondrial respiration on in vitro ATP production of structurally different monosaccharides, more specific glucose, fructose, galactose and L-arabinose, and ATP rates were determined with respirometry. In HepG2 cells long-term pre-treated with 5.5 mM glucose or galactose, injection of 10 mM glucose or fructose resulted in an acute and direct increase in ECAR and glycolytic ATP production, which was not observed upon mannitol, galactose or L-arabinose injection (Figure 7a,b,e,f). Injection of glucose, but not fructose, also resulted in a significant increase in the total ATP production rate (Figure 7e,f). Galactose and L-arabinose did not alter any of the ATP rates in either glucose or galactose pre-treated HepG2 cells, suggesting that these sugars have little impact on the liver metabolism. Oligomycin further increased anaerobic glycolysis slightly in cells exposed to glucose, but had a significantly different impact on cells exposed to fructose, in which it reduced anaerobic glycolysis (Figure 7a). HepG2 cells pre-treated with galactose had higher basal OCR and lower ECAR levels before injection of the sugars, suggesting that these cells are in a more aerobic state. Upon injection of glucose, these cells also provided more potent ECAR responses and experienced a more pronounced decrease in OCR (Figure 7).

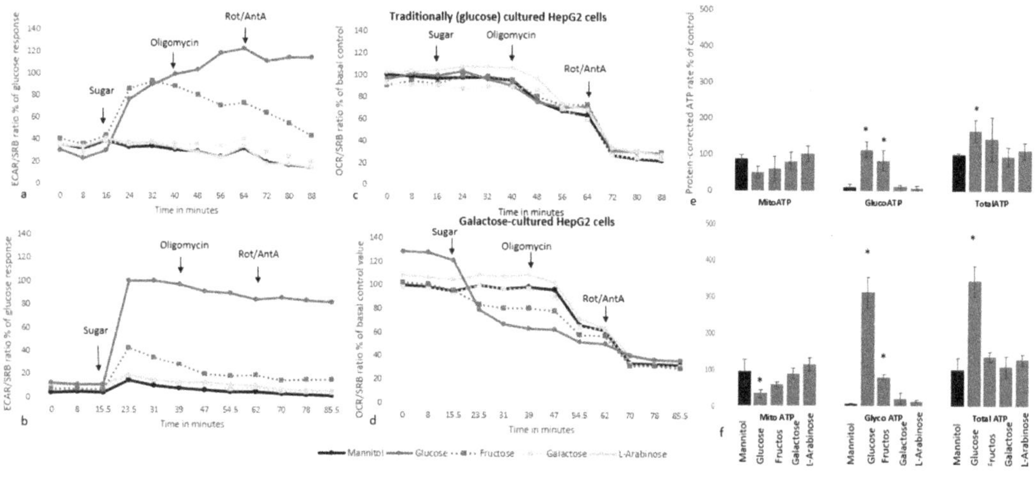

Figure 7. The acute response to different sugars at 10 mM in HepG2 cells cultured under standard conditions (5.5 mM glucose, (**a**,**c**,**e**)) and after chronic galactose (5.5 mM, (**b**,**d**,**f**)) pre-treatment. Subfigures show the protein-corrected effects on the ECAR profile (**a**,**b**), the OCR profile normalized to the basal level in control cells (**c**,**d**) and ATP rates as % of total ATP rate during mannitol exposure (**e**,**f**). The assay medium at the start of the assay consisted of XF base medium, without sugars, L-glutamine or pyruvate. Data were generated from a total of eight wells per condition spread over two plates (**a**,**c**,**e**), or from four wells of a single plate (**b**,**d**,**f**)). * indicates statistical significance ($p < 0.05$) compared to the mannitol response.

3.3. *Insulin Sensitivity and Glucose Uptake in L6 Muscle Cells*

To investigate how sugars impact insulin sensitivity in skeletal muscle cells, insulin-stimulated glucose uptake was determined, starting with a L6 monoculture and gradually building a Caco-2/HepG2/L6 triple coculture model.

3.3.1. Insulin-Mediated Glucose Uptake Determined with the GOD-POD Assay

Insulin (24 h exposure) increased glucose uptake in L6 muscle cells, starting from concentrations of 1.67 μIU/mL (Figure 8a). In an experiment to test the impact of a hyperglycemic environment on insulin sensitivity in L6 cells, exposure to 25 mM glucose for 24 h significantly reduced (±23%) insulin-stimulated glucose uptake, compared to cells exposed to 5.5 mM glucose (Figure 8b). In a HepG2/L6 coculture using the same exposure regimen, 24 h exposure to glucose or fructose at 28 mM reduced glucose uptake in L6 cells by 25 to 35%, while L-arabinose did not have an effect (Figure 8c). These findings suggest that L-arabinose, unlike glucose and fructose, does not induce skeletal muscle insulin resistance in the HepG2/L6 coculture model. In the triple coculture with intestinal, liver and skeletal muscle cells, no significant differences in insulin-mediated glucose uptake were observed in L6 cells following a 24 h exposure (at Caco-2 level) to different disaccharides at 14 mM (Figure 8d).

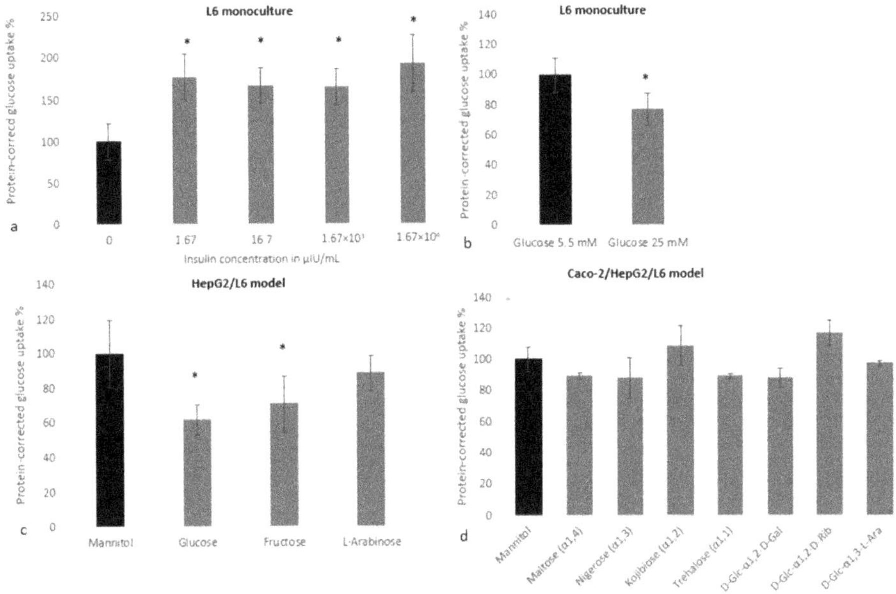

Figure 8. The effect of sugars and insulin on insulin-mediated glucose uptake in L6 cells determined with the GOD-POD assay, using models with different complexity. (**a**) Shows the effect of insulin on glucose uptake from an experiment with six replicates from a single plate. The other subfigures show how 24 h pre-treatment with monosaccharides (28 mM) or disaccharides (14 mM) impact insulin-mediated (50 µIU/mL) glucose uptake in a L6 monoculture model ((**b**): three independent plates for a total of nine replicates), HepG2/L6 coculture model ((**c**): two independent plates for a total of eight replicates) and Caco-2/HepG2/L6 triple coculture model ((**d**): at least three replicates per condition). Data were corrected for SRB (**a**–**c**) or average protein concentration within conditions measured with the Bio-Rad DC protein assay (**d**), and are presented as mean ± standard deviation with * indicating significant differences ($p < 0.05$) compared to the mannitol control.

3.3.2. Insulin Sensitivity Determined with Cellular Bioenergetics

In an experiment to test the effects of insulin pre-treatment on glucose-induced ECAR in L6 cells, the physiological insulin concentration of 50 µIU/mL was found to significantly increase the glucose-induced ECAR response by 36%, whereas the glucose-induced response was only 16% and not significantly higher after 1.67×10^4 µIU/mL insulin pre-treatment (Figure 9a,b). High glucose (28 mM) pre-treatment for 24 h did not significantly affect the insulin-stimulated glucose-induced ECAR response, as this response was only lowered in one of the repetitions. Insulin did not impact the glucose-induced OCR response or the ECAR and OCR responses to the mitochondrial stressors oligomycin, FCCP and rotenone/antimycinA (Figure A1).

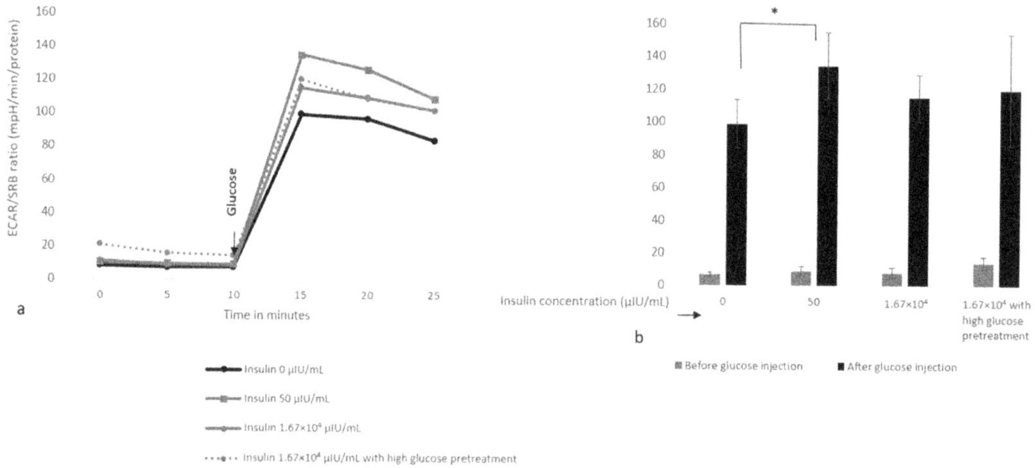

Figure 9. The effect of insulin concentrations and high glucose pre-treatment on the glucose-induced ECAR response in L6 cells. Subfigures show the protein-corrected glucose-induced ECAR response as a Seahorse profile (**a**) and a bar graph. (**b**) The assay medium at the start of the assay consisted of XF base medium, without sugars, glutamine or pyruvate. (**b**) Shows bars for the 10 min (gray before bar) and 15 (black after bar) minute timepoint. Data were generated from 2 independent plates for a total of 10 wells per condition, and are presented as mean ± standard deviation with * indicating significant differences ($p < 0.05$).

4. Discussion

We investigated how (rare) sugars impact physiological processes in the gut–liver–muscle axis controlling the glycemic index, more specifically, (I) their brush border digestion, (II) their impact on acute and adaptive metabolic responses to the monosaccharides entering the liver and (III) their effect on skeletal muscle insulin sensitivity. Hereby, the research builds further upon the current knowledge on rare sugars and the existing in vitro models for testing cellular impact of nutrients. Currently, there is only a small number of rare sugars for which the metabolic health impact is known, as mentioned in recent reviews [35,36], and this is the first study to investigate metabolic health effects of the rare sugars D-Glc-α1,2-D-Gal, D-Glc-α1,2-D-Rib and D-Glc-α1,3-L-Ara. In addition, this study has used cellular energetics to obtain knowledge on cellular effects of disaccharides, whereas previous research has mainly focused on conventional monosaccharides [25,37]. Lastly, this study introduces a new triple coculture model and new assay combinations to evaluate skeletal insulin muscle sensitivity in vitro, along with suggestions to improve the model further. Using these approaches, we achieved (I) the identification of differential effects of both conventional and rare disaccharides on the energy metabolism, and (II) improved understanding of how specific monosaccharides impact energy metabolism following acute and chronic exposures, demonstrating differences in metabolic flexibility upon chronic exposure to structurally different monosaccharides.

4.1. *Rare Disaccharides Are More Slowly Digested than Maltose*

Respirometry and HPAEC-PAD glucose quantification indicate differences in digestion between disaccharides in Caco-2 cells, reflected by ECAR responses and changes in glucose concentrations over time. The ECAR responses highlight the importance of the glycosidic bond for digestibility, explaining the more delayed digestion of kojibiose and especially trehalose. This is confirmed by the HPAEC-PAD findings, while these analyses also showed that the monosaccharide composition has an impact. Although the decreased glucose levels upon exposure to D-Glc-α1,2-D-Gal, D-Glc-α1,2-D-Rib and D-Glc-α1,3-L-Ara could be related to the fact that only one molecule of glucose is released during their digestion, we have previously demonstrated that cellular ATP production and the change in disaccharide concentration over time is similar for these sugars in comparison to a mannitol control [38]. Therefore, the decrease in glucose concentration during exposure to Glc-α1,2-D-Gal, D-Glc-α1,2-D-Rib and D-Glc-α1,3-L-Ara is most likely a result of both a lower glucose content of the sugar and delayed digestion. Our findings for maltose, kojibiose and trehalose support the ranking of these glucobioses based on their digestibility with rat intestinal extract [39]. These differences may translate to a reduced glycemic effect of kojibiose and trehalose, relevant in the diabetes context. However, inducibility of brush border enzymes may impact digestion, and it is not yet known whether the tested rare sugars can induce expression of brush border enzymes as described for maltose at 12.5 mM [40]. Diet composition with different macronutrients, fibers and bioactives may impact glycemic responses to specific sugars in vivo, which is an aspect that is not included in this study [41–43]. Furthermore, inter-personal differences in brush border expression impact glycemic responses in vivo, and are highly relevant for trehalose digestion as intestinal trehalase expression differs considerably between individuals [44]. As a consequence, blood glucose responses to trehalose are small in individuals with low trehalase activity and significantly higher in individuals with high trehalase activity [44].

4.2. Different Sugars Influence Short-Term Aerobic and Anaerobic Hepatic Metabolism

Differences in disaccharide digestion result in a different flow of monosaccharides to the liver. Our results suggest that these monosaccharides provide different acute effects on cellular metabolism in the liver, with increased ATP production from glucose and fructose mediated via anaerobic glycolysis. These findings can be linked to studies reporting that glucose and fructose contribute to lactate production, which is less the case with galactose [45–48]. The absence of an acute sugar-induced increase in mitochondrial ATP production may be explained by the cancerous nature of the cells with a tendency to be more dependent on glycolysis [49], access to alternative (non-sugar) substrates that are depleted after longer exposures and in the case of galactose, an inefficient galactose metabolism in HepG2 cells [48]. Sugars with a larger acute impact on liver metabolism such as glucose and fructose may not be problematic in moderate amounts, but should not exceed amounts that can be metabolized for energy production, as excessive hepatic acetyl-CoA beyond energy may be used for synthesis of cholesterol and triglycerides with adverse influences on metabolic health [50]. This has been observed in interventions with high concentrations of fructose [51], and upon skeletal muscle insulin resistance when accumulation of blood glucose increases the flow of glucose to the liver [16]. It should be mentioned that ATP production measured with the Seahorse is different from the intracellular ATP content, which can be depleted following high-fructose exposures, potentially resulting in excessive uric acid production as metabolite contributing to oxidative stress and endothelial dysfunction [52,53]. Measurement for these adverse effects are not included in our setup.

4.3. Chronic Replacement of Glucose in the Culture Stage Alters the Hepatic Energy Metabolism

Using different applications of cellular bioenergetics, we observed that chronic exposure to specific monosaccharides alters basal metabolism and sugar-specific responses, indicative of altered capacities to metabolize metabolic substrates. Although more relevant in the context of metabolic diseases than a single acute dose response, chronic exposure is rarely investigated using cellular in vitro models. The few previous studies on chronic exposure to particular monosaccharides showed that monocultures of L6 and HepG2 cells become more responsive to mitochondrial toxicants following chronic pre-treatment with galactose [48,54]. Our observations suggest that chronic monosaccharide exposures also impact other aspects (e.g., anaerobic glycolysis) of the cellular energy metabolism and could be relevant in the metabolic health context.

A major finding from both the respirometry and resazurin experiments is that HepG2 cells always (independent of the monosaccharide pre-treatment) respond significantly to glucose and fructose, though not to L-arabinose. These findings confirm the non-metabolizable nature of L-arabinose [55] and show that metabolic conversion of this sugar cannot be induced by chronic exposure to L-arabinose. However, multiple other responses are altered upon chronic exposures and can be divided in general metabolic adaptations and sugar-specific responses.

General adaptations include the increase in basal aerobic metabolism (cellular bioenergetics) in cells pre-treated with galactose, as well as the higher basal energy production (resazurin assay) in cells chronically exposed to galactose or L-arabinose. These changes suggest that chronic exposure to galactose and L-arabinose skews cells towards a more efficient aerobic metabolism. The adaptation towards a more aerobic metabolism could be the result of an improved mitochondrial function and may either be an effect of the sugars themselves or an effect caused by glucose retraction, thereby forcing cells to use other substrates present in the medium [48] (such as glutamine and other amino acids) or energy stores inside the cells (such as glycogen, fatty acids and proteins). In this context, mitochondrial function is an important indicator for metabolic health that is improved by insulin [22]. Improved mitochondrial function may facilitate fatty acid oxidation and thereby prevent accumulation of intrahepatic and intramuscular fatty acids capable of inducing local insulin resistance [56,57]. In contrast, hyperglycemia impairs mitochondrial function in cardiomyocytes and rabbit liver [58,59]. However, it should be mentioned that cancerous nature of our cell lines may be partly responsible for the relatively inactive mitochondrial metabolism in cells pre-treated with glucose, since cancer cells rely more on anaerobic glycolysis even when oxygen is present [49].

Interesting sugar-specific responses include (I) altered glucose-induced ECAR and OCR response in galactose pre-treated cells (cellular bioenergetics), (II) a large response to galactose upon galactose pre-treatment (resazurin assay) and (III) an increase in energy generated from glucose in cells pre-treated with fructose (resazurin assay).

Chronic exposure to galactose induced a number of changes related to the glucose metabolism, which was visible as a (I) higher glucose-induced ECAR response, (II) a larger glucose-induced decrease in OCR and (III) a larger difference between the acute glucose-induced and fructose-induced ECAR response. Together, these findings suggest that chronic galactose exposure improves the capacity of cells to metabolize glucose, which could be explained by an increase in glucose uptake and mitochondrial capacity, and contributes to improved glucose handling [60]. Improved handling of glucose may prevent hyperglycemia, as illustrated in vivo by rapid glucose clearance in athletes (who generally have a high mitochondrial capacity, large glycogen stores and above average muscle mass) during hyperinsulinemic-euglycemic clamps [61]. Enhanced hepatic glycolysis as we observed following chronic galactose pre-treatment can result in lower blood glucose levels via consumption of glucose and inhibition of hepatic gluconeogenesis via the enzyme fructose-2,6-biphosphate [62]. This finding may also be an indication of improved metabolic flexibility following chronic galactose pre-treatment, considering the enhanced switch from aerobic metabolism to anaerobic glycolysis. Collectively, cellular alterations upon chronic glucose replacement can impact health by improving metabolism of metabolic substrates, most importantly glucose and fatty acids. However, it should be mentioned that substrate availability will not be affected as much in vivo, even upon permanent replacement of dietary sugars, considering glucose present in the circulation and a variety of nutrients provided by the diet.

The alterations of sugar-specific responses observed with the resazurin assay cannot be coupled directly to metabolic health, but can potentially be explained by changes in enzymatic activity. Chronic exposure to galactose resulted in 75% more resazurin conversion than chronic exposure to glucose when extra galactose is added to the cells, which may, along with a higher basal metabolism, be the result of the upregulation of enzymes involved in the Leloir pathway, as reported in yeast when galactose is present as the only carbohydrate source for an extended period [63]. Davit–Spraul et al. reported that a similar adaptation occurs in HepG2 cells, resulting in enhanced conversion of galactose to glucose as well as an increased activity of galactose-1-phosphate-uridyltransferase (GALT) and glucose-6-phosphate-dehydrogenase (G6PDH) as important enzymes within galactose metabolism [64]. Likewise, the increased glucose response following chronic fructose exposure—despite the initially similar basal energy levels—could be explained by the stimulation of hepatic glucokinase, as it has been reported that this enzyme is stimulated even by small amounts of fructose/fructose-1-phosphate, resulting in an increased formation of glycolytic intermediates [65].

4.4. *Sugars Differentially Impact Skeletal Muscle Insulin Sensitivity?*

Insulin treatment reduces glucose concentrations in the cell medium, and hence enhances glucose uptake in muscle cells, which is in line with the well-described cellular role of insulin [66]. Cellular respirometry with L6 cells confirms that insulin successfully increased cellular uptake and metabolic conversion of glucose, as insulin pre-treatment increased the glucose-stimulated ECAR response in L6 cells. Whereas insulin was effective in stimulating glucose uptake, pre-treatment with 25 mM glucose resulted in a reduced glucose uptake of ±25% in the presence of insulin compared to cells pre-treated with 5.5 mM glucose, suggesting that high glucose exposure was able to trigger insulin resistance, in line with previous publications using L6 muscle cells or adipocytes [67,68]. Although these monoculture effects on glucose uptake and the insulin signaling pathway are a confirmation of previous knowledge, we also studied the impact of different sugars on insulin sensitivity using a coculture model of intestinal, hepatic and skeletal muscle cells.

Glucose and fructose exposure at the HepG2 level in a HepG2/L6 coculture had a similar inhibitory effect on glucose uptake in L6 cells despite the indirect exposure, showing that the effect is not specific for glucose and that the hepatic metabolism may contribute to peripheral insulin resistance as well. It is unclear which mechanism would mediate skeletal muscle insulin resistance in our setup, but it is known that interruption of the hepatic parasympathetic reflex (preventing release of hepatic insulin-sensitizing substance) and antagonism of hepatic nitric oxide synthase can both cause skeletal muscle insulin resistance [69]. In addition, inflammatory cytokines and saturated fatty acids can contribute insulin resistance in skeletal muscle [70], and may be produced by the liver upon high sugar exposures [71]. The absence of an insulin resistance response to L-arabinose exposure suggests that this sugar and sugars containing L-arabinose may be healthier alternatives.

In the triple coculture, however, an effect of disaccharides (including maltose) on insulin-mediated glucose uptake was not visible, which may be explained by higher background concentrations of glucose in this setup. This may mask effects in the triple coculture models compared to the other models, since disaccharides provide more subtle effects due to the delayed release of monosaccharides. Furthermore, the absence of an effect could be due to the intestinal layer reducing the impact of the sugars on the skeletal muscle cells, for example, by metabolizing part of the monosaccharides and thereby reducing the impact of sugars on cocultured liver and skeletal muscle cells. In this latter scenario, the in vivo impact of all the tested disaccharides may be limited as well. Other experimental setups in which longer and/or repeated exposures are applied in media with a small background of glucose may then be required to detect relatively subtle differences between disaccharides. It is important to note that the physiology of Caco-2, HepG2 and L6 cells was not affected by coculture, which potentially allows those extended exposures.

In this context, it is important to realize that glucose homeostasis in vivo is impacted by the intestinal microbiota as well [72]. It has been shown that certain rare sugars, such as D-Glc-α1,3-L-Ara and kojibiose can have prebiotic effects [73], thereby further enlarging their potential as sugar replacers with a more beneficial impact on glucose homeostasis. In contrast, trehalose as one of the more beneficial sugars regarding delayed glucose release in our study was previously found to promote growth of pathogenic gut bacteria and adversely impact alpha diversity of the gut microbiota [73,74]. The microbial aspects should be part of the final evaluation of the health potential of a specific rare sugar. Likewise, other safety aspects have to be taken into account. Rare sugars such as trehalose can impair cellular uptake of conventional sugars [75], which theoretically could both be beneficial in decreasing their adverse metabolic impact and be problematic by reducing glucose availability for glucose-dependent tissues (such as erythrocytes). Currently, there is mostly scientific evidence for the beneficial impact of reducing cellular uptake of conventional sugars such as the glucose-lowering effect of allulose and the anti-cancer effect of allose via interference with the tumor glucose metabolism [76,77].

4.5. Rare Sugars with the Largest Health Potential

Our findings on disaccharide digestion, energy metabolism and insulin sensitivity have contributed to some insights in the health perspectives of rare sugars. Firstly, L-arabinose showed no adverse impact on energy metabolism or insulin sensitivity in our models, and although they are not per se rare in nature, we demonstrated that a nigerose analogue with L-arabinose also lacks the adverse metabolic effects associated with conventional sugars. Furthermore, we have shown that multiple rare disaccharides display considerably delayed digestion rates, amongst which D-Glc-α1,2-D-Rib and D-Glc-α1,3-L-Ara may be especially promising because of their delayed digestion and unconventional monosaccharide composition. Nevertheless, the full potential of these sugars remains to be determined in in vitro insulin sensitivity experiments with repeated exposures and finally in vivo.

4.6. Model Suitability and Future Perspectives

Coculture models were used to mimic inter-organ crosstalk and the complexity of metabolic health. Although our cell models cannot simulate whole-body metabolism, the intestine, liver and skeletal muscle are arguably the most important tissues for sugar metabolism and digestion, as well as the organs necessary to simulate hepatic and peripheral insulin resistance as the primary diagnostic criteria for diabetes [6,78,79]. The addition of adipocytes, immune cells, pancreatic β-cells and glucose-dependent tissues (e.g., erythrocytes) may strengthen the model further. Furthermore, chronic (as performed in HepG2 monocultures) and repeated sugar exposures in coculture models may provide additional insight in metabolic health.

L6 myotubes of rat origin are a frequently used model to study insulin sensitivity [80,81]. Although rats differ from humans, Wistar rats can develop insulin resistance following exposures that impact insulin sensitivity in humans [82]. Although L6 myotubes have increased expressions of glucose transporters compared to primary human myocytes [83], insulin resistance in skeletal muscle cells is more closely related to alterations in GLUT4 translocation than GLUT4 expression [67]. Nevertheless, comparative studies reported that L6 rat myotubes also experience a larger insulin-mediated effect on glucose uptake than primary human skeletal muscle cells [83,84]. However, GLUT4 expression necessary for the pronounced insulin-dependent glucose uptake in L6 myotubes is obtained during differentiation of the cells [85], whereas our cells were likely not fully differentiated. Our L6 cells have been exposed under serum-free conditions, but were not pre-treated with standardized low serum conditions that induce differentiation [85]. Therefore, our L6 were still in a myoblast stage and this model will underestimate rather than overestimate the insulin-dependent glucose uptake in primary human skeletal muscle cells, as confirmed by our finding that insulin increased glucose uptake only two-fold. This reduced window of effect may partly explain why no effects of disaccharides on insulin sensitivity were observed in the triple coculture model, and highlights the importance of future research with models having a confirmed potent GLUT4 expression.

Final experiments were performed under physiological conditions which include culturing (I) HepG2 and L6 cells on 5.5 mM glucose as normal glucose concentration in the bloodstream [86]; (II) using insulin concentrations within the normal range for fasted (±8.3 μIU/mL) and stimulated (67 μIU/mL after an oral glucose challenge) conditions [87]; and

(III) performing intestinal exposures with disaccharides at 14 mM, which is well below luminal glucose peak concentration observed after a meal [88]. In contrast, experiments building up to the triple coculture involving direct exposures to L6 and HepG2 cells were performed with a sugar concentration (28 mM) that exceeds portal vein glucose concentrations found in feeding trials (8 mM) [89]. Moreover, this concentration is higher than the cut-off value for the diagnosis of diabetes determined 2 h after an oral glucose challenge (11.1 mM) [86], which may explain the rapid development of metabolic complications in these cell models. These high monosaccharide exposures were used to test if metabolic complications could be induced in a worst-case scenario and serve as preparations for the disaccharide comparison in coculture models with an intestinal compartment.

5. Conclusions

Insulin resistance in skeletal muscles contributes to type 2 diabetes development by interfering with peripheral glucose uptake, although it is not completely understood whether and how different sugars impact insulin resistance. Our results demonstrate that glucose and fructose, unlike L-arabinose, are metabolized efficiently in an anaerobic manner in HepG2 cells and are capable of interfering with skeletal muscle insulin resistance when exposed for 24 h at high concentrations in a muscle or liver/muscle model. The rare disaccharides trehalose, D-Glc-α1,2-D-Gal, D-Glc-α1,2-D-Rib and D-Glc-α1,3-L-Ara are slowly digested in comparison to maltose, and the glucose released upon their digestion is unlikely to adversely impact insulin sensitivity. However, the direct link between intestinal disaccharide digestion rate and insulin sensitivity could not be made as none of the tested disaccharides induced skeletal muscle insulin resistance in the gut/liver/muscle model. Therefore, repeated exposures in realistic in vitro models and finally in vivo experiments are needed to further investigate this hypothesis. The most promising rare sugars would need to be subjected to well-designed clinical trials, and their integration in the diet needs to take both multi-endpoint health aspects among different target groups of consumers, as well as food production aspects, into account.

Author Contributions: Conceptualization research article, A.v.L., C.G. and J.V.C.; formal analysis, A.v.L. and K.B.; investigation, A.v.L.; resources, A.R., J.V.C., K.B. and T.D.; writing—original draft preparation, A.v.L., C.G. and J.V.C.; writing—review and editing, A.R., K.B. and T.D.; supervision, C.G. and J.V.C.; project administration, T.D., K.B. and J.V.C.; funding acquisition, T.D., K.B. and J.V.C. All authors have read and agreed to the published version of the manuscript.

Funding: This research was funded by Research Foundation Flanders (FWO-Vlaanderen) as part of the 'GlycoProFit project', under Strategic Basic Research Grant S003617N. The Seahorse XF24 analyser was funded by the BOF Special Research Fund from Ghent University, GOA project no. 01G02213. The TEER equipment was funded by the Special Research Fund from Ghent University, 01B04212.

Institutional Review Board Statement: This study did not involve humans or animals. Caco-2, HepG2 and L6 cells were obtained from ATCC.

Informed Consent Statement: Not applicable.

Data Availability Statement: Data is included in the article submission to Nutrients. Samples generated from our cell lines are registered in a database: 'Biobank Vakgroep Levensmiddelentechnology, Voedselveiligheid en Gezondheid' with accesscode BB190156.

Acknowledgments: We would like to thank Shari Dhaene, Tim Dedroog, Diete Verfaillie and Lenina Adriaenssens for their contribution to the work as project partners or master students.

Conflicts of Interest: The authors declare no conflict of interest. The funders had no role in the design of the study; in the collection, analyses, or interpretation of data; in the writing of the manuscript; or in the decision to publish the results.

48. Xu, Q.; Liu, L.; Vu, H.; Kuhls, M.; Aslamkhan, A.G.; Liaw, A.; Yu, Y.; Kaczor, A.; Ruth, M.; Wei, C.; et al. Can galactose be converted to glucose in HepG2 cells? Improving the in vitro mitochondrial toxicity assay for the assessment of drug induced liver injury. *Chem. Res. Toxicol.* **2019**, *32*, 1528–1544. [CrossRef] [PubMed]
49. Gogvadze, V.; Orrenius, S.; Zhivotovsky, B. Mitochondria in cancer cells: What is so special about them? *Trends Cell Biol.* **2008**, *18*, 165–173. [CrossRef] [PubMed]
50. Kawano, Y.; Cohen, D.E. Mechanisms of hepatic triglyceride accumulation in non-alcoholic fatty liver disease. *J. Gastroenterol.* **2013**, *48*, 434–441. [CrossRef]
51. Schaefer, E.J.; Gleason, J.A.; Dansinger, M.L. Dietary fructose and glucose differentially affect lipid and glucose homeostasis. *J. Nutr.* **2009**, *139*, 1257S–1262S. [CrossRef]
52. Jia, G.; Aroor, A.R.; Whaley-Connell, A.T.; Sowers, J.R. Fructose and uric acid: Is there a role in endothelial function? *Curr. Hypertens. Rep.* **2014**, *16*, 434. [CrossRef]
53. Lanaspa, M.A.; Tapia, E.; Soto, V.; Sautin, Y.; Sánchez-Lozada, L.G. Uric acid and fructose: Potential biological mechanisms. *Semin. Nephrol.* **2011**, *31*, 426–432. [CrossRef]
54. Dott, W.; Mistry, P.; Wright, J.; Cain, K.; Herbert, K.E. Modulation of mitochondrial bioenergetics in a skeletal muscle cell line model of mitochondrial toxicity. *Redox Biol.* **2014**, *2*, 224–233. [CrossRef]
55. Pol, K.; de Graaf, K.; Bruin, M.D.-D.; Balvers, M.; Mars, M. The effect of replacing sucrose with L-arabinose in drinks and cereal foods on blood glucose and plasma insulin responses in healthy adults. *J. Funct. Foods* **2020**, *73*, 104114. [CrossRef]
56. Monsénégo, J.; Mansouri, A.; Akkaoui, M.; Lenoir, V.; Esnous, C.; Fauveau, V.; Tavernier, V.; Girard, J.; Prip-Buus, C. Enhancing liver mitochondrial fatty acid oxidation capacity in obese mice improves insulin sensitivity independently of hepatic steatosis. *J. Hepatol.* **2011**, *56*, 632–639. [CrossRef]
57. Hegarty, B.D.; Furler, S.M.; Ye, J.; Cooney, G.J.; Kraegen, E.W. The role of intramuscular lipid in insulin resistance. *Acta Physiol. Scand.* **2003**, *178*, 373–383. [CrossRef]
58. Dassanayaka, S.; Readnower, R.D.; Salabei, J.K.; Long, B.W.; Aird, A.L.; Zheng, Y.-T.; Muthusamy, S.; Facundo, H.T.; Hill, B.G.; Jones, S.P. High glucose induces mitochondrial dysfunction independently of protein O-GlcNAcylation. *Biochem. J.* **2015**, *467*, 115–126. [CrossRef]
59. Vanhorebeek, I.; Ellger, B.; De Vos, R.; Boussemaere, M.; Debaveye, Y.; Perre, S.V.; Rabbani, N.; Thornalley, P.J.; Berghe, G.V.D. Tissue-specific glucose toxicity induces mitochondrial damage in a burn injury model of critical illness. *Crit. Care Med.* **2009**, *37*, 1355–1364. [CrossRef]
60. Kase, E.T.; Nikolić, N.; Bakke, S.S.; Bogen, K.K.; Aas, V.; Thoresen, G.H.; Rustan, A.C. Remodeling of oxidative energy metabolism by galactose improves glucose handling and metabolic switching in human skeletal muscle cells. *PLoS ONE* **2013**, *8*, e59972. [CrossRef]
61. Pratley, R.E.; Hagberg, J.M.; Rogus, E.M.; Goldberg, A.P. Enhanced insulin sensitivity and lower waist-to-hip ratio in master athletes. *Am. J. Physiol. Metab.* **1995**, *268*, E484–E490. [CrossRef]
62. Guo, X.; Li, H.; Xu, H.; Woo, S.; Dong, H.; Lu, F.; Lange, A.J.; Wu, C. Glycolysis in the control of blood glucose homeostasis. *Acta Pharm. Sin. B* **2012**, *2*, 358–367. [CrossRef]
63. Sellick, C.A.; Campbell, R.N.; Reece, R.J. Galactose Metabolism in yeast—Structure and regulation of the leloir pathway enzymes and the genes encoding them. *Int. Rev. Cell Mol. Biol.* **2008**, *269*, 111–150.
64. Davit-Spraul, A.; Pourci, M.; Soni, T.; Lemonnier, A. Metabolic effects of galactose on human HepG2 hepatoblastoma cells. *Metabolism* **1994**, *43*, 945–952. [CrossRef]
65. Feinman, R.D.; Fine, E.J. Fructose in perspective. *Nutr. Metab.* **2013**, *10*, 45. [CrossRef]
66. Hespel, P.; Vergauwen, L.; Vandenberghe, K.; Richter, E.A. Important role of insulin and flow in stimulating glucose uptake in contracting skeletal muscle. *Diabetes* **1995**, *44*, 210–215. [CrossRef]
67. Huang, C.; Somwar, R.; Patel, N.; Niu, W.; Török, D.; Klip, A. Sustained exposure of L6 myotubes to high glucose and insulin decreases insulin-stimulated GLUT4 translocation but upregulates GLUT4 activity. *Diabetes* **2002**, *51*, 2090–2098. [CrossRef] [PubMed]
68. Tomás, E.; Lin, Y.-S.; Dagher, Z.; Saha, A.; Luo, Z.; Ido, Y.; Ruderman, N.B. Hyperglycemia and insulin resistance: Possible mechanisms. *Ann. N. Y. Acad. Sci.* **2006**, *967*, 43–51. [CrossRef] [PubMed]
69. Sadri, P.; Lautt, W.W. Blockade of hepatic nitric oxide synthase causes insulin resistance. *Am. J. Physiol.-Gastrointest. Liver Physiol.* **1999**, *277*, G101–G108. [CrossRef] [PubMed]
70. Martins, A.R.; Nachbar, R.T.; Gorjao, R.; Vinolo, M.A.; Festuccia, W.T.; Lambertucci, R.H.; Cury-Boaventura, M.F.; Silveira, L.R.; Curi, R.; Hirabara, S.M. Mechanisms underlying skeletal muscle insulin resistance induced by fatty acids: Importance of the mitochondrial function. *Lipids Health Dis.* **2012**, *11*, 30. [CrossRef] [PubMed]
71. Pagliassotti, M.J.; Shahrokhi, K.A.; Moscarello, M. Involvement of liver and skeletal muscle in sucrose-induced insulin resistance: Dose-response studies. *Am. J. Physiol. Integr. Comp. Physiol.* **1994**, *266*, R1637–R1644. [CrossRef]
72. Portincasa, P.; Bonfrate, L.; Vacca, M.; De Angelis, M.; Farella, I.; Lanza, E.; Khalil, M.; Wang, D.Q.-H.; Sperandio, M.; Di Ciaula, A. Gut microbiota and short chain fatty acids: Implications in glucose homeostasis. *Int. J. Mol. Sci.* **2022**, *23*, 1105. [CrossRef]
73. Onyango, S.O.; Beerens, K.; Li, Q.; Van Camp, J.; Desmet, T.; Van de Wiele, T. Glycosidic linkage of rare and new-to-nature disaccharides reshapes gut microbiota in vitro. *Food Chem.* **2023**, *411*, 135440. [CrossRef]
74. Vanaporn, M.; Titball, R.W. Trehalose and bacterial virulence. *Virulence* **2020**, *11*, 1192–1202. [CrossRef]

75. DeBosch, B.J.; Heitmeier, M.R.; Mayer, A.L.; Higgins, C.B.; Crowley, J.R.; Kraft, T.E.; Chi, M.; Newberry, E.P.; Chen, Z.; Finck, B.N.; et al. Trehalose inhibits solute carrier 2A (SLC2A) proteins to induce autophagy and prevent hepatic steatosis. *Sci. Signal.* **2016**, *9*, ra21. [CrossRef]
76. Hossain, A.; Yamaguchi, F.; Matsuo, T.; Tsukamoto, I.; Toyoda, Y.; Ogawa, M.; Nagata, Y.; Tokuda, M. Rare sugar d-allulose: Potential role and therapeutic monitoring in maintaining obesity and type 2 diabetes mellitus. *Pharmacol. Ther.* **2015**, *155*, 49–59. [CrossRef]
77. Razban, V.; Khajeh, S.; Ganjavi, M.; Panahi, G.; Zare, M.; Zare, M.; Tahami, S.M. D-allose: Molecular pathways and therapeutic capacity in cancer. *Curr. Mol. Pharmacol.* **2022**, *in press*. [CrossRef]
78. Perry, R.J.; Samuel, V.T.; Petersen, K.F.; Shulman, G.I. The role of hepatic lipids in hepatic insulin resistance and type 2 diabetes. *Nature* **2014**, *510*, 84–91. [CrossRef]
79. Woolnough, J.W.; Monro, J.A.; Brennan, C.S.; Bird, A.R. Simulating human carbohydrate digestion in vitro: A review of methods and the need for standardisation. *Int. J. Food Sci. Technol.* **2008**, *43*, 2245–2256. [CrossRef]
80. Sinha, S.; Perdomo, G.; Brown, N.F.; O'Doherty, R.M. Fatty acid-induced insulin resistance in L6 myotubes is prevented by inhibition of activation and nuclear localization of nuclear factor κB. *J. Biol. Chem.* **2004**, *279*, 41294–41301. [CrossRef]
81. Bailey, C.J.; Turner, S.L. Glucosamine-induced insulin resistance in L6 muscle cells. *Diabetes Obes. Metab.* **2004**, *6*, 293–298. [CrossRef]
82. Ai, J.; Wang, N.; Yang, M.; Du, Z.-M.; Zhang, Y.-C.; Yang, B.-F. Development of wistar rat model of insulin resistance. *World J. Gastroenterol.* **2005**, *11*, 3675–3679. [CrossRef]
83. Abdelmoez, A.M.; Puig, L.S.; Smith, J.; Gabriel, B.M.; Savikj, M.; Dollet, L.; Chibalin, A.V.; Krook, A.; Zierath, J.R.; Pillon, N.J. Comparative profiling of skeletal muscle models reveals heterogeneity of transcriptome and metabolism. *Am. J. Physiol. Cell Physiol.* **2020**, *318*, C615–C626. [CrossRef]
84. Sarabia, V.; Ramlal, T.; Klip, A. Glucose uptake in human and animal muscle cells in culture. *Biochem. Cell Biol.* **1990**, *68*, 536–542. [CrossRef]
85. Mitsumoto, Y.; Burdett, E.; Grant, A.; Klip, A. Differential expression of the GLUT1 and GLUT4 glucose transporters during differentiation of L6 muscle cells. *Biochem. Biophys. Res. Commun.* **1991**, *175*, 652–659. [CrossRef]
86. Evans, P.L.; McMillin, S.L.; Weyrauch, L.A.; Witczak, C.A. Regulation of skeletal muscle glucose transport and glucose metabolism by exercise training. *Nutrients* **2019**, *11*, 2432. [CrossRef]
87. Metter, E.J.; Windham, B.G.; Maggio, M.; Simonsick, E.M.; Ling, S.M.; Egan, J.M.; Ferrucci, L. Glucose and insulin measurements from the oral glucose tolerance test and mortality prediction. *Diabetes Care* **2008**, *31*, 1026–1030. [CrossRef] [PubMed]
88. Kellett, G.L.; Helliwell, P.A. The diffusive component of intestinal glucose absorption is mediated by glucose-induced recruitment of GLUT2 to the brush-border membrane. *Biochem. J.* **2000**, *350*, 155–162. [CrossRef] [PubMed]
89. Strubbe, J.; Steffens, A. Blood glucose levels in portal and peripheral circulation and their relation to food intake in the rat. *Physiol. Behav.* **1977**, *19*, 303–307. [CrossRef] [PubMed]

nutrients

MDPI

Article

Exposure to Obesogenic Environments during Perinatal Development Modulates Offspring Energy Balance Pathways in Adipose Tissue and Liver of Rodent Models

Diana Sousa [1,2,3], Mariana Rocha [1,2,3], Andreia Amaro [1,2,3], Marcos Divino Ferreira-Junior [4], Keilah Valéria Naves Cavalcante [4], Tamaeh Monteiro-Alfredo [1,2,3], Cátia Barra [1,2,3,5], Daniela Rosendo-Silva [1,2,3], Lucas Paulo Jacinto Saavedra [6], José Magalhães [7], Armando Caseiro [8,9], Paulo Cezar de Freitas Mathias [6], Susana P. Pereira [2,10], Paulo J. Oliveira [2,10], Rodrigo Mello Gomes [4] and Paulo Matafome [1,2,3,8,*]

1 Coimbra Institute for Clinical and Biomedical Research (iCBR) and Institute of Physiology, Faculty of Medicine, University of Coimbra, 3000-548 Coimbra, Portugal
2 Center for Innovative Biomedicine and Biotechnology (CIBB), University of Coimbra, 3000-548 Coimbra, Portugal
3 Clinical Academic Center of Coimbra (CACC), 3000-061 Coimbra, Portugal
4 Department of Physiological Sciences, Institute of Biological Sciences, University Federal of Goiás, Goiânia 74690-900, Brazil
5 Hospitals of the University of Coimbra, 3004-561 Coimbra, Portugal
6 Laboratory of Secretion Cell Biology, Department of Biotechnology, Genetics and Cell Biology, State University of Maringa, Maringa 87020-900, Brazil
7 Laboratory of Metabolism and Exercise (LaMetEx), Research Centre in Physical Activity, Health and Leisure (CIAFEL) and Laboratory for Integrative and Translational Research in Population Health (ITR), Faculty of Sports, University of Porto, 4200-450 Porto, Portugal
8 Polytechnic Institute of Coimbra, Coimbra Health School (ESTeSC), 3046-854 Coimbra, Portugal
9 Molecular Physical-Chemistry R&D Unit, Department of Chemistry, University of Coimbra, 3004-535 Coimbra, Portugal
10 CNC—Centre for Neuroscience and Cell Biology, University of Coimbra, 3004-504 Coimbra, Portugal
* Correspondence: paulo.matafome@uc.pt; Tel.: +351-239480014; Fax: +351-239480034

Abstract: Obesogenic environments such as Westernized diets, overnutrition, and exposure to glycation during gestation and lactation can alter peripheral neuroendocrine factors in offspring, predisposing for metabolic diseases in adulthood. Thus, we hypothesized that exposure to obesogenic environments during the perinatal period reprograms offspring energy balance mechanisms. Four rat obesogenic models were studied: maternal diet-induced obesity (DIO); early-life obesity induced by postnatal overfeeding; maternal glycation; and postnatal overfeeding combined with maternal glycation. Metabolic parameters, energy expenditure, and storage pathways in visceral adipose tissue (VAT) and the liver were analyzed. Maternal DIO increased VAT lipogenic [NPY receptor-1 (NPY1R), NPY receptor-2 (NPY2R), and ghrelin receptor], but also lipolytic/catabolic mechanisms [dopamine-1 receptor (D1R) and p-AMP-activated protein kinase (AMPK)] in male offspring, while reducing NPY1R in females. Postnatally overfed male animals only exhibited higher NPY2R levels in VAT, while females also presented NPY1R and NPY2R downregulation. Maternal glycation reduces VAT expandability by decreasing NPY2R in overfed animals. Regarding the liver, D1R was decreased in all obesogenic models, while overfeeding induced fat accumulation in both sexes and glycation the inflammatory infiltration. The VAT response to maternal DIO and overfeeding showed a sexual dysmorphism, and exposure to glycotoxins led to a thin-outside-fat-inside phenotype in overfeeding conditions and impaired energy balance, increasing the metabolic risk in adulthood.

Keywords: metabolic diseases; energy balance; metabolic programming; sugars and AGEs; obesity/adipose tissue

1. Introduction

Since the 1980s, the incidence and prevalence of obesity and type 2 diabetes (T2D) have been escalating worldwide, being associated with Westernized diet intake and a sedentary lifestyle [1–3]. Abnormal body fat accumulation in obesity can promote the development of other diseases such as T2D, characterized by pancreatic β-cells dysfunction and insulin resistance in target organs [4,5]. Insulin resistance in peripheral organs stimulates insulin's continuous release, leading to hyperinsulinemia and the exhaustion of pancreatic β-cells [4,6,7]. Furthermore, increased levels of free fatty acids (FFAs) in obesity contribute to lipotoxicity, one of the main factors for insulin resistance [8]. Overfeeding and energy balance dysregulation are among the main factors contributing to the development of obesity [9]. Under obesogenic conditions, energy balance regulators such as ghrelin, neuropeptide Y (NPY), leptin, glucagon-like peptide-1 (GLP-1), and dopamine levels are altered, disrupting the mechanisms involved with insulin secretion and energy storage and expenditure [10–14]. Furthermore, westernized diets are rich in saturated and monounsaturated fats, simple carbohydrates, and poor in fibers, making them a major source of advanced glycation end-products (AGEs) and their precursors—glycotoxins (reviewed by [15–19]). Previous reports from our laboratory showed that adipose tissue glycation in high-fat diet-fed rats impairs its expandability—which may be related to energy balance mechanisms dysregulation—leading to insulin resistance [20,21]. Moreover, glycation reduction through pharmacological strategies prevents such harmful effects [21,22].

Neuroendocrine pathways such as NPY and dopamine act in peripheral tissues to regulate lipid and glucose metabolism [23–25]. However, the effects mediated by dopamine and NPY depend on their receptor subtype. Different dopamine receptors trigger opposite effects: while dopamine receptor 1 (D1R) induces lipolysis and catabolic activity through AMP-activated protein kinase (AMPK) activation [26,27], dopamine receptor 2 (D2R) inhibits lipolysis by decreasing AMPK activity, hormone-sensitive lipase (HSL), and ATP citrate lyase (ACL), and it also induces lipogenesis by increasing Acetyl-CoA carboxylase (ACC) activity [25,26]. Regarding NPY receptors in white adipose tissue (WAT), NPY receptor 1 (NPY1R) induces lipogenic effects, while NPY receptor 2 (NPY2R) is associated with adipogenic and angiogenic processes [23,28–31]. Known as the stomach-derived hunger hormone, ghrelin binding to growth hormone secretagogue receptor 1α (GHS-R1α) in the hypothalamus regulates energy balance by NPY/Agouti-related protein (AgRP) neurons activation [32,33]. In WAT, the direct activation of GHS-R1α on adipocytes decreases insulin sensitivity and stimulates adiposity [12,34,35]. Acyl-ghrelin in retroperitoneal adipose tissue (AT) increases sterol regulatory element-binding transcription factor 1 (SREBP1C), a master regulator of lipogenesis, while decreasing fatty acid (FA) transport, which contributes to fat accumulation [35,36]. Ghrelin acts as an anti-inflammatory agent in the liver and promotes hepatic lipogenesis by activating the mTOR-PPARγ signaling pathway, although its role on glucose and lipid metabolism remains unknown [34,37]. Overall, both NPY and acyl ghrelin levels are increased in patients with obesity and T2D, contributing to adiposity and reduced insulin sensitivity [11,12,38,39], whereas dopamine action is dependent on subtype receptors.

Maternal nutrition impacts offspring gene expression and epigenome, metabolism, and cellular function, affecting organ development and later newborns' lives [40]. Gestation and lactation play a major role in programming windows. Maternal nutrition and metabolic status induce alterations in the intrauterine environment and breastmilk composition that determine offspring adiposity [41,42]. Worldwide, studies demonstrated a higher risk of obesity development when the organism was exposed to maternal overnutrition and obesity [43–45]. At the central level, it has already been demonstrated that both maternal obesity and postnatal overfeeding, besides leading to overweight, induce NPY hypothalamic changes in offspring [40,46]. Moreover, maternal obesogenic diets, such as westernized diets, can induce obesity and contribute to metabolic complications during lactation [47,48], showing the importance of maternal diet/lifestyle during lactation. Thus, it is necessary to understand if insulin-sensitive tissues such as the liver and AT undergo alterations in the pathways that regulate energy expenditure after exposure to an unhealthy maternal lifestyle and postnatal overfeeding.

In the present study, we compared the role of different obesogenic environments, namely maternal hypercaloric diets (gestation and lactation) and postnatal overfeeding, in energy balance mechanisms, particularly ghrelin, NPY, and dopamine signaling in insulin-sensitive tissues of young animals. Additionality, in order to disclose the role of maternal diet in the metabolic state, we addressed the impact of glycotoxins, common in Western diets, in postnatal overfed rats. We hypothesize that animals with postnatal overweight induced by both maternal obesity and overfeeding show an adaptation of the mechanisms of energy storage and expenditure to unhealthy motherhood and the larger amount of food available. However, exposure to maternal glycation in postnatal overfed rats may hamper these compensatory mechanisms, aggravating the risk for obesity, T2D, and other metabolic diseases that are major healthcare concerns worldwide.

2. Materials and Methods

2.1. *In Vivo Models of Obesogenic Metabolic Programming*

2.1.1. Maternal Diet-Induced Obesity (DIO) during Gestation and Lactation

Female Sprague-Dawley rats were fed an HFHS diet (containing 42% metabolizable energy from fat, 27% from proteins, and 31% from carbohydrates) before pregnancy until lactation. At PND 21, newborns were weaned and fed with chow. At PND 42, male and female offspring were euthanized, and peripheral tissues were collected for molecular and cellular analysis. In this study, liver and visceral AT (VAT) samples were used from a previously published study, where the experimental design (Figure 1A) and all the biochemical profiles of the rats with 42 days were already described (control litters = 6; HFHS litters = 6; number of males from control dams = 4; number of males from dams submitted to HFHS diet = 5; number of females from control dams = 4; number of females from dams submitted to HFHS diet = 5). More information about this animal model is available in Stevanović-Silva et al. [49,50]. The results of this animal model are depicted in Figures 1 and 2.

2.1.2. Postnatal Overfeeding and Glycation Models

The procedures were approved by the Animal Welfare Committee (ORBEA) of the Coimbra Institute for Clinical and Biomedical Research (iCBR), Faculty of Medicine, University of Coimbra. Animal experimentation was performed following the European Community directive guidelines for the use of laboratory animals (2010/63/EU), transposed into Portuguese law in 2013 (Decreto-Lei 113/2013). Wistar rats were housed under standard conditions (ventilation; 22 °C temperature; 55% humidity; 12 h/12 h light/dark cycle) with ad libitum access to food and water. This work used three animal models of young Wistar rats (postnatal overfeeding model, a maternal glycation model, and postnatal overfed rats exposed to Maternal Glycotoxins). After delivery, all litter sizes were reduced to 8 pups for standardization. After birth, body weight was monitored at postnatal day (PND) 0, PND 4, PND 7, PND 14, PND 21, PND 35, and PND 45. On PND 21, newborns were weaned and separated from their mothers until PND 45 and fed a standard diet. During this period, food consumption was weekly monitored. At PND 45, triglyceride levels were measured in the tail vein (Accutrend, Roche, Mannheim, Germany), and an insulin tolerance test was performed. After blood collection, animals were anesthetized with an IP injection of ketamine/chlorpromazine and euthanized by cervical displacement, and the VAT and liver were collected for molecular analysis. The dams were anesthetized with an IP injection of ketamine/chlorpromazine, milk samples were collected, and the females were euthanized by cervical displacement at 21 PND for liver collection and tissue morphology analyses.

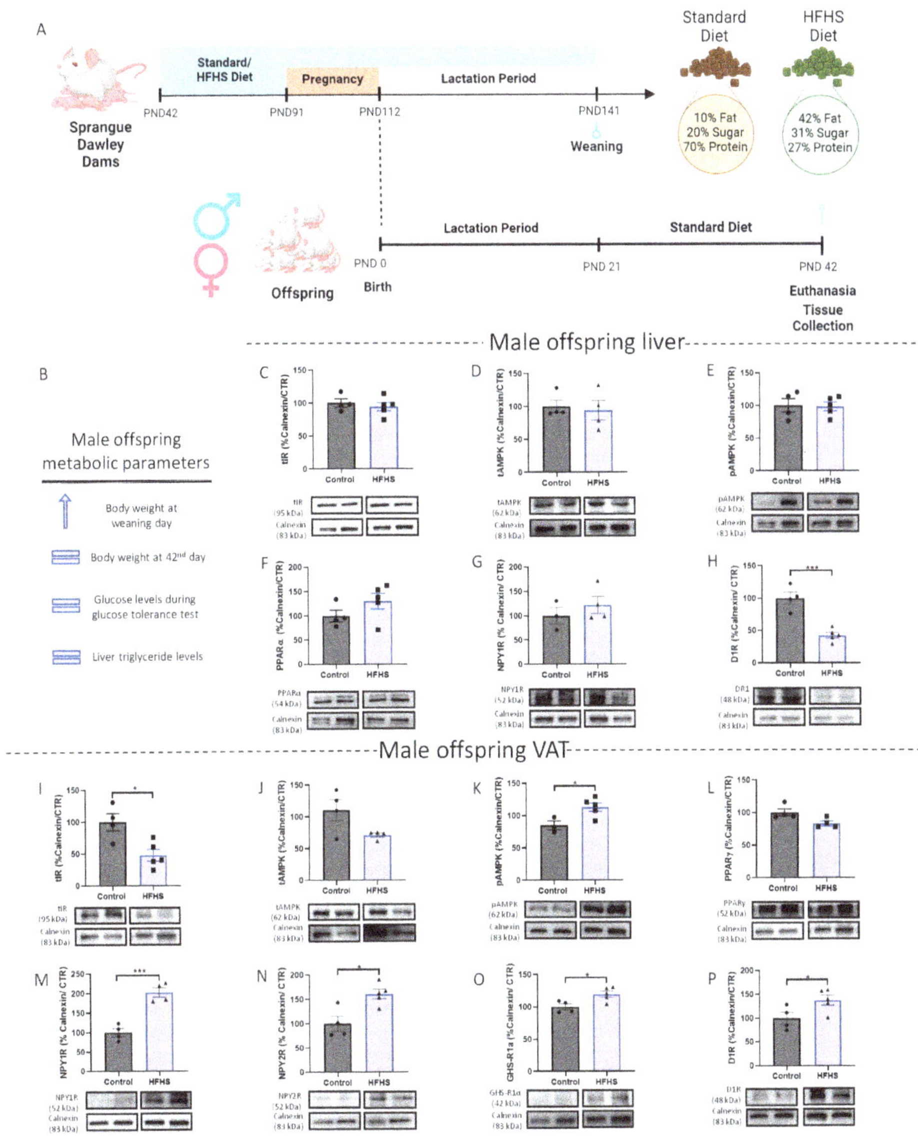

Figure 1. Maternal obesity increases both lipogenic and catabolic mechanisms in WAT from male offspring at PND 42. Maternal DIO experimental design (Created with BioRender.com, accessed on 25 January 2023) (**A**). Summary of metabolic parameters of male offspring after the data from [50] (**B**). Total liver IR (**C**), total AMPK (**D**), p-AMPK (**E**), PPARα (**F**), and NPY1R (**G**) levels remain similar to the control, while D1R levels are reduced in male offspring (**H**). Maternal obesity reduces total IR levels (**I**) in the WAT of male offspring. Total levels of AMPK (**J**) and PPARγ (**L**) were maintained in WAT, while D1R (**P**) AMPK phosphorylation (**K**), NPY1R (**M**), NPY2R (**N**), and ghrelin receptor (GHS-R1α) (**O**) levels increase in WAT of male offspring. Representative images of Western blot proteins of interest and loading controls (Calnexin) are shown at the bottom. Control: 42-day-old Sprague-Dawley males from dams fed a standard diet; HFHS: male offspring with 42-day-old Sprague-Dawley dams fed an HFHS diet. Bars represent mean ± SEM of 3, 4, or 5 animals per group, and unpaired t-tests were performed to compare the groups. * $p < 0.05$; *** $p < 0.001$.

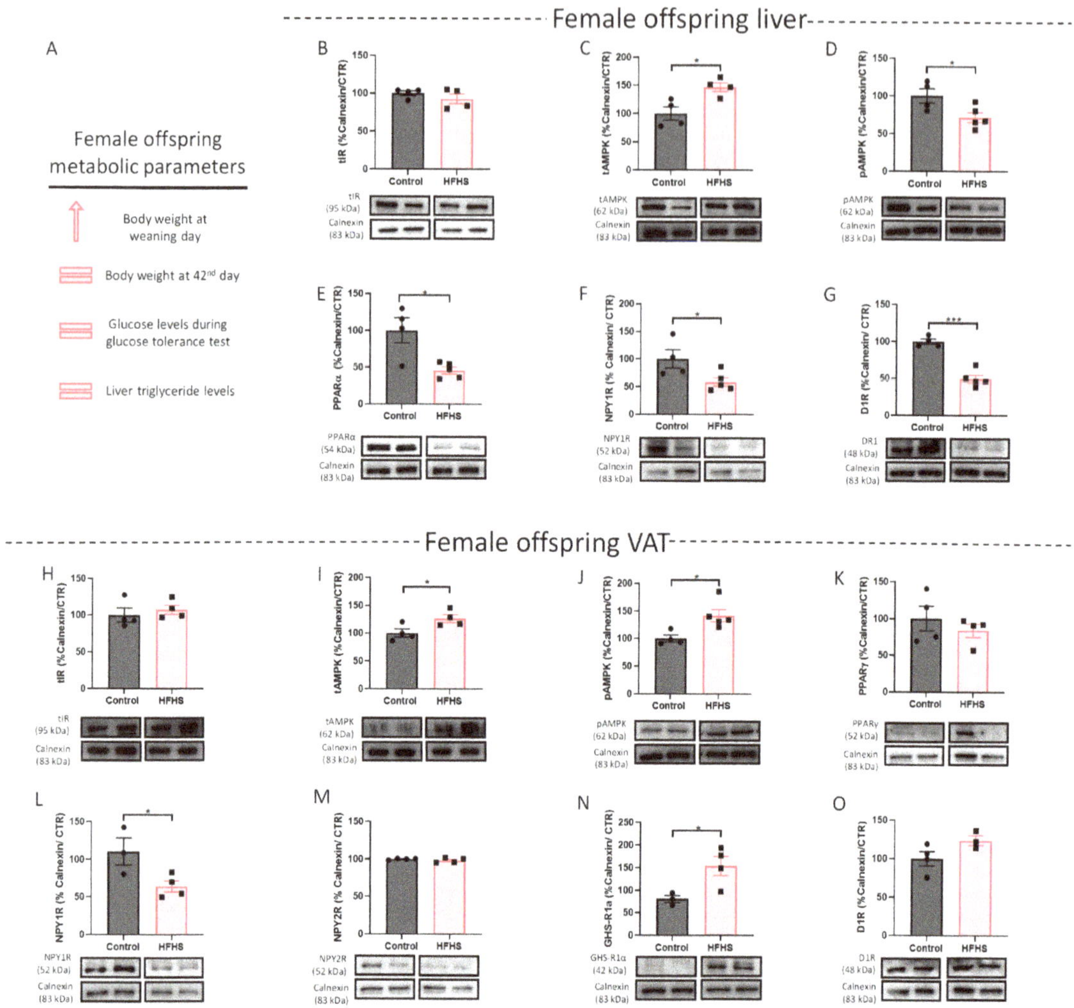

Figure 2. Effects of maternal DIO in liver and VAT of female offspring at PND 42. Summary of metabolic parameters of female offspring, after the date from [49] (**A**). Total liver IR levels were maintained (**B**), while PPARα (**E**), NPY1R (**F**), and D1R (**G**) levels are reduced in the liver of female offspring. Total AMPK levels increases (**C**) while AMPK phosphorylation decreased (**D**) in the liver of female rats from DIO dams. On WAT, total IR (**H**), PPARγ (**K**), NPY2R (**M**) and D1R (**O**) levels remain similar to the control. Maternal HFHS diet increases both total (**I**) and phosphorylated (**J**) forms of AMPK in female offspring. NPY1R levels were decreased (**L**) while GHS-R1α levels increased (**N**) in VAT of female offspring of DIO dams. Representative images of Western blot proteins of interest and loading controls (Calnexin) are shown at the bottom. Control—42-day-old Sprague-Dawley females from dams fed a standard diet; HFHS—female offspring with 42-day-old of Sprague-Dawley dams fed an HFHS diet. p-AMPK and D1R in liver samples were detected in the same membrane, as were PPARγ and NPY2R in VAT samples, which had the same loading control. Bars represent mean ± SEM of 3, 4, or 5 animals per group, and unpaired *t*-tests were performed to compare the groups. * $p < 0.05$; *** $p < 0.001$.

Postnatal Overfeeding Model

On the third day after the birth, a small litter (SL) protocol was implemented by reducing the litter size to 3 pups per dam to induce postnatal overfeeding and overweight (Figure 3A) (control litters = 7; SL litters = 5; number of males from normal litters (NL) = 35; number of males from SL = 9; number of females from NL = 12; number of females from SL = 6). The results of this animal model are depicted in Figures 3 and 4.

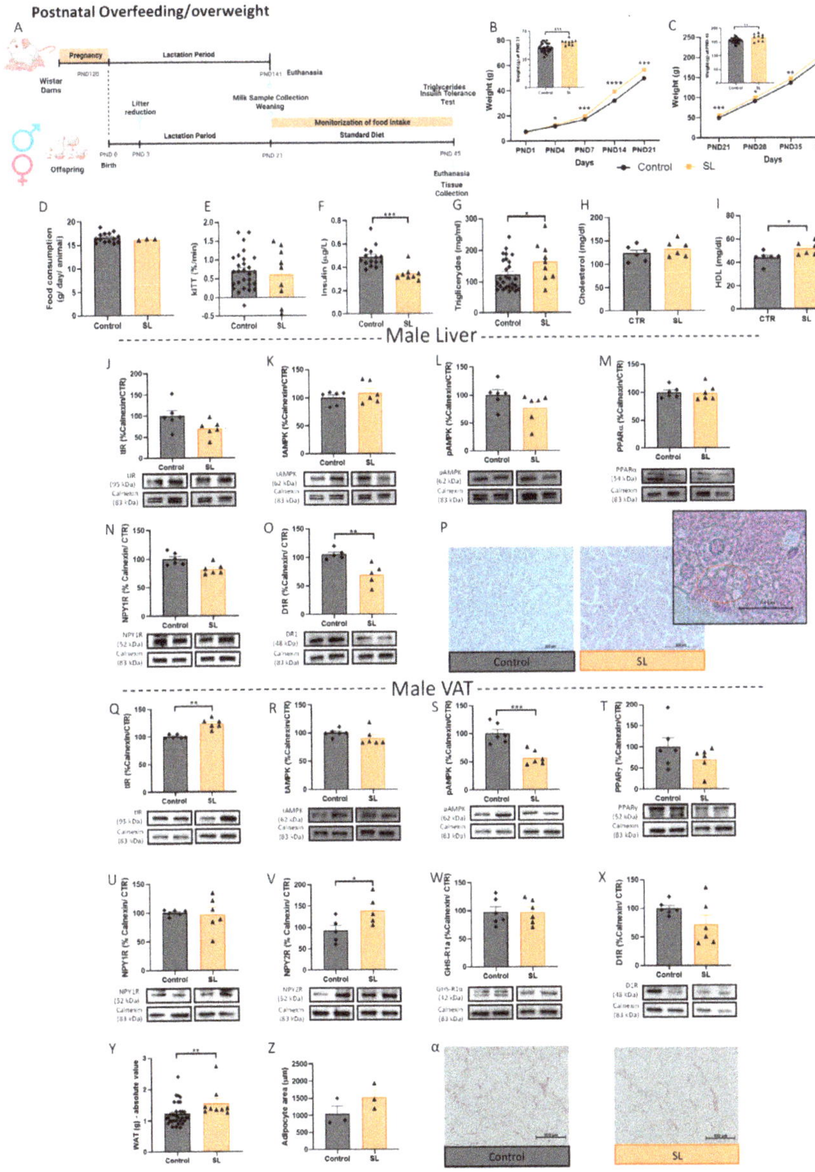

Figure 3. Postnatal overfeeding (SL) enhances VAT adipogenesis regulators and fat mass in male offspring. Postnatal overfeeding-induced early life obesity experimental design (created with BioRender.com, accessed on 25 January 2023) (**A**). Weight gain curves during the first 21 PND

(weaning day) (**B**) and between PND 21 to PND 45 (period after weaning) (**C**). Postnatal overfeeding did not alter food intake in SL male animals after weaning (n = average per cage) (**D**). The kITT of overfed male animals on PND 45 (**E**). Postnatal overfeeding reduced plasma insulin levels (**F**) and increased plasma triglycerides (**G**) and HDL (**I**) in male SL animals without altering cholesterol plasma levels (**H**). Total IR (**J**), total AMPK (**K**), p-AMPK (**L**), PPARα (**M**), and NPY1R (**N**) levels remained similar to the control while D1R levels were reduced (**O**) in the liver of male overfed animals. Hematoxylin–eosin staining (100×) of liver from overfed male rats (**P**). Early life obesity increased total IR (**Q**) and NPY2R (**V**) levels in visceral AT, without affecting PPARγ (**T**), NPY1R (**U**), GHS-R1α (**W**), and D1R (**X**) levels in male SL animals. Overfeeding induced lower AMPK phosphorylation (**S**) but maintained total AMPK levels (**R**) in male SL animals. The absolute value of perigonal fat mass (mg) at PND 45 (**Y**), adipocytes area (**Z**), and representative images of perigonadal AT stained with hematoxylin–eosin (100×) (**α**). Control—45-day-old Wistar males from normal litters; SL—45-day-old Wistar males from SLs. IR and AMPK in liver samples were marked in the same membrane as well as IR and p-AMPK in VAT samples, having the same loading control. Bars represent the mean ± SEM of 35 (or 6 in WB data or 3 in adipocyte area) animals in the control and 9 (or 6 in WB data or 3 in adipocyte area) in the SL group, and unpaired t-tests were conducted to compare the groups. * $p < 0.05$; ** $p < 0.01$; *** $p < 0.001$.

Maternal Glycation Model

Wistar dams were injected via intraperitoneal (IP) with S-P-Bromobenzylgutathione cyclopentyl diester—BBGC (5 mg/kg)—a selective inhibitor of Glyoxalase 1 (GLO1), during the first six days post-partum, whereas vehicle dams were injected with the vehicle dimethyl sulfoxide—DMSO (60 µL) (Figure 5A) (control litters = 7; vehicle litters = 5; BBGC group = 5; number of male pups from control dams = 35; number of male pups from dams treated with vehicle = 22; number of male pups from dams treated with BBGC = 29; number of female pups from control dams = 12; number of female pups from dams treated with vehicle = 9; number of female pups from dams treated with BBGC = 9). The results of this animal model are depicted in Figures 5–7.

Postnatal Overfed Rats Exposed to Maternal Glycotoxins Model

Male offspring from dams treated with BBGC as described in the maternal glycation model were submitted to an SL reduction to 3 pups for litter at PND 3 (Figure 8A) (SL litters = 5; BBGC + SL litters = 3; number of SL = 9; number of BBGC + SL = 9). The results of this animal model are depicted in Figure 8.

2.2. Milk Sample Collection and Determination of Total Antioxidant Capacity and Triglycerides

The Wistar female dams were anesthetized and injected with oxytocin (Facilpart) at a concentration of 10 UI/mL after 6 h of fasting, on day 21 postpartum. Milk samples were collected, milk triglycerides were determined using the Accutrend, Roche, Germany, and milk total antioxidant capacity was assessed with an assay kit (ab65329) according to the manufacturer's instructions.

2.3. Plasma Determinations

Wistar rats' blood samples were collected by cardiac puncture under anesthesia and immediately before sacrifice in Vacuette K3EDTA tubes (Greiner Bio-one, Kremsmunster, Austria) at PND 45. Blood samples were immediately centrifuged (2200× g, 4 °C, 15′) and the plasma fraction was stored at −80 °C until performing the Rat Insulin ELISA Kit (Mercodia, Uppsala, Sweden), according to the manufacturer's instructions. Total and HDL cholesterol were determined using the Prestige 24i Tokyo Boeky system with reagents from Cormay, Poland.

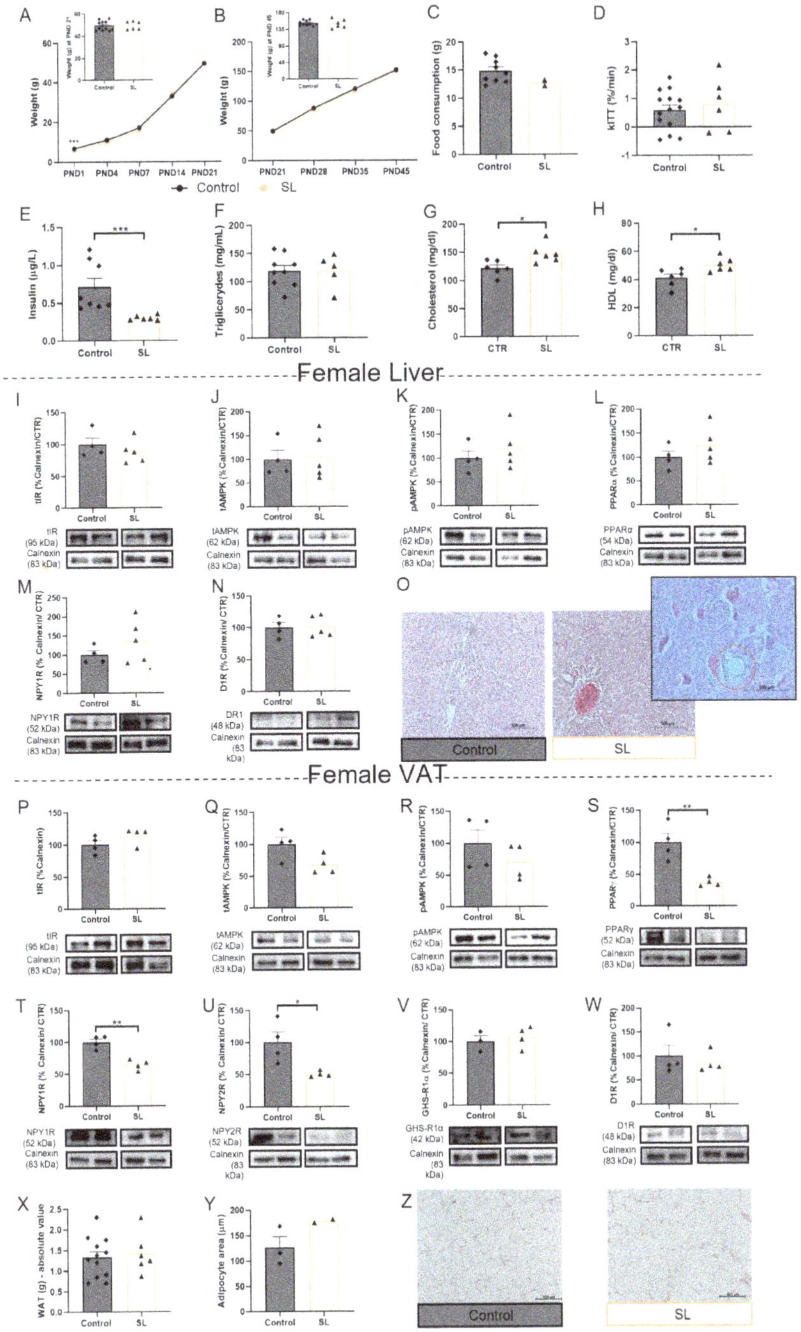

Figure 4. Adipogenic and lipogenic regulators were reduced in VAT female overfed rats. Weight gain curves during the first PND 21 (weaning day) (**A**) and between PND 21 to PND 45 (period after weaning) (**B**). Postnatal overfeeding did not alter food intake in obese female animals after weaning (**C**). The kITT

of female overfed animals on PND 45 (**D**). Postnatal overfeeding induced by SL reduced plasma insulin levels (**E**), maintained plasma triglycerides (**F**), and increased both cholesterol (**G**) and HDL (**H**) levels. Total IR (**I**), total AMPK (**J**), p-AMPK (**K**), PPARα (**L**), NPY1R (**M**), D1R (**N**) levels remained similar to the control in the liver of SL females. Representative liver images stained with hematoxylin–eosin (100×) (**O**). Overfeeding during lactation did not affect total IR (**P**) total AMPK (**Q**), p-AMPK (**R**), GHS-R1α (**V**), and D1R (**W**) levels while impairing adipogenesis and lipogenesis-associated mechanisms NPY1R (**T**), NPY2R (**U**), and PPARγ (**S**) in VAT. The absolute value of fat mass (mg) at PND 45 (**X**), adipocytes area (**Y**), and representative images of perigonadal AT stained with hematoxylin–eosin (100×) (**Z**). Control—45-day-old Wistar females from normal litters; SL—45-day-old Wistar females from SLs. AMPK and PPARα, IR, and D1R in liver samples were marked in the same membrane as well as AMPK and NPY1R in VAT samples. Bars represent the mean ± SEM of 12 (or 4 in WB data or 3 in adipocyte area) animals in the control and 9 (or 4/5 in WB data or 2 in adipocyte area) in the SL group, and unpaired t-tests were conducted to compare among the groups. * $p < 0.05$; ** $p < 0.01$; *** $p < 0.001$.

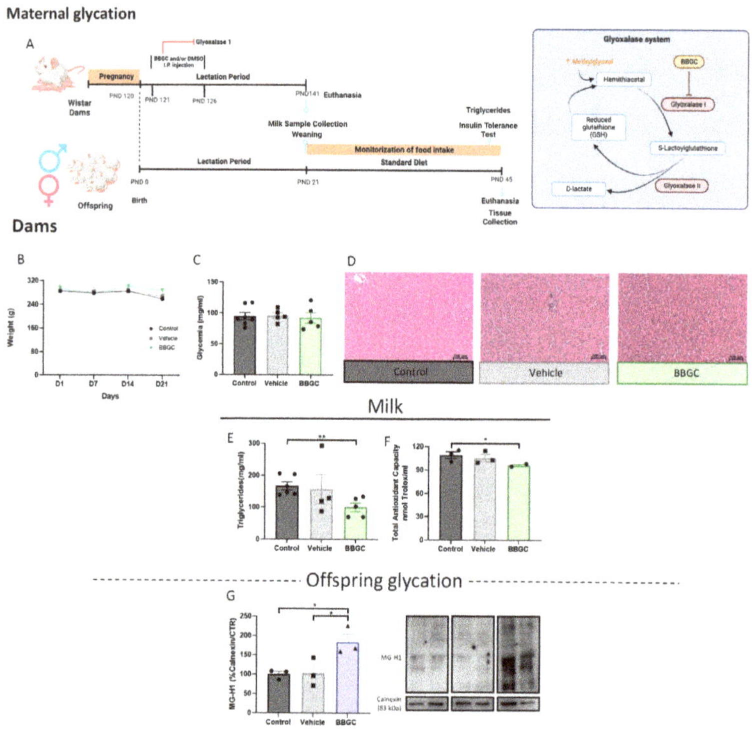

Figure 5. Maternal glycation alters breastmilk composition: Maternal glycation experimental design (created with BioRender.com, accessed on 25 January 2023) (**A**). Maternal weight gain curves during the first 21 days postpartum (**B**). Fasting glycemia of dams on the 21st postpartum day (**C**). Representative liver images from dams stained with hematoxylin–eosin staining (100×) (**D**). Glycation induced by BBGC reduced triglyceride levels (**E**) and total antioxidant capacity (**F**) content in breastmilk. Exposure to maternal glycation during lactation increases MG-H1 levels in VAT from male offspring (**G**). Control: Wistar control dams (or their offspring in **G**); vehicle: Wistar dams treated with DMSO (or their offspring in **G**); BBGC: Wistar dams treated with BBGC (5 mg/kg) through i.p. (or their offspring in **G**). Bars represent mean ± SEM of 2–5 dams per group or 3 rat offspring, and the Kruskal–Wallis test or one-way ANOVA was conducted to compare the groups. * $p < 0.05$; ** $p < 0.01$.

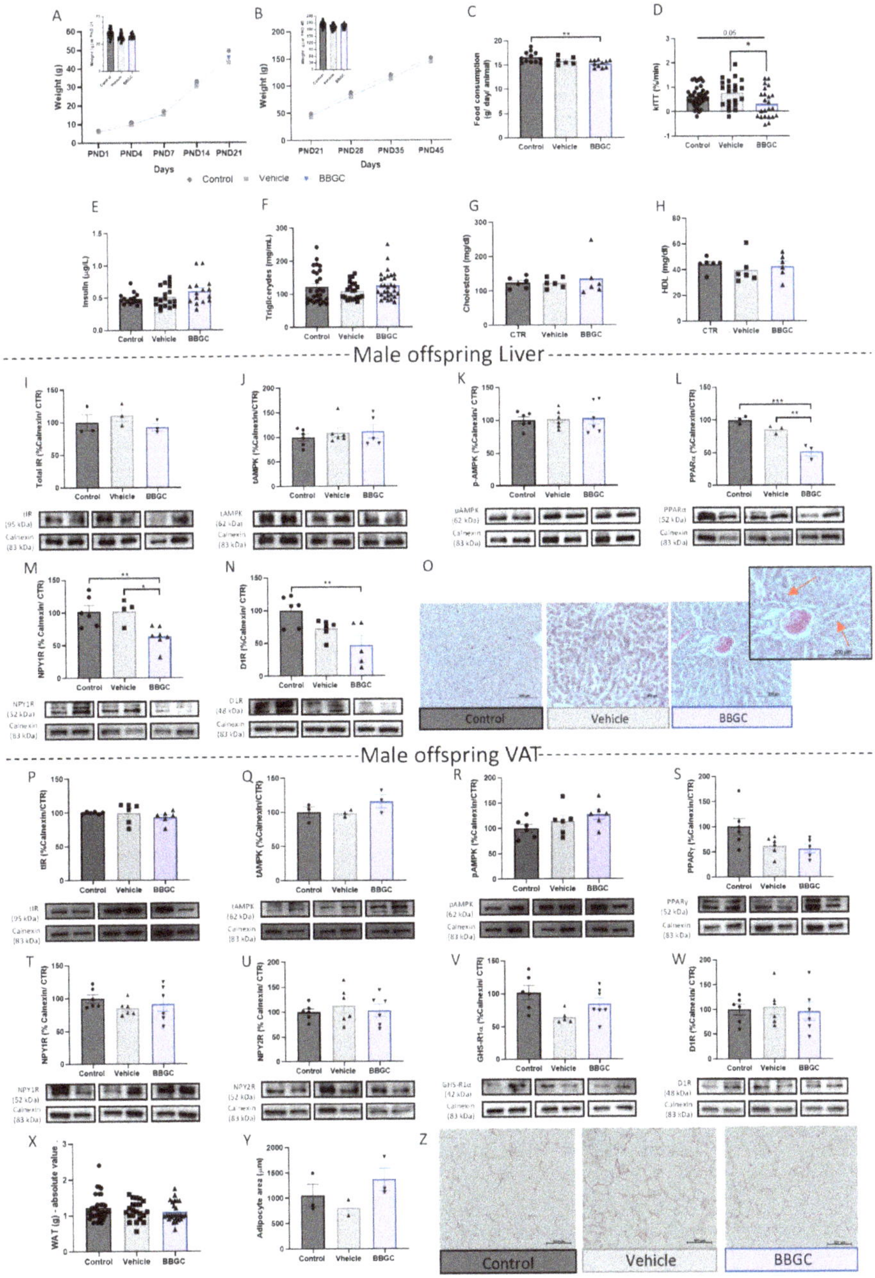

Figure 6. Maternal glycation reduced insulin sensitivity and liver NPY and D1R levels, not affecting VAT energy balance mechanisms in lean male offspring. Weight gain curves during the first PND 21 (weaning day) (**A**). Weight gain curves between PND 21 to PND 45 (period after weaning) (**B**). Maternal glycation reduced food intake in male offspring after weaning (**C**). Maternal glycation reduced the kITT (**D**) without affecting plasma insulin levels (**E**). Plasma triglyceride levels (**F**), plasma cholesterol levels (**G**) and plasma HDL levels (**H**) do not change. Offspring exposed to glycotoxins during lactation presented lower levels of PPARα (**L**), NPY1R (**M**), and D1R (**N**) while total IR (**I**), total

AMPK (**J**), and p-AMPK (**K**) levels were conserved. Representative image of inflammatory infiltration in the liver of male offspring exposed to maternal glycation (**O**) (100×). Levels of total IR (**P**), total AMPK (**Q**), p-AMPK (**R**), PPARγ (**S**), NPY1R (**T**), NPY2R (**U**), GHS-R1α (**V**), and D1R (**W**) in VAT of male offspring submitted to glycotoxins exposure during lactation. The absolute value of fat mass (mg) at PND 45 (**X**), adipocytes area (**Y**), and representative images of perigonadal AT stained with hematoxylin–eosin (100×) (**Z**). Representative images of Western blot proteins of interest and loading controls (Calnexin) are shown at the bottom. Control: male offspring of control dams; vehicle: male offspring of dams treated with DMSO; BBGC- Wistar male offspring of dams treated with BBGC (5 mg/kg) through IP. Bars represent the mean ± SEM of 35 animals (or 6 in WB data or 3 in adipocyte area) in control group, 22 (or 6 in WB data or 2 in adipocyte area) in vehicle group, and 29 in BBGC group (or 6 in WB data or 3 in adipocyte area) and one-way ANOVAs were performed to compare the groups. * $p < 0.05$; ** $p < 0.01$; *** $p < 0.001$.

2.4. *Histology—Haematoxylin-Eosin*

Livers and VAT from dams and offspring of exclusive metabolic programming during lactation models were fixed in formalin solution (10%), dehydrated in an increasing series of alcohol concentrations (70% to 100%), cleared in xylene, and then embedded in histological paraffin. The livers were sectioned in a microtome, on a non-serial section of 4 μm thickness (n = 3/group) and subsequently dried overnight at room temperature (RT). The paraffin-embedded liver sections were submitted to paraffin-removing protocols, using xylol, progressive hydration (EtOH 100%/70%/30% 3′/each and Milli-Q water during 3′ at RT), and stained with hematoxylin and eosin (H&E). Then, the liver sections were washed again, and coverslips were mounted using a mounting medium (DAKO, Kyoto, JAPAN). Lastly, images (100×) were captured in a Zeiss microscope with an incorporated camera (Zeiss, Jena, Germany). All full-size representative images in the manuscript are presented in the Supplementary Data.

2.5. *Western Blot*

Hepatic and VAT samples were collected and washed with PBS and disrupted in lysis buffer (0.25 M Tris-HCl, 125 mM NaCl, 1% TritonX-100, 0.5% SDS, 1 mM EGTA, 1 mM EDTA, 20 mM NaF, 2 mM Na3VO4, 10 mM βglycerophosphate, 2.5 mM sodium pyrophosphate, 10 mM PMSF, 40 μL of protease inhibitor) using the TissueLyser system (Quiagen, Hilden, Germany). The bicinchoninic acid (BCA) Protein Assay Kit was carried out on the supernatant (14,000 rpm for 20 min at 4 °C, followed by the addition of Laemmeli buffer (62.5 mM Tris-HCl, 10% glycerol, 2% SDS, 5% β-mercaptoethanol, and 0.01% bromophenol blue). Tissue samples (20 μg) were loaded onto SDS-PAGE and electroblotted into polyvinylidene difluoride (PVDF) membrane (Advansta, San Jose, CA, USA). Tris-buffered saline-tween (TBS-T) 0.01% and bovine serum albumin (BSA) 5% were used to block the membranes, which were then incubated with primary (overnight, 4 °C) and secondary antibodies (2 h RT), following the dilutions listed in the Supplementary Table S1. The proteins of interest were detected using enhanced chemiluminescence (ECL) substrate with the LAS 500 system (GE Healthcare, Chicago, IL, USA). The bands of interest were quantified with Image Quant 5.0 software (Molecular Dynamics). The results were expressed as a percentage of control and normalized for the loading control (calnexin, 83 kDa).

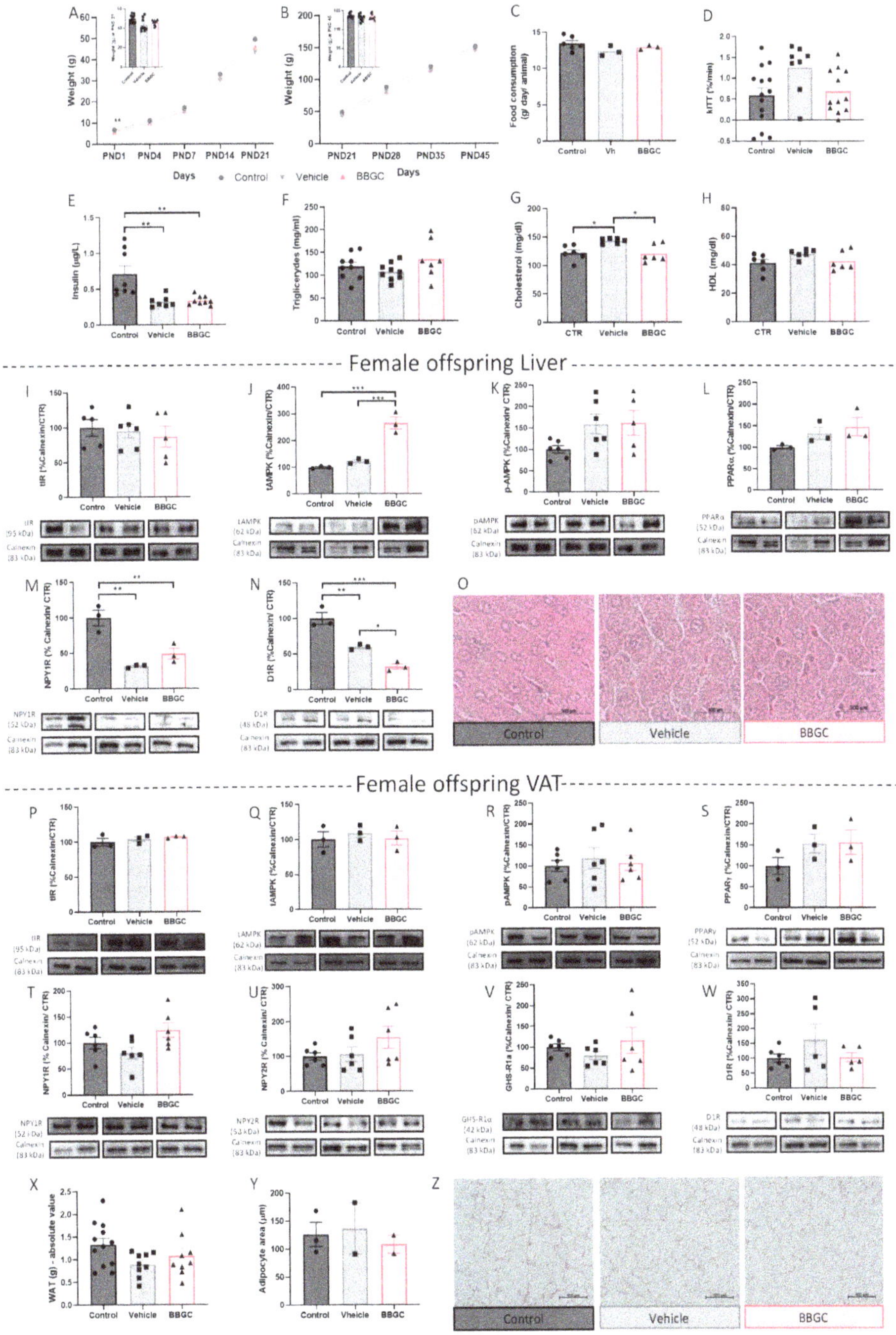

Figure 7. Maternal glycation does not affect metabolic profile and VAT energy balance in BBGC female offspring. Weight gain curves until 21 PND (weaning day) (**A**) and between PND 21 to PND 45 of female offspring (**B**). Maternal glycation did not alter food intake after weaning (**C**), the kITT (**D**), and plasma levels of insulin (**E**), triglyceride levels (**F**), total cholesterol (**G**), and HDL-cholesterol (**H**) in

female offspring exposed to maternal glycotoxins. Maternal glycation increased AMPK levels (**J**) in the liver of female offspring without affecting AMPK phosphorylation (**K**), total IR (**I**), and PPARα levels (**L**). Both vehicle and BBGC treatment in dams decreased NPY1R (**M**) and D1R (**N**) levels in the liver of female offspring. Representative images of liver from female offspring stained with hematoxylin–eosin staining (100×) (**O**). Levels of total IR (**P**), total AMPK (**Q**), p-AMPK (**R**), PPARγ (**S**), NPY1R (**T**), NPY2R (**U**), GHS-R1α (**V**), and D1R (**W**) in WAT of female offspring submitted to maternal glycation. The absolute value of fat mass (mg) at PND 45 (**X**), adipocytes area (**Y**), and representative images of perigonadal AT stained with hematoxylin–eosin (100×) (**Z**). Representative images of Western blot proteins of interest and loading controls (Calnexin) are shown at the bottom. Control: female offspring of control dams; vehicle: female offspring of dams treated with DMSO; BBGC: Wistar female offspring of dams treated with BBGC. AMPK and PPARα in liver samples were marked in the same membrane. Bars represent the mean ± SEM of 12 (or 3/6 in WB data or 3 in adipocyte area) animals in control group, 9 (or 3/6 in WB data or 2 in adipocyte area) in vehicle group, and 9 (or 3/6 in WB data or 2 in adipocyte area) in BBGC group, and one-way ANOVAs were conducted to compare the groups. * $p < 0.05$; ** $p < 0.01$; *** $p < 0.001$.

2.6. Statistical Analyses

The results are presented as the mean ± standard error of the mean (SEM). Statistical analysis was performed with GraphPad Prism 8 (GraphPad Software, Inc., San Diego, CA, USA). The normality of the data was assessed with the Shapiro–Wilk normality test. Accordingly, data with two conditions were analyzed with a nonpaired *t*-test or Mann–Whitney test, and data with more than two conditions were analyzed with the Kruskal–Wallis test or with a one-way ANOVA followed by Tukey's post hoc test. Differences were considered for $p < 0.05$.

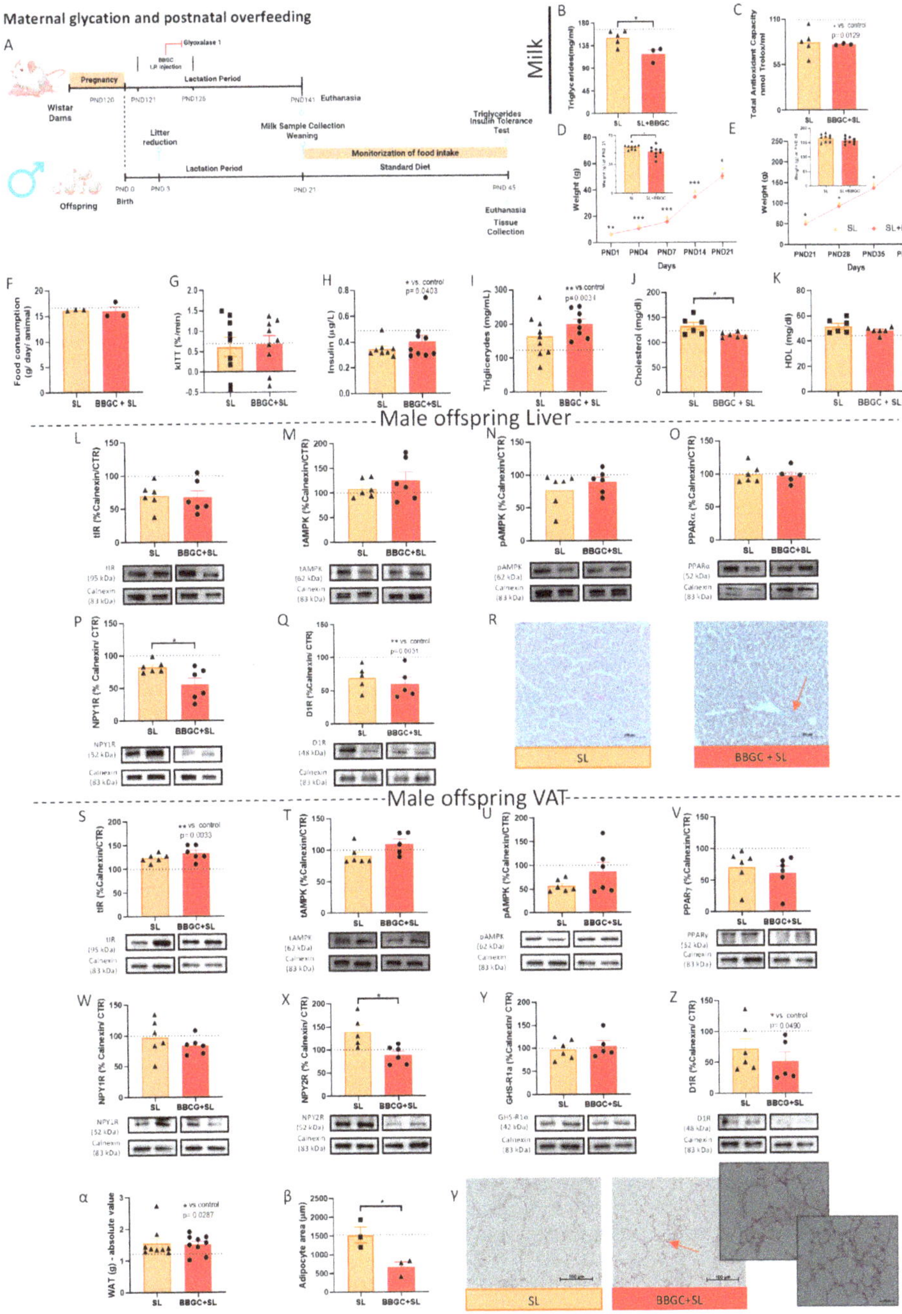

Figure 8. Maternal glycation impairs energy storage preventing weight gain in male offspring overfed induced by SL. Maternal glycation on SL-induced overfed rats experimental design (created with BioRender.com, accessed on 25 January 2023) (**A**). The triglyceride levels were decreased in breastmilk of BBGC dams (**B**). The total antioxidant capacity in breastmilk was affected by both the SL process and BBGC treatment (**C**). Weight gain curves during the first PND 21 (weaning day) (**D**) and between

PND 21 to PND 45 (period after weaning) (**E**). Postnatal overfeeding did not alter food intake in male SL offspring exposed to maternal glycation after weaning (**F**). The kITT of SL offspring from dams treated with BBGC on the PND 45 (**G**). Plasma insulin levels (**H**), plasma triglyceride levels (**I**), plasma cholesterol levels (**J**), and plasma HDL levels (**K**) of overfed offspring exposed to maternal glycation. The levels of total IR (**L**), total AMPK (**M**), p-AMPK (**N**), PPARα (**O**), and D1R (**Q**) in the liver of overfed offspring exposed to maternal glycation remained similar to the animals with early-life obesity. NPY1R levels (**P**) in the liver of overfed animals exposed to glycotoxins during lactation were decreased compared to the postnatal overfed ones. Representative images of liver stained with hematoxylin–eosin (100×) (**R**). Compared to obese animals, the levels of total IR (**S**), total AMPK (**T**), p-AMPK (**U**), PPARγ (**V**), NPY1R (**W**), GHS-R1α (**Y**), and D1R (**Z**) were maintained in WAT from postnatal overfed offspring exposed to maternal glycation, while NPY2R (**X**) levels and adipocytes area (**β**) were decreased. The absolute value of fat mass (mg) at PND 45 (**α**) and representative images of perigonadal AT stained with hematoxylin–eosin (100×) (**γ**). SL: 45-day-old Wistar males from SLs; BBGC +SL: 45-day-old male obese offspring of dams treated with BBGC. Bars represent the mean ± SEM of 9 (or 6 in WB data or 3 in adipocyte area) animals in the SL group and 9 (or 6 in WB data or 3 in adipocyte area) animals in the BBGC + SL group, and unpaired t-tests were performed to compare the groups. * $p < 0.05$. Dashed lines represent the value of control animals.

3. Results

3.1. *Liver and VAT Sexual Dysmorphism in the Response to Different Perinatal Obesogenic Environments*

3.1.1. Maternal DIO Modulates Energy Balance Mechanisms in the VAT and Liver of Male Offspring

Exposure to maternal obesity increased male offspring body weight until weaning day, as previously published by Stevanović-Silva et al. [50] (summary presented in Figure 1B). Maternal obesity did not cause alterations in hepatic levels of IR (Figure 1C), AMPK (Figure 1D), phosphorylated AMPK (Figure 1E), and PPARα in male offspring (Figure 1F). Exposure to the maternal high-fat high-sugar (HFHS) diet induced a drastic reduction in D1R levels in the male offspring liver ($p < 0.001$ vs. control) (Figure 1H) without affecting the levels of NPY1R (Figure 1G).

Regarding VAT, maternal obesity induced by the HFHS diet decreased total IR levels ($p < 0.05$ vs. control) (Figure 1I) in male offspring without changing AMPK (Figure 1J) or PPARγ levels (Figure 1L). NPY1R (Figure 1M), NPY2R (Figure 1N), and GHS-R1α (Figure 1O) were significantly increased in male offspring VAT ($p < 0.001$ vs. control, $p < 0.05$ vs. control, and $p < 0.05$ vs. control, respectively), suggesting adipogenesis and lipogenesis upregulation. Moreover, the maternal HFHS diet increased male offspring D1R levels ($p < 0.05$ vs. control) (Figure 1P) and AMPK phosphorylation ($p < 0.05$ vs. control) (Figure 1K) in VAT compared to offspring from dams fed a standard diet.

3.1.2. Maternal Obesity Alters Levels of Energy Balance-Regulating Receptors in the VAT of Female Offspring

Maternal DIO female offspring were also studied to assess a potential sexual dimorphism in the liver and VAT adaptation to the maternal metabolic state. As occurs in males, maternal obesity increased female offspring body weight until weaning day (in summary in Figure 2A) (previously published [49]). Similar to male livers, there were no alterations in total IR levels (Figure 2B), and a significant reduction in D1R (Figure 2G) in female offspring exposed to the maternal HFHS diet during the perinatal period was observed ($p < 0.001$ vs. control). Nevertheless, in female offspring livers, NPY1R, PPARα, and p-AMPK levels also decreased ($p < 0.05$ vs. control, $p < 0.05$ vs. control, and $p < 0.05$ vs. control, respectively) (Figure 2D–F, respectively), while total levels of AMPK increased ($p < 0.05$ vs. control) (Figure 2C), showcasing the disparity between sexes.

Regarding VAT, besides the increased p-AMPK in female offspring ($p < 0.05$ vs. control) (Figure 2J), the total levels of AMPK also augmented ($p < 0.05$ vs. control) (Figure 2I), suggesting that the higher activation of AMPK is a consequence of additional AMPK being available. The changes in the NPY2R, D1R, and total IR levels observed in males were not noticed in the VAT of female offspring (Figure 2H,M,O). Surprisingly, the NPY1R levels were significantly decreased in the female offspring VAT ($p < 0.05$ vs. control) (Figure 2L), demonstrating the sexual dysmorphism of such mechanisms. In the case of the ghrelin receptor, as well as in males, an increase in VAT was observed in female offspring ($p < 0.05$ vs. control) (Figure 2N).

3.1.3. Postnatal Overfeeding Modulates Energy Balance Mechanisms in the VAT and Liver of Male Offspring

In this study, we used males from SL as a model of postnatal overfeeding without modification in the maternal diet, shown to induce obesity early in life [46,51]. Indeed, animals from SL had significantly higher body weight than the control group at the weaning PND 21 ($p < 0.001$ vs. control) (Figure 3B). The higher weight gain was maintained over time until the day of euthanasia—PND 45 ($p < 0.01$ vs. control) (Figure 3C), despite no changes in food intake being observed (Figure 3D). The decay of the glucose rate during the insulin tolerance test per minute (kITT) revealed that insulin sensitivity was not affected in SL rats (Figure 3E), although plasma insulin levels were decreased ($p < 0.001$ vs. control) (Figure 3F). Triglyceride ($p < 0.05$ vs. control) (Figure 3G) and HDL cholesterol levels were increased in male overfed rats ($p < 0.05$ vs. control) (Figure 4I) without affecting cholesterol total level (Figure 3H).

Similar to the maternal HFHS diet model (Figure 1), total liver levels of IR (Figure 3J), PPARα (Figure 3M), NPY1R (Figure 3N), AMPK (Figure 3K), and phosphor-AMPK (Figure 3L) were not affected, and D1R levels were decreased ($p < 0.05$ vs. control) (Figure 3O) in the SL animals with higher consumption of breastmilk during the perinatal period. Moreover, histology showed dispersed fat accumulation zones in the liver of overfed male animals (Figure 3P), while no changes were observed in the liver weight (supplementary data—Figure S1A).

The alterations observed in VAT in postnatal overweight offspring induced by overfeeding were different from rats exposed to maternal DIO. Interestingly, the total IR levels in the VAT were increased ($p < 0.01$ vs. control) (Figure 3Q) suggesting an adaptation for the reduced levels of insulin in the plasma. AMPK activation was lower in male overfed rats ($p < 0.001$ vs. control) (Figure 3S), while the total AMPK (Figure 3R) and PPARγ (Figure 3T) levels were maintained. Regarding neuroendocrine mechanisms controlling adipogenic and lipogenic processes, NPY1R, GHS-R1α, and D1R levels in WAT were maintained (Figure 3U,W,X, respectively), while an increase in the adipogenic-related NPY2R levels ($p < 0.05$ vs. control) was also observed (Figure 3V). This was consistent with an increase in fat mass in overfed animals ($p < 0.01$) (Figure 3Y), without adipocyte hypertrophy (Figure 3Z,α).

3.1.4. Female Postnatal Overfed Rats Exhibit Similar Metabolic Adaptations to Male Offspring except the Downregulation of VAT Adipogenic Mechanisms

Food intake (Figure 4C) and liver weight (supplementary data—Figure S1B) were not altered in SL female animals, and contrary to SL males, postnatal overfeeding did not induce overweight in female rats over the 45 days (Figure 4A,B), nor did it increase triglyceride levels (Figure 4F). Regarding insulin levels, the outcome of postnatal overfeeding was the same in both sexes: insulin plasma levels were reduced ($p < 0.001$ vs. control) (Figure 4E), while insulin sensitivity did not alter according to kITT (Figure 4D). Cholesterol levels in the plasma of overfed females increased ($p < 0.05$ vs. control) (Figure 4G), accompanied by an increase in HDL levels ($p < 0.05$ vs. control) (Figure 4H).

No changes in the studied mechanisms were observed in the postnatally overfed females' livers: total IR, total AMPK, p-AMPK, PPARα, NPY1R, and D1R levels (Figure 3I–N, respectively). Histology demonstrated few areas of lipid droplet accumulation in the liver (Figure 3O), as occurs in male overfed animals. Regarding AT, total IR levels are maintained in overfed females (Figure 4P), contrary to overfed male rats, as well as total AMPK and p-AMPK levels (Figure 4Q,R, respectively). Regarding energy storage mechanisms, both NPY1R and NPY2R were reduced in the VAT of postnatal overfed females ($p < 0.01$ vs. control and $p < 0.05$ vs. control, respectively) (Figure 4T,U, respectively), which was consistent with decreased PPARγ levels ($p < 0.01$ vs. control) (Figure 4S), unaffected fat mass (Figure 4X) and adipocyte size (Figure 4Y,Z), indicating that lipogenesis and adipogenesis may be reduced in postnatally overfed females. Ghrelin receptor (Figure 4V) and D1R levels were not altered in the VAT of SL females (Figure 4W).

3.2. *The Role of Maternal Glycation in Impairing the Neuroendocrine Mechanism of Adaptation to Obesogenic Environments*

3.2.1. Glycation Changes Breastmilk Composition

The glycation induced by BBGC (5 mg/kg) through IP did not alter the body weight or glycemic profile of dams (Figure 5B,C, respectively). Furthermore, BBGC did not cause toxicity in the liver (Figure 5D). However, glycation changed breastmilk quality, decreasing triglyceride levels ($p < 0.01$ vs. control) and total antioxidant capacity ($p < 0.05$ vs. control) (Figure 5E,F, respectively). A known methylgyoxal-derived AGE is N-(5-hydro-5-methyl-4-imidazolon-2-yl)-ornithine (MG-H1) [52]. This compound was increased in VAT from male offspring exposed to maternal glycotoxins ($p < 0.05$ vs. control; $p < 0.05$ vs. vehicle) (Figure 5G), showing that inhibition of glyoxalse I in dams leads to AGEs accumulation in VAT offspring.

3.2.2. Exposure to Glycotoxins during Lactation Decreases Offspring Food Intake and Impairs Insulin-Dependent Glucose Uptake without Affecting the Insulin Levels of Male Offspring

Obesogenic environments contribute to body weight alteration in early life, which is not often observed in lean or adult experimental models of glycation [53]. Here, we observed that maternal glycation did not affect the body weight of male offspring either during the breastfeeding period or after weaning (Figure 6A,B, respectively), although lower food consumption was observed ($p < 0.01$ vs. control) (Figure 6C). Despite no alterations In insulin plasma levels (Figure 6E), the kITT was reduced in male offspring exposed to maternal glycation ($p < 0.05$ vs. control) (Figure 6D), suggesting lower insulin sensitivity. Exposure to glycotoxins reduced triglyceride levels in breastmilk without affecting the plasma triglyceride, total, and HDL cholesterol levels in male offspring (Figure 6F–H).

3.2.3. Exposure to Maternal Glycation Reduces Both NPY and Dopamine Signalling in the Liver from Male Offspring without Affecting the VAT

Although insulin sensitivity was reduced in male offspring exposed to glycotoxins during lactation, total IR levels in the liver and WAT were not altered (Figure 6I,P, respectively). Total and activated AMPK also remained similar between groups in the hepatic tissue (Figure 6J,K, respectively).

In the liver, maternal glycation induced a reduction in NPY1R ($p < 0.01$ vs. control; $p < 0.05$ vs. vehicle), D1R ($p < 0.01$ vs. control), and PPARα levels ($p < 0.001$ vs. control; $p < 0.01$ vs. vehicle) (Figure 6M,N,L, respectively). Furthermore, maternal glycation appeared to induce portal inflammatory infiltration at the hepatic level (Figure 6O), while no changes were observed in the liver weight (Supplementary Data—Figure S1C).

Adult models exposed to glycated products showed that glycation does not cause significant changes in VAT function in the lean phenotype [53]. Here, we demonstrate that male offspring from dams treated with glyoxalase 1 inhibitor did not present changes in energy expenditure or storage mechanisms in VAT. The receptors of NPY, ghrelin, and dopamine evaluated (NPY1R, NPY2R, GHS-R1α, and D1R, respectively) display preserved levels when compared with the vehicle and with the control (Figure 6T–W, respectively) as well as the downstream proteins AMPK and PPARγ (Figure 6Q–S). Furthermore, neither fat mass (Figure 6X) nor adipocyte size was altered in male offspring exposed to maternal glycation at PND 45 (Figure 6Y,Z).

3.2.4. BBGC-Induced Maternal Glycation Does Not Affect Food intake, Metabolic Profile and VAT Mechanisms of Lipid Storage and Energy Expenditure in Female Offspring

As well as in the male offspring, during the lactation period and after weaning, no alteration in body weight of females exposed to maternal glycation (Figure 7A,B) or liver weight (supplementary data—Figure S1D) was observed. Contrary to what was observed in male offspring, food intake was not affected in BBGC female offspring (Figure 7C). Furthermore, no changes in kITT were observed in female offspring (Figure 7D), suggesting that male offspring are more susceptible to alteration in insulin sensitivity and feeding regulation when exposed to the same maternal condition. Although plasma insulin levels in BBGC female offspring were decreased ($p < 0.01$ vs. control), this was also observed in the vehicle group ($p < 0.01$ vs. control) (Figure 7E). Triglyceride (Figure 7F) and HDL (Figure 7H) levels were also unaltered between groups, while cholesterol levels were affected by vehicle ($p < 0.05$ vs. control; $p < 0.05$ vs. BBGC) (Figure 7G). Total IR levels were also maintained in both the liver and WAT of female offspring (Figure 7I,P, respectively).

The total levels of AMPK were increased in the liver of female offspring exposed to glycotoxins ($p < 0.001$ vs. control; $p < 0.001$ vs. vehicle) (Figure 7J) while its activity remained similar to the other groups (Figure 7K). PPARα levels also remained similar between all conditions (Figure 7L), and no inflammatory infiltration was observed in the livers of female rats exposed to maternal glycation (Figure 7O). NPY1R levels were decreased in females from dams treated with DMSO ($p < 0.01$ vs. control) and from dams subjected to glyoxalase 1 inhibition ($p < 0.01$ vs. control) (Figure 7M). However, the effect on D1R observed in the vehicle group ($p < 0.01$ vs. control) seems to be partially independent of the reduction induced by BBGC ($p < 0.001$ vs. control; $p < 0.05$ vs. vehicle) (Figure 7N), which is in accordance with the results observed in male offspring.

As occurs in male offspring exposed to maternal glycation during lactation, in BBGC female offspring, the WAT mechanisms associated with energy expenditure and storage were not affected by maternal glycation alone (Figure 7P–W) as were the fat mass and the adipocyte size (Figure 7X–Z). These results demonstrated once again that glycation alone may not be sufficient to alter the energy balance in the VAT of lean animals at PND 45.

3.2.5. Maternal Glycation Inhibits SL-Induced Weight Gain in Male and Impairs VAT Mechanisms of Adaptation to Overfeeding

BBGC also reduced milk triglyceride levels ($p < 0.05$ vs. SL) and total antioxidant capacity ($p < 0.05$ vs. control) in SL dams (Figure 8B,C, respectively). Exposure to maternal glycation during lactation prevented male offspring weight gain induced by postnatal overfeeding from PND14 to PND35 ($p < 0.05$ vs. SL) (Figure 8D,E). No alterations were observed in food intake or the kITT (Figure 8F,G, respectively) or in the liver weight (supplementary data—Figure S1E) in the BBGC + SL group as occurs in SL animals with healthy motherhood. The alterations in plasma insulin ($p < 0.05$ vs. control) (Figure 8H) and triglyceride ($p < 0.05$ vs. control) (Figure 8I) levels observed in SL males were maintained in SL + BBGC offspring. Cholesterol levels were reduced in SL rats exposed to maternal glycation ($p < 0.05$ vs. SL) (Figure 8J), while the increase in HDL-cholesterol levels observed in male SL was not noticed when the animals were exposed to glycotoxins (Figure 8K). Total IR levels were maintained in the liver and VAT of male SL + BBGC offspring compared to SL (Figure 8L,S, respectively), and its upregulation in the liver of SL males was maintained in relation to control rats.

AMPK levels and activity in the liver did not alter in BBGC + SL rats (Figure 8M,N) as occurs in animals exposed to both conditions independently. Surprisingly, the PPARα levels were not altered upon exposure to maternal glycation in animals overfed and induced by SL (Figure 8O) as occurred in lean offspring. As well as observed in the liver from lean male offspring exposed to maternal glycation, in obese conditions, the decrease in NPY1R levels was also verified ($p < 0.01$ vs. SL) (Figure 8P), while the downregulation of D1R levels observed with both conditions per se was maintained ($p < 0.001$ vs. control) (Figure 8Q). These changes were associated with liver inflammatory infiltration, which occurs in lean male offspring from dams treated with BBGC (Figure 8R). Importantly, this male offspring did not show visible liver fat accumulation, contrary to the overfed ones. The decrease in p-AMPK levels induced by postnatal overfeeding in VAT was lost (Figure 8U), while the total levels of AMPK were maintained (Figure 8T) in male offspring exposed to glycotoxins. As already mentioned, postnatal overfeeding causes adaptations in VAT, namely NPY2R overexpression. Exposure to glycated products during lactation induced VAT energy balance dysregulation in postnatal overfed offspring. Despite no changes in PPARγ, NPY1R, or GHS-R1α levels in the VAT of overfed offspring from dams treated with BBGC (Figure 8V,W,Y, respectively), NPY2R levels significantly decreased ($p < 0.05$ vs. SL) (Figure 8X) and D1R further decreased compared to the control ($p < 0.05$ vs. control) (Figure 8Z). Interestingly, although the visceral fat mass did not change in these animals (Figure 8α), their adipocyte size was significantly reduced ($p < 0.05$ vs. SL), following reduced NPY2R levels, and several multivesicular adipocytes were observed, a marker of adipocyte dysfunction (Figure 8β,γ).

4. Discussion

Exposure to unhealthy motherhood during the perinatal period has been suggested as a risk factor for the rise of metabolic diseases worldwide. Herein, we describe: (1) the distinct metabolic and neuroendocrine effects in the liver and VAT caused by two obesogenic conditions during developmental programming phases, exposure to perinatal maternal HFHS diet and postnatal overfeeding; (2) the sexual dysmorphism of such responses; and (3) the impact of maternal glycation in disrupting mechanisms of adaptation to postnatal overfeeding (Figure 9).

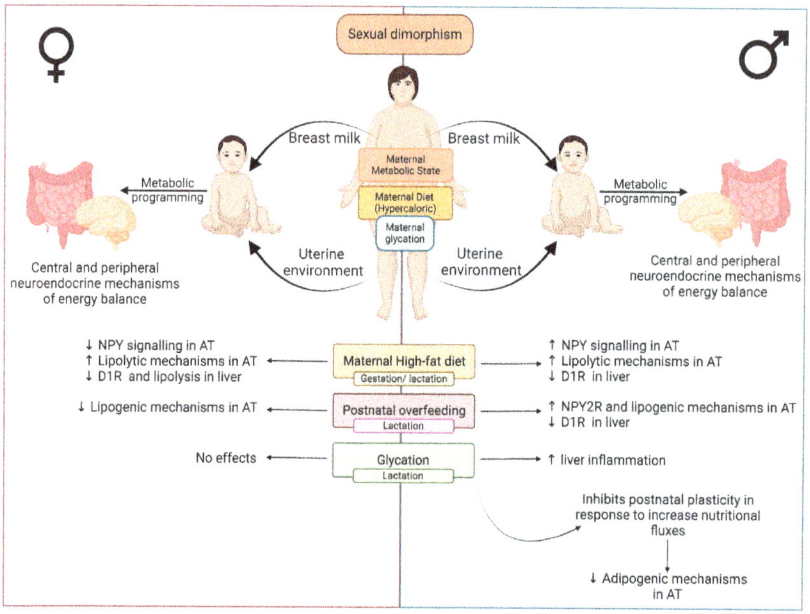

Figure 9. Hypothetical effects of obesogenic environments (hypercaloric diet, postnatal overfeeding, and maternal glycation) on human energy balance mechanisms depending on sexual dysmorphism observed in rat models (created with BioRender.com, accessed on 25 January 2023).

Maternal gestational obesity induced by the HFHS diet in the same model causes postnatal overweight and hepatic lipid accumulation in male offspring [49,50]. Here we show that this maternal diet decreases liver D1R levels in both sexes. In adult diabetic rats, lower D1R levels were also observed. Its upregulation by bromocriptine was associated with lipid mobilization from the liver and hepatic steatosis improvement [54]. However, Stevanović-Silva et al. showed no lipid accumulation in female offspring [49,50]. Interestingly, the liver of female offspring presented lower levels of NPY1R, which suggests less inhibition of CTP-1 and lipid accumulation [55], regardless of PPARα level reduction.

In male offspring VAT, maternal DIO decreased IR levels, which may impact glucose uptake and triglyceride formation [56]. However, p-AMPK was upregulated, which is known to promote HSL activity and lipid oxidation. This upregulation was associated with increased D1R levels, in accordance with previous reports showing its role in lipid oxidation [27]. Moreover, the receptors associated with adipogenic and lipogenic processes (NPY1R, NPY2R, and GHS-R1α) levels were also increased, suggesting an adiposity effect of maternal DIO in the offspring. These results suggest that maternal obesity may trigger compensatory mechanisms to enhance lipid storage in male offspring VAT while also counteracting this by simultaneously improving energy expenditure in VAT. Such VAT alterations were sex specific since NPY1R levels were decreased in female offspring, suggesting a reduction in lipogenic/adipogenic processes, contrary to what occurs in males. Indeed, fat distribution depends on sex, with females more prone to accumulate subcutaneous AT (SAT), while males store lipids predominantly in VAT [56,57]. SAT has a lower lipolysis rate but is more efficient on the FFAs uptake than VAT [56,57]. On the other hand, VAT has a lower capacity to store lipids and a more inflammatory environment [56,57]. One limitation of our study is the lack of SAT data to compare the same mechanisms in male and female SAT. Thus, the discrepancy between sex may be associated with the fatty liver presented by the male offspring. Males have a larger VAT area than females, exposing the liver to a higher amount of FFA via the portal vein, which induces fat ectopic deposition in the liver [58].

Postnatal overweight was also induced by neonatal overfeeding, although body weight gain did not occur in females and no increased food consumption was observed after weaning. Male overfed rats have lower plasma insulin levels, suggesting insulin release impairment. Accordingly, Robert A. et al. (2002) demonstrated that the SL animal model at two different ages (PND 26 and PND 110) presents an impairment in the release of insulin from β-cells [59]. Interestingly, IR total levels were increased in male VAT, which may be a compensatory response to hypoinsulinemia. In male rats, overfeeding also promoted an impairment of D1R hepatic levels, as occurs in maternal HFHS-induced early-life obesity. These findings suggest that this may be an important mechanism of lipid retention in the liver, being associated with the presence of lipid droplets in this tissue (Figure 3N). In VAT, neonatal overfeeding in male rats also upregulates NPY2R signaling, potentiating AT expansion and preventing hypertrophy, while no other mechanisms were changed. Thus, the impact of postnatal overweight on VAT energy balance mechanisms apparently depends on the nature of the obesogenic environment. In fact, contrarily to rats exposed to maternal DIO, p-AMPK levels were reduced in VAT from overfed male animals. This suggests that, besides higher caloric intake from milk, nutritional cues may have a direct impact on VAT mechanisms of energy expenditure. It is possible that maternal consumption of fats and sugars stimulates not only energy storage but also energy dissipation pathways in the male offspring. Again, sex was shown to be a crucial factor in energy balance mechanisms. Neonatal overfeeding in females reduces lipogenic and adipogenic processes by decreasing NPY/PPARγ signaling, which may be the reason for the absence of weight gain and may be related to the distinct stimulation of different fat pads, as already discussed.

Given that maternal diets are often rich in sugars, we also intended to study the effects of exposure to maternal glycotoxins on the mechanisms of adaptation to obesogenic environments. Previous studies have used direct administration of methylglyoxal (MG)—an AGEs precursor—in dams [48], and we aimed to use a more physiological model using BGGC injection, inhibiting the enzyme responsible for MG detoxification—GLO1. In the study by Francisco et al. (2018), breastmilk triglyceride levels were increased upon maternal MG ingestion. However, we verified the opposite and a lower antioxidant capacity, suggesting that maternal glycation alters breastmilk composition. As described before, AGEs formation debilitates antioxidant defenses (reviewed by [53]), and we have here shown that maternal glycation has also an impact on breastmilk antioxidant capacity and possibly on the detoxification mechanisms of the offspring. As occurs in adults exposed to glycotoxins, BBGC did not affect maternal or male and female offspring body weight [20]. Nevertheless, Francisco et al. (2018) showed that offspring from MG-treated dams present excessive weight gain only after PND 77, which may be a consequence of metabolic dysregulation secondary to the dose used [48]. As observed in adult models, animals exposed to maternal glycation do not develop insulin resistance, glucose dysmetabolism, or changes in insulinemia [20], although a lower kITT indicates an insufficiency in insulin action, suggesting a predisposition to insulin resistance. Regarding female offspring, plasma insulin levels were decreased, although female offspring of DMSO-treated dams also present this deficiency. Thus, there is no evidence of insulin resistance, consistent with several studies that have reported that females are less prone to develop insulin resistance due to hormonal sex differences [57].

Other studies indicate that glycation plays a more prevailing role in obesity than in normal conditions [57]. Here we observed that maternal glycation does not alter the molecular mechanisms associated with energy balance in the WAT of lean female and male offspring. However, we detected that impairment of D1R levels in the livers of both sexes, as observed in the overfeeding models. These results suggest that changes in milk composition induced by glycation also impact the hepatic level, decreasing D1R and PPARα, a transcription factor known to be activated by FA, and thus possibly reducing lipid oxidation [60]. In addition, NPY1R levels are also diminished in the liver of male offspring exposed to maternal glycotoxins. Moreover, maternal glycation induced inflammatory infiltration in the liver of male offspring. Interestingly, NPY1R has been linked with an anti-inflammatory action associated with a pro-inflammatory M1-like phenotype in other regions, namely the kidney and cardiac system [61,62]. Regarding female livers, no major changes were observed besides those associated with the vehicle, DMSO, and the stress of the daily IP injection. Diverse studies have shown that females are more prone to develop obesity when exposed to maternal stress than males [63]. Moreover, the effects of maternal prenatal stress on gut microbiota colonization are sex-dependent [64]. As a regulator of gut hormone release [65], alterations induced by maternal alterations in the microbiota may change the release of gut-derived hormones.

Previous reports from our group showed that weight gain induced by the HF diet in adult Wistar rats is lost when MG is administered, suggesting that glycation impairs the necessary tissue plasticity to respond to higher nutritional fluxes [20,53]. Indeed, the weight gain provoked by postnatal overfeeding is lost when the young animals are exposed to maternal glycotoxins during lactation. However, the metabolic effects of overfeeding in male offspring regarding plasma triglycerides and insulin levels were maintained when exposed to maternal glycation. Such inhibition of weight gain was associated with the loss of compensatory NPY2R upregulation in overfeeding-induced overweight rats, suggesting a decrease in adipogenesis rates promoting a lower energy storage capacity. Indeed, despite the fact that maternal glycation did not change fat mass, the adipocytes were smaller when the animals experienced maternal glycation. Moreover, several regions show multivesicular adipocytes, a marker of impaired lipid storage and adipocyte dysfunction. Thus, maternal glycation may modify the AT environment, especially in conditions of overfeeding/WAT expansion. So, exposure to glycotoxins during lactation in obesogenic conditions apparently inhibits the development of storage mechanisms to compensate for higher food consumption. These alterations predispose to the development of metabolic syndrome due to their contribution to adipocyte hypertrophy, lipotoxicity, hypoxia in the AT, and insulin resistance, hallmarks of an unhealthy WAT phenotype.

In the liver, maternal glycation impairs NPY1R in lean and overfed animals. Moreover, the hepatic inflammatory infiltration provoked by glycotoxins exposure during lactation was also noticed. These results suggest that alterations in NPY1R signaling and inflammatory infiltration are an exclusive consequence of maternal glycation exposure, as in the livers of overfed animals, NPY1R levels were preserved.

5. Conclusions

Dysregulation of energy balance mechanisms is one of the major factors contributing to the development of metabolic diseases. Insults such as obesogenic environments during the perinatal period program adaptations of nutrient-sensing mechanisms, readjusting anabolic/catabolic processes. Here, we characterized, for the first time to our knowledge, the changes in VAT and liver energy balance pathways caused by maternal obesity during pregnancy and postnatal overfeeding, although different rat strains were used. Besides the weight gain, young male animals exposed to the maternal HFHS diet developed compensatory mechanisms regarding energy storage and expenditure, allowing a higher AT plasticity. Postnatal overfeeding apparently protects from AT dysfunction by enhancing energy storage. So, the different consequences in both models suggest that maternal nutritional cues during lactation have a crucial role in their modulation. Moreover, females showed distinct mechanisms in both models, suggesting that lipid partitioning may be different in both sexes since the early stages of development and the energy balance mechanisms in SAT should be of future interest. Nevertheless, despite developing these compensatory peripheral mechanisms, modulation of such pathways can become unhealthy since these compensatory actions may be lost later in life. This may disturb energy balance and therefore contribute to predisposing to the development of metabolic complications.

To characterize the role of Westernized diet consumption during gestation and lactation, we studied the role of maternal glycation under normal and obese conditions on VAT and liver energy balance mechanisms. Although glycated products do not have major effects on a lean phenotype, they are related to a compromise of AT plasticity, which may lead to AT dysfunction and lipotoxicity when combined with an obesogenic environment (overfeeding). We followed a protocol based on GLO-1 inhibition in order to avoid possible supraphysiological doses after MG administration. However, it is possible that exposure to MG in high-sugar diets may be even higher than in our protocol. In the future, both protocols may be compared for such effects.

Our work points out the relevance of better characterizing these mechanisms in later stages of development and how they predispose to insulin resistance and metabolic diseases at a more advanced age. The role played by pregnancy and lactation should also be disclosed using cross-fostering. Another concern in the future is understanding the importance of offspring environment upon unhealthy motherhood and how the offspring can change its fate and decrease the predisposition of metabolic disease development to break the intergenerational cycle of obesity.

Supplementary Materials: The following supporting information can be downloaded at: https://www.mdpi.com/article/10.3390/nu15051281/s1, Figure S1: The liver mass per body weight at PND 45: Male control and SL (A), female control and SL (B), male control, vehicle and BBGC (C), female control, vehicle and BBGC (D) and male SL and BBGC + SL (E). Representative images of the liver and VAT stained with Haematoxylin-eosin (100×) – liver and VAT from male SL (F and G, respectively) and liver and VAT from female SL (H and I, respectively); Figure S2: Representative images of the liver and VAT stained with Haematoxylin-eosin (100×) – liver from dams treated with BBGC (A); liver and VAT from male BBGC (B and C, respectively) and liver and VAT from female BBGC (D and E, respectively); Figure S3: Representative images of the liver and VAT stained with Haematoxylin-eosin (100×) – liver and VAT from BBGC-offspring male SL (A and B, respectively); Table S1: Primary antibodies used in Western Blotting.

Author Contributions: Conceptualization, J.M., P.C.d.F.M., S.P.P., P.J.O., R.M.G. and P.M.; methodology, D.S., M.R., A.A., M.D.F.-J., K.V.N.C., T.M.-A., C.B., D.R.-S., L.P.J.S. and P.M.; software, D.S. and P.M.; validation, D.S. and P.M.; formal analysis, P.M.; investigation, D.S., M.R., A.A., M.D.F.-J., K.V.N.C., T.M.-A., C.B., D.R.-S., L.P.J.S., A.C. and P.M.; resources, J.M., P.C.d.F.M., S.P.P., P.J.O., R.M.G. and P.M.; data curation, P.M.; writing—original draft preparation, D.S. and P.M.; writing—review and editing, P.M.; visualization, P.M.; supervision, P.M.; project administration, P.M.; funding acquisition, P.M. All authors have read and agreed to the published version of the manuscript.

Funding: This work was supported by Portugal Foundation for Science and Technology (strategic projects UIDB/04539/2020, UIDP/04539/2020, and LA/P/0058/2020: CIBB), PTDC/DTP-DES/1082/2014 (POCI-01-0145-FEDER-016657), CENTRO-01-0246-FEDER-000010 (Multidisciplinary Institute of Ageing in Coimbra), and FCT-Post-doctoral Fellowship (SPP, SFRH/BPD/116061/2016).

Institutional Review Board Statement: The Ethical Committee of the Institute for Research and Innovation in Health—i3S, University of Porto, and National Government Authority (Direção Geral de Alimentação e Veterinánia—No.0421/000/000/2018) approved the experimental protocol of Maternal diet-induced obesity (DIO) during gestation and lactation, which was in compliance with the Guidelines for Care and Use of Laboratory Animals in Research advised by the Federation of European Laboratory Animal Science Associations (FELASA), being all the authors involved in animal manipulation accredited by FELASA to perform animal experiments. The experiments involving Wistar animals were approved by the local Institutional Animal Care and Use Committee (ORBEA UC 04-2015).

Informed Consent Statement: Not applicable.

Data Availability Statement: The datasets generated during and/or analyzed during the current study are available from the corresponding author upon reasonable request.

Acknowledgments: Biorender.

Conflicts of Interest: The authors declare no conflict of interest.

References

1. Ng, M.; Fleming, T.; Robinson, M.; Thomson, B.; Graetz, N.; Margono, C.; Mullany, E.C.; Biryukov, S.; Abbafati, C.; Abera, S.F.; et al. Global, regional, and national prevalence of overweight and obesity in children and adults during 1980–2013: A systematic analysis for the Global Burden of Disease Study 2013. *Lancet Lond. Engl.* **2014**, *384*, 766–781. [CrossRef]
2. GBD 2015 Obesity Collaborators; Afshin, A.; Forouzanfar, M.H.; Reitsma, M.B.; Sur, P.; Estep, K.; Lee, A.; Marczak, L.; Mokdad, A.H.; Moradi-Lakeh, M.; et al. Health Effects of Overweight and Obesity in 195 Countries over 25 Years. *N. Engl. J. Med.* **2017**, *377*, 13–27. [CrossRef] [PubMed]
3. Khanna, D.; Khanna, S.; Khanna, P.; Kahar, P.; Patel, B.M. Obesity: A Chronic Low-Grade Inflammation and Its Markers. *Cureus* **2022**, *14*, e22711. [CrossRef]
4. Chatterjee, S.; Khunti, K.; Davies, M.J. Type 2 diabetes. *Lancet* **2017**, *389*, 2239–2251. [CrossRef]
5. Gravina, G.; Ferrari, F.; Nebbiai, G. The obesity paradox and diabetes. *Eat. Weight Disord.-Stud. Anorex. Bulim. Obes.* **2021**, *26*, 1057–1068. [CrossRef] [PubMed]
6. Hudish, L.I.; Reusch, J.E.; Sussel, L. β Cell dysfunction during progression of metabolic syndrome to type 2 diabetes. *J. Clin. Invest.* **2019**, *129*, 4001–4008. [CrossRef]
7. Melo, B.F.; Sacramento, J.F.; Ribeiro, M.J.; Prego, C.S.; Correia, M.C.; Coelho, J.C.; Cunha-Guimaraes, J.P.; Rodrigues, T.; Martins, I.B.; Guarino, M.P.; et al. Evaluating the Impact of Different Hypercaloric Diets on Weight Gain, Insulin Resistance, Glucose Intolerance, and its Comorbidities in Rats. *Nutrients* **2019**, *11*, 1197. [CrossRef]
8. Lai, M.; Chandrasekera, P.C.; Barnard, N.D. You are what you eat, or are you? The challenges of translating high-fat-fed rodents to human obesity and diabetes. *Nutr. Diabetes* **2014**, *4*, e135. [CrossRef] [PubMed]
9. Rohde, K.; Schamarek, I.; Blüher, M. Consequences of Obesity on the Sense of Taste: Taste Buds as Treatment Targets? *Diabetes Metab. J.* **2020**, *44*, 509. [CrossRef]
10. Alamri, B.N.; Shin, K.; Chappe, V.; Anini, Y. The role of ghrelin in the regulation of glucose homeostasis. *Horm. Mol. Biol. Clin. Investig.* **2016**, *26*, 3–11. [CrossRef]
11. Li, R.; Guan, H.; Yang, K. Neuropeptide Y potentiates beta-adrenergic stimulation of lipolysis in 3T3-L1 adipocytes. *Regul. Pept.* **2012**, *178*, 16–20. [CrossRef] [PubMed]
12. Barazzoni, R.; Zanetti, M.; Ferreira, C.; Vinci, P.; Pirulli, A.; Mucci, M.; Dore, F.; Fonda, M.; Ciocchi, B.; Cattin, L.; et al. Relationships between Desacylated and Acylated Ghrelin and Insulin Sensitivity in the Metabolic Syndrome. *J. Clin. Endocrinol. Metab.* **2007**, *92*, 3935–3940. [CrossRef]
13. Loh, K.; Shi, Y.-C.; Bensellam, M.; Lee, K.; Laybutt, D.R.; Herzog, H. Y1 receptor deficiency in β-cells leads to increased adiposity and impaired glucose metabolism. *Sci. Rep.* **2018**, *8*, 11835. [CrossRef]
14. Maffei, A.; Segal, A.M.; Alvarez-Perez, J.C.; Garcia-Ocaña, A.; Harris, P.E. Anti-incretin, Anti-Proliferative action of dopamine on B-Cells. *Mol. Endocrinol.* **2015**, *29*, 542–557. [CrossRef] [PubMed]

15. Bettiga, A.; Fiorio, F.; Di Marco, F.; Trevisani, F.; Romani, A.; Porrini, E.; Salonia, A.; Montorsi, F.; Vago, R. The Modern Western Diet Rich in Advanced Glycation End-Products (AGEs): An Overview of Its Impact on Obesity and Early Progression of Renal Pathology. *Nutrients* **2019**, *11*, 1748. [CrossRef]
16. Matafome, P.; Rodrigues, T.; Sena, C.; Seiça, R. Methylglyoxal in Metabolic Disorders: Facts, Myths, and Promises. *Med. Res. Rev.* **2017**, *37*, 368–403. [CrossRef]
17. Martinez, K.B.; Leone, V.; Chang, E.B. Western diets, gut dysbiosis, and metabolic diseases: Are they linked? *Gut Microbes* **2017**, *8*, 130–142. [CrossRef] [PubMed]
18. Varlamov, O. Western-style diet, sex steroids and metabolism. *Biochim. Biophys. Acta BBA-Mol. Basis Dis.* **2017**, *1863*, 1147–1155. [CrossRef]
19. Monteiro-Alfredo, T.; Matafome, P. Gut Metabolism of Sugars: Formation of Glycotoxins and Their Intestinal Absorption. *Diabetology* **2022**, *3*, 596–605. [CrossRef]
20. Rodrigues, T.; Matafome, P.; Sereno, J.; Almeida, J.; Castelhano, J.; Gamas, L.; Neves, C.; Gonçalves, S.; Carvalho, C.; Arslanagic, A.; et al. Methylglyoxal-induced glycation changes adipose tissue vascular architecture, flow and expansion, leading to insulin resistance. *Sci. Rep.* **2017**, *7*, 1698. [CrossRef]
21. Rodrigues, T.; Borges, P.; Mar, L.; Marques, D.; Albano, M.; Eickhoff, H.; Carrêlo, C.; Almeida, B.; Pires, S.; Abrantes, M.; et al. GLP-1 improves adipose tissue glyoxalase activity and capillarization improving insulin sensitivity in type 2 diabetes. *Pharmacol. Res.* **2020**, *161*, 105198. [CrossRef] [PubMed]
22. Rodrigues, T.; Matafome, P.; Santos-Silva, D.; Sena, C.; Seiça, R. Reduction of Methylglyoxal-Induced Glycation by Pyridoxamine Improves Adipose Tissue Microvascular Lesions. *J. Diabetes Res.* **2013**, *2013*, 690650. [CrossRef] [PubMed]
23. Shi, Y.C.; Baldock, P.A. Central and peripheral mechanisms of the NPY system in the regulation of bone and adipose tissue. *Bone* **2012**, *50*, 430–436. [CrossRef] [PubMed]
24. Rubí, B.; Maechler, P. Minireview: New Roles for Peripheral Dopamine on Metabolic Control and Tumor Growth: Let's Seek the Balance. *Endocrinology* **2010**, *151*, 5570–5581. [CrossRef]
25. Borcherding, D.C.; Hugo, E.R.; Idelman, G.; De Silva, A.; Richtand, N.W.; Loftus, J.; Ben-Jonathan, N. Dopamine Receptors in Human Adipocytes: Expression and Functions. *PLoS ONE* **2011**, *6*, e25537. [CrossRef] [PubMed]
26. Tavares, G.; Martins, F.O.; Melo, B.F.; Matafome, P.; Conde, S.V. Peripheral Dopamine Directly Acts on Insulin-Sensitive Tissues to Regulate Insulin Signaling and Metabolic Function. *Front. Pharmacol.* **2021**, *12*, 713418. [CrossRef]
27. Yu, J.; Zhu, J.; Deng, J.; Shen, J.; Du, F.; Wu, X.; Chen, Y.; Li, M.; Wen, Q.; Xiao, Z.; et al. Dopamine receptor D1 signaling stimulates lipolysis and browning of white adipocytes. *Biochem. Biophys. Res. Commun.* **2022**, *588*, 83–89. [CrossRef]
28. Park, S.; Fujishita, C.; Komatsu, T.; Kim, S.E.; Chiba, T.; Mori, R.; Shimokawa, I. NPY antagonism reduces adiposity and attenuates age-related imbalance of adipose tissue metabolism. *FASEB J.* **2014**, *28*, 5337–5348. [CrossRef]
29. Sainsbury, A.; Schwarzer, C.; Couzens, M.; Herzog, H. Y2 Receptor Deletion Attenuates the Type 2 Diabetic Syndrome of *ob/ob* Mice. *Diabetes* **2002**, *51*, 3420–3427. [CrossRef]
30. Yang, K.; Guan, H.; Arany, E.; Hill, D.J.; Cao, X. Neuropeptide Y is produced in visceral adipose tissue and promotes proliferation of adipocyte precursor cells via the Y1 receptor. *FASEB J. Off. Publ. Fed. Am. Soc. Exp. Biol.* **2008**, *22*, 2452–2464. [CrossRef]
31. Kuo, L.E.; Kitlinska, J.B.; Tilan, J.U.; Li, L.; Baker, S.B.; Johnson, M.D.; Lee, E.W.; Burnett, M.S.; Fricke, S.T.; Kvetnansky, R.; et al. Neuropeptide Y acts directly in the periphery on fat tissue and mediates stress-induced obesity and metabolic syndrome. *Nat. Med.* **2007**, *13*, 803–811. [CrossRef] [PubMed]
32. Andrews, Z.B.; Erion, D.M.; Beiler, R.; Choi, C.S.; Shulman, G.I.; Horvath, T.L. Uncoupling protein-2 decreases the lipogenic actions of ghrelin. *Endocrinology* **2010**, *151*, 2078–2086. [CrossRef]
33. Muhammad, H.F.L. Obesity as the Sequel of Childhood Stunting: Ghrelin and GHSR Gene Polymorphism Explained. *Acta Medica Indones.* **2018**, *50*, 159–164.
34. Poher, A.-L.; Tschöp, M.H.; Müller, T.D. Ghrelin regulation of glucose metabolism. *Peptides* **2018**, *100*, 236–242. [CrossRef] [PubMed]
35. Lin, L.; Saha, P.K.; Ma, X.; Henshaw, I.O.; Shao, L.; Chang, B.H.J.; Buras, E.D.; Tong, Q.; Chan, L.; McGuinness, O.P.; et al. Ablation of ghrelin receptor reduces adiposity and improves insulin sensitivity during aging by regulating fat metabolism in white and brown adipose tissues: GHS-R ablation improves insulin resistance. *Aging Cell* **2011**, *10*, 996–1010. [CrossRef] [PubMed]
36. Davies, J.S.; Kotokorpi, P.; Eccles, S.R.; Barnes, S.K.; Tokarczuk, P.F.; Allen, S.K.; Whitworth, H.S.; Guschina, I.A.; Evans, B.A.J.; Mode, A.; et al. Ghrelin Induces Abdominal Obesity Via GHS-R-Dependent Lipid Retention. *Mol. Endocrinol.* **2009**, *23*, 914–924. [CrossRef]
37. Li, Z.; Xu, G.; Qin, Y.; Zhang, C.; Tang, H.; Yin, Y.; Xiang, X.; Li, Y.; Zhao, J.; Mulholland, M.; et al. Ghrelin promotes hepatic lipogenesis by activation of mTOR-PPARγ signaling pathway. *Proc. Natl. Acad. Sci. USA* **2014**, *111*, 13163–13168. [CrossRef] [PubMed]
38. Mercer, R.E.; Chee, M.J.S.; Colmers, W.F. The role of NPY in hypothalamic mediated food intake. *Front. Neuroendocrinol.* **2011**, *32*, 398–415. [CrossRef]
39. Cho, Y.K.; Lee, Y.L.; Jung, C.H. Pathogenesis, Murine Models, and Clinical Implications of Metabolically Healthy Obesity. *Int. J. Mol. Sci.* **2022**, *23*, 9614. [CrossRef]
40. Desai, M.; Ross, M.G. Maternal-infant nutrition and development programming of offspring appetite and obesity. *Nutr. Rev.* **2020**, *78*, 25–31. [CrossRef]

41. Grilo, L.F.; Tocantins, C.; Diniz, M.S.; Gomes, R.M.; Oliveira, P.J.; Matafome, P.; Pereira, S.P. Metabolic Disease Programming: From Mitochondria to Epigenetics, Glucocorticoid Signalling and Beyond. *Eur. J. Clin. Invest.* **2021**, *51*, e13625. [CrossRef] [PubMed]
42. Marousez, L.; Lesage, J.; Eberlé, D. Epigenetics: Linking Early Postnatal Nutrition to Obesity Programming? *Nutrients* **2019**, *11*, 2966. [CrossRef]
43. Ehrenthal, D.B.; Maiden, K.; Rao, A.; West, D.W.; Gidding, S.S.; Bartoshesky, L.; Carterette, B.; Ross, J.; Strobino, D. Independent relation of maternal prenatal factors to early childhood obesity in the offspring. *Obstet. Gynecol.* **2013**, *121*, 115–121. [CrossRef] [PubMed]
44. Whitaker, R.C. Predicting preschooler obesity at birth: The role of maternal obesity in early pregnancy. *Pediatrics* **2004**, *114*, e29–e36. [CrossRef]
45. Sridhar, S.B.; Darbinian, J.; Ehrlich, S.F.; Markman, M.A.; Gunderson, E.P.; Ferrara, A.; Hedderson, M.M. Maternal gestational weight gain and offspring risk for childhood overweight or obesity. *Am. J. Obstet. Gynecol.* **2014**, *211*, 259.e1–8. [CrossRef] [PubMed]
46. Habbout, A.; Li, N.; Rochette, L.; Vergely, C. Postnatal overfeeding in rodents by litter size reduction induces major short- and long-term pathophysiological consequences. *J. Nutr.* **2013**, *143*, 553–562. [CrossRef]
47. Bayol, S.A.; Farrington, S.J.; Stickland, N.C. A maternal "junk food" diet in pregnancy and lactation promotes an exacerbated taste for "junk food" and a greater propensity for obesity in rat offspring. *Br. J. Nutr.* **2007**, *98*, 843–851. [CrossRef]
48. Francisco, F.A.; Barella, L.F.; Silveira, S.d.S.; Saavedra, L.P.J.; Prates, K.V.; Alves, V.S.; Franco, C.C.d.S.; Miranda, R.A.; Ribeiro, T.A.; Tófolo, L.P.; et al. Methylglyoxal treatment in lactating mothers leads to type 2 diabetes phenotype in male rat offspring at adulthood. *Eur. J. Nutr.* **2018**, *57*, 477–486. [CrossRef]
49. Stevanović-Silva, J.; Beleza, J.; Coxito, P.; Rocha, H.; Gaspar, T.B.; Gärtner, F.; Correia, R.; Fernandes, R.; Oliveira, P.J.; Ascensão, A.; et al. Exercise performed during pregnancy positively modulates liver metabolism and promotes mitochondrial biogenesis of female offspring in a rat model of diet-induced gestational diabetes. *Biochim. Biophys. Acta Mol. Basis Dis.* **2022**, *1868*, 166526. [CrossRef]
50. Stevanović-Silva, J.; Beleza, J.; Coxito, P.; Pereira, S.; Rocha, H.; Gaspar, T.B.; Gärtner, F.; Correia, R.; Martins, M.J.; Guimarães, T.; et al. Maternal high-fat high-sucrose diet and gestational exercise modulate hepatic fat accumulation and liver mitochondrial respiratory capacity in mothers and male offspring. *Metabolism.* **2021**, *116*, 154704. [CrossRef]
51. Lutz, T.A. Considering our methods: Methodological issues with rodent models of appetite and obesity research. *Physiol. Behav.* **2018**, *192*, 182–187. [CrossRef]
52. Matafome, P. Another Player in the Field: Involvement of Glycotoxins and Glycosative Stress in Insulin Secretion and Resistance. *Diabetology* **2020**, *1*, 24–36. [CrossRef]
53. Neves, C.; Rodrigues, T.; Sereno, J.; Simões, C.; Castelhano, J.; Gonçalves, J.; Bento, G.; Gonçalves, S.; Seiça, R.; Domingues, M.R.; et al. Dietary Glycotoxins Impair Hepatic Lipidemic Profile in Diet-Induced Obese Rats Causing Hepatic Oxidative Stress and Insulin Resistance. *Oxid. Med. Cell. Longev.* **2019**, *2019*, 6362910. [CrossRef] [PubMed]
54. Tavares, G.; Marques, D.; Barra, C.; Rosendo-Silva, D.; Costa, A.; Rodrigues, T.; Gasparini, P.; Melo, B.F.; Sacramento, J.F.; Seiça, R.; et al. Dopamine D2 receptor agonist, bromocriptine, remodels adipose tissue dopaminergic signalling and upregulates catabolic pathways, improving metabolic profile in type 2 diabetes. *Mol. Metab.* **2021**, *51*, 101241. [CrossRef]
55. Zhang, L.; Macia, L.; Turner, N.; Enriquez, R.F.; Riepler, S.J.; Nguyen, A.D.; Lin, S.; Lee, N.J.; Shi, Y.C.; Yulyaningsih, E.; et al. Peripheral neuropeptide Y Y1 receptors regulate lipid oxidation and fat accretion. *Int. J. Obes.* **2010**, *34*, 357–373. [CrossRef] [PubMed]
56. Lustig, R.H.; Collier, D.; Kassotis, C.; Roepke, T.A.; Kim, M.J.; Blanc, E.; Barouki, R.; Bansal, A.; Cave, M.C.; Chatterjee, S.; et al. Obesity I: Overview and molecular and biochemical mechanisms. *Biochem. Pharmacol.* **2022**, *199*, 115012. [CrossRef]
57. Chang, E.; Varghese, M.; Singer, K. Gender and Sex Differences in Adipose Tissue. *Curr. Diab. Rep.* **2018**, *18*, 69. [CrossRef] [PubMed]
58. Björntorp, P. How should obesity be defined? *J. Intern. Med.* **1990**, *227*, 147–149. [CrossRef] [PubMed]
59. Waterland, R.A.; Garza, C. Early Postnatal Nutrition Determines Adult Pancreatic Glucose-Responsive Insulin Secretion and Islet Gene Expression in Rats. *J. Nutr.* **2002**, *132*, 357–364. [CrossRef]
60. Wang, Y.; Nakajima, T.; Gonzalez, F.J.; Tanaka, N. PPARs as Metabolic Regulators in the Liver: Lessons from Liver-Specific PPAR-Null Mice. *Int. J. Mol. Sci.* **2020**, *21*, 2061. [CrossRef]
61. Qin, Y.-Y.; Huang, X.-R.; Zhang, J.; Wu, W.; Chen, J.; Wan, S.; Yu, X.-Y.; Lan, H.-Y. Neuropeptide Y attenuates cardiac remodeling and deterioration of function following myocardial infarction. *Mol. Ther.* **2022**, *30*, 881–897. [CrossRef]
62. Tan, R.-Z.; Li, J.-C.; Zhu, B.-W.; Huang, X.-R.; Wang, H.-L.; Jia, J.; Zhong, X.; Liu, J.; Wang, L.; Lan, H.-Y. Neuropeptide Y protects kidney from acute kidney injury by inactivating M1 macrophages via the Y1R-NF-κB-Mincle-dependent mechanism. *Int. J. Biol. Sci.* **2023**, *19*, 521–536. [CrossRef] [PubMed]
63. Leppert, B.; Junge, K.M.; Röder, S.; Borte, M.; Stangl, G.I.; Wright, R.J.; Hilbert, A.; Lehmann, I.; Trump, S. Early maternal perceived stress and children's BMI: Longitudinal impact and influencing factors. *BMC Public Health* **2018**, *18*, 1211. [CrossRef] [PubMed]

64. Jašarević, E.; Howard, C.D.; Misic, A.M.; Beiting, D.P.; Bale, T.L. Stress during pregnancy alters temporal and spatial dynamics of the maternal and offspring microbiome in a sex-specific manner. *Sci. Rep.* **2017**, *7*, 44182. [CrossRef] [PubMed]
65. El-Salhy, M.; Hatlebakk, J.G.; Hausken, T. Diet in Irritable Bowel Syndrome (IBS): Interaction with Gut Microbiota and Gut Hormones. *Nutrients* **2019**, *11*, 1824. [CrossRef]

nutrients

MDPI

Article

Association of Healthy Eating Index-2015 and Dietary Approaches to Stop Hypertension Patterns with Insulin Resistance in Schoolchildren

María Dolores Salas-González [1,*], Aranzazu Aparicio [2], Viviana Loria-Kohen [1], Rosa M. Ortega [2] and Ana M. López-Sobaler [2]

[1] VALORNUT Research Group, Department of Nutrition and Food Science, Faculty of Pharmacy, Complutense University of Madrid, 28040 Madrid, Spain
[2] VALORNUT Research Group, Department of Nutrition and Food Science, Faculty of Pharmacy, Complutense University of Madrid, IdISSC, 28040 Madrid, Spain
* Correspondence: masala06@ucm.es

Abstract: Background: Diet quality patterns are associated with a lower incidence of insulin resistance (IR) in adults. The aim of this study was to investigate the association between two diet quality indices and IR in schoolchildren and to identify the best diet quality index associated with a lower risk of IR. Methods: A total of 854 schoolchildren (8–13 years) were included in a cross-sectional study, who completed a three-day dietary record to assess their diet. Fasting plasma glucose and insulin were also measured, and anthropometric data were collected. Healthy Eating Index-2015 (HEI-2015), Dietary Approaches to Stop Hypertension (DASH), and adjusted DASH (aDASH) were calculated as diet quality indices. The homeostasis model assessment of insulin resistance (HOMA-IR) was used, and IR was defined as HOMA-IR > 3.16. Results: The prevalence of IR was 5.5%, and it was higher in girls. The mean HEI-2015 and DASH scores were 59.3 and 23.4, respectively, and boys scored lower in both indices. In girls, having a HEI-2015 score above the 33rd percentile was associated with a lower risk of IR (odds ratio [95% CI]: 0.43 [0.19–0.96], $p = 0.020$). Conclusion: Greater adherence to a healthy dietary pattern, as assessed by a higher HEI-2015 score, was associated with a lower risk of IR in schoolchildren, especially in girls.

Keywords: children; insulin resistance; HEI; DASH

1. Introduction

With the growing epidemic of obesity, incidence of insulin resistance (IR) in childhood and adolescence has increased worldwide. Moreover, IR is accompanied by other components of metabolic syndrome and continues into adulthood [1].

IR is a common manifestation of obesity. Initially, pancreatic beta cells are able to compensate for IR by increasing insulin secretion in the pathogenesis of glucose intolerance. Compensatory hyperinsulinemia induces an increase in appetite, and consequently, weight gain. After pancreatic beta cell function decreases, insufficient insulin secretion occurs, leading to a transition from IR to impaired glucose tolerance, followed by type 2 diabetes [2,3].

Diet is one of the main modifiable factors that can help control and prevent obesity-associated disorders such as IR. Several scores and indices are available to assess the diet quality or the adherence to a healthy dietary pattern. They are a multidimensional representation of diet and do not focus on specific nutrients or foods. Adherence to healthier dietary patterns has been associated with lower risk of several diseases, such as cardiovascular disease, type 2 diabetes, metabolic syndrome, and some of its characteristics, such as IR, in both children and adults. For example, a healthier diet was associated with lower mortality [4], better cardiovascular health [5], and lower risk of metabolic syndrome [6], type 2 diabetes [7], hypertension [8] or IR in adults [9,10]. Moreover, there are some studies in children that have observed that a higher adherence to a healthy dietary pattern was associated with better cardiovascular health [11] and lower risk of metabolic syndrome [12,13], hypertension [14], or IR [15,16], although in children, the evidence is still very scarce.

A number of indices have been proposed to assess adherence to a Mediterranean dietary pattern. However, some have been designed for application in the adult population, such as the Mediterranean Diet Adherence Screener (MEDAS) of the PREDIMED study [17] and are not applicable in children. Others, such as the Mediterranean Diet Score (MDS) developed by Trichopoulou et al. [18], have been subsequently modified to include fish consumption [19] and/or has been adapted to children [20,21]. However, the MDS has a small range (0–9 points), and more than half of the population scored between 3 and 5, which suggests that MDS may not be able to distinguish between individuals with different patterns of dietary intake [22]. On the other hand, in Spain, there is no consensus on dietary recommendations or guidelines for Spain, so there is no validated diet quality index for the Spanish population.

The Dietary Approaches to Stop Hypertension (DASH) [23] and the Healthy Eating Index (HEI), updated in 2015 (HEI-2015) [24,25] are some of the most widely used diet quality indices to assess adherence to dietary guidelines in Americans. However, many studies worldwide have used it [14,26,27]. Spain is one of the countries in which the DASH and HEI have been used in both adult and child populations [28–30]. The use of diet quality indices worldwide is feasible due to the similarity of recommendations and allows comparison of the results of the relationship between insulin resistance and diet.

According to a systematic review [31] on the epidemiology of IR in children, most studies focus on adolescence, and few studies (four out of 18) focused on children younger than 10 years. This could be due to the fact that the insulin resistance peaks at mid-puberty [32] and most studies have focused on this age group. Nevertheless, Chiarelli et al. [33] point out that studying the IR in prepubertal children is especially important, as IR and related complications could be exacerbated by the influence of puberty, due to the physiological decrease in insulin sensitivity associated with normal pubertal development.

However, there are few studies on diet quality indices and IR in children, and, therefore, the possible relationship between them needs to be investigated. This study aimed to investigate the association of different diet quality indices with IR in schoolchildren and to identify the diet quality index associated with a lower risk of IR. This study could help identify appropriate measures for use in epidemiological studies and improve the success of nutritional measures and interventions by targeting significant dietary components in at-risk groups.

2. Materials and Methods

2.1. Study Design and Sample

The cross-sectional observational study included a convenience sample of schoolchildren aged 8–13 years from five Spanish provinces (A Coruña, Barcelona, Madrid, Seville, and Valencia). In each province, schools were randomly selected from a list of primary schools with at least two classrooms per grade. The principals of 55 schools were contacted by telephone to explain the characteristics and importance of the study, and to request the school's permission to participate in the survey. Thirteen principals did not want to collaborate. The principals of the remaining 42 schools showed interest in the study and forwarded the information to the school council for approval. Twenty-two schools gave their permission to conduct the survey on their premises and to organize a meeting with the parents of the children in the target age range. At this meeting, the study was explained in detail, questions raised by the parents were answered, and signed parental authorization for the children to take part was requested.

The potential initial sample size was approximately 3850 participants. These children were studied previously in another context [34]. It was calculated taking into account the number of schools that agreed to participate (n = 22), the number of students per classroom (n = 25) and the number of classrooms per grade (between 2 and 3). Subsequently, 1035 schoolchildren obtained written consent from their parents or guardians to participate in the study, so the approximate acceptance rate was 27%. Finally, 854 students completed all the dietary questionnaires and the anthropometric and biochemical study. Figure 1 shows a flow chart of the sample selection process.

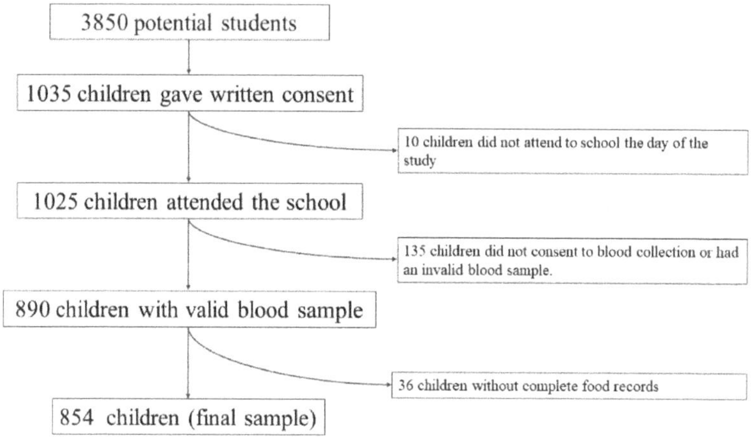

Figure 1. Flow chart of the selection process.

The inclusion criteria were as follows: boys and girls aged 8–13 years who were in 4th, 5th, and 6th grades of primary education; written informed consent signed by the child's parents and/or guardians; and acceptance of all the study conditions.

The following exclusion criteria were applied: the presence of any disease, such as metabolic or chronic diseases (diabetes, renal diseases, etc.) that could affect the study outcomes, inability to visit the study centre on the agreed days, and pharmacological treatment, such as corticosteroids, insulin, and levothyroxine, that could interfere with study outcomes.

The participating schoolchildren underwent a dietary, anthropometric, and biochemical study at school by trained investigators. Measurements were made between February 2005 and June 2009.

This study was conducted in accordance with the guidelines of the Declaration of Helsinki. This study was approved by the Human Research Review Committee of the Faculty of Pharmacy of the Complutense University of Madrid (PI060318 approved at 17 March 2006).

2.2. Anthropometric Study

All measurements were performed at schools in the morning according to the World Health Organization (WHO) standards [35].

Weight and height were measured with the participants in underwear and barefoot using a digital electronic scale (model SECA ALPHA, GMBH & Co., Igny, France) (range: 0.1–150 kg, accuracy 100 g) and a Harpenden digital stadiometer (Pfifter, Carlstadt, NJ, USA) (range: 70–205 cm, accuracy 1 mm), respectively. Body mass index (BMI) was calculated by dividing the weight (kg) by height squared (m^2). Subsequently, the BMI z-score (z-BMI) was calculated. The cut-off points were established according to the standard deviation of the z-score established by the WHO [36] ($z < -2$: underweight, z: -2–1: normal weight z: 1–2: overweight; y $z > 2$: obesity) [37].

The tricipital skinfold was measured on the right side of the body using a Holtain skinfold caliper (constant pressure of 10 g/mm^2 (range: 0–39 mm) and 0.1 mm accuracy,

Holtain Ltd., Crymych, Wales) following the recommendations by the ISAK [38]. Fat mass (FM) was obtained using the formula by Dezemberg et al. in 1999 [39]. FM was classified according to McCarthy's cut-off points as insufficient fat, normal fat, excessive fat, and obesity [40].

2.3. Physical Activity

Parents completed an adapted daily physical activity questionnaire for their children [41], which has been used in other studies [34,42–44].

The questionnaire asked about the time dedicated to different activities usually carried out during the day: sleeping, being in class, study time, time dedicated to different meals, time dedicated to sedentary play (with PC, video consoles or watching television); time dedicated to playing actively in the street, play time at home; time dedicated to gymnastics or sports activities at school, at recess and in extracurricular activities, time and form of travel between home and school and to other activities.

The time spent on the different activities was then grouped into 4 categories: sleep, very light activities (activities done lying, sitting or standing -painting, playing an instrument, cooking...), light activities (equivalent to walking on a flat surface at 4–5 km/h, cleaning the house, golf, table tennis...) and moderate and/or vigorous activities (physical activities requiring greater energy expenditure such as cycling, skiing, tennis, dancing, basketball, soccer, rugby, running...).

An activity coefficient was established for each subject by multiplying the time dedicated to each activity by the established coefficients [45,46]: 1 for sleeping and resting, 1.5 for very light activities, 2.5 for light activities, 5 for moderate activities, and 7 for intense activities, and then dividing by 24 h. Once the individual physical activity coefficient was obtained, it was used to establish the physical activity category for the calculation of energy expenditure (EE) according to IOM equations [47].

2.4. Dietetic Study

For dietary intake, a 'food record' was filled out for three consecutive days, including one weekend day (Sunday to Tuesday) [48]. Parents were asked to record the weights, if possible, or alternatively, to record home measurements of all foods consumed outside of school by their children. On Mondays and Tuesdays, the trained staff visited the school canteen, recorded the food on the menu, and weighed the amount of food served to each child who stayed to eat in the canteen and what each child left on the plate. DIAL software (Alce Ingeniería, Madrid, Spain) was used to process the dietary surveys [49]. The food composition tables from Ortega et al. [50] were used. Energy intake (EI) and macro- and micronutrients were calculated. Energy expenditure (EE) was calculated according to the IOM equations [47].

The possible underestimation or overestimation of EI was determined as the discrepancy between EI and EE, measured using the formula: (EE-EI) $\times$ 100/GE. This formula provides the percentage of possible underestimation (if values are positive) or overestimation of intake (when values are negative) [51].

The HEI-2015 was used to assess diet quality according to the recommendations of the U.S. Dietary Guidelines. It is composed of 13 components: 9 for adequacy and 4 for moderation. Whole grains, dairy, and fatty acid ratio items were scored 0–10; total fruits, whole fruits, vegetables, seafood and vegetable protein, green vegetables and legumes, and protein food items were scored 0–5; and moderation, refined grains, sodium, added sugars, and saturated fat items were scored 0–10 [24].

The DASH index was also calculated [23], It includes eight components that are relevant to the diet used for hypertension management. For the DASH index, scoring is based on intake quintiles created within the full data set; participants in the lowest quintile received 1 point, and individuals in quintile five received 5 points. Red and processed meat, sugar-sweetened beverages, and sodium were reverse coded.

Finally, adjusted DASH (aDASH) was calculated in the same way as DASH after adjusting the grams of the different food groups according to the energy consumed [16].

2.5. Biochemical Studies

Participants' blood samples were drawn by venepuncture between 08.00 and 09.00 h after 12 h of fasting. The nurses verified the adequacy of the fasting period before sample collection. All extractions were performed at the respective centers.

Plasma glucose was determined calorimetrically using the glucose oxidase-peroxidase method [52] (Vitros GLU Slides, Rochester, NY, USA; CV = 2.8%). Fasting insulin was measured by immunochemiluminometric assay [53] (Abbott Diagnostics, Madrid, Spain; CV = 4.8%). The QUICKI index was evaluated as1/[log (basal insulin) + log (basal glucose)] [54].

The homeostasis model assessment of insulin resistance (HOMA-IR) was used to reflect the degree of IR [55,56]: HOMA-IR = [basal glucose (mmol/L) × basal insulin (μU/mL)]/22.5. In our study IR was defined as HOMA-IR > 3.16, as proposed by Keskin et al. [57] for children and adolescents, since this cut-off point showed the highest sensitivity and specificity for the diagnosis of IR assessed by an oral glucose tolerance test. This cut-off point has also been used in other studies in schoolchildren of similar age to our study [16,58–61].

2.6. Statistical Analysis

Descriptive data were expressed as means and standard deviations. Diets were compared according to sex and presence or absence of IR. For comparison of means, the Mann–Whitney U test was used if the distribution of the variables was not homogeneous, the Student's *t*-test for homogeneous distributions, and the two-way ANOVA test. The χ^2 test was used to determine the significance of the differences between proportions. Logistic regression analysis was performed to identify risk or protective factors associated with IR. Significance was set at $p < 0.05$. All calculations were performed using IBM SPSS Statistics for Windows, Version 25.0. Armonk, NY, USA: IBM Corp. Released 2017.

3. Results

3.1. Sample Characteristics

A total of 1035 schoolchildren (49.2% boys) were initially included. Ten participants did not attend school on the day of the anthropometric study and were, therefore, excluded from the analysis. Valid dietary data were only obtained from 965 schoolchildren (48.8% boys), valid blood samples from 890 participants (48.4% boys), and 854 (48.4% boys) who have both blood and dietary samples to constitute the sample of the present study.

Table 1 shows the age and anthropometric, biochemical, and lifestyle data of the total sample classified by sex. Approximately one third of the schoolchildren (27.4%) were overweight and 12.4% were obese, mostly comprising boys. In addition, 31.6% of the participants had excessive fat and 31.5% had obesity according to their fat mass, with boys having higher obesity percentage.

Regarding biochemical data, boys had higher values of glucose and QUICKI index but lower values of insulin and HOMA-IR index. A 5.3% of the sample presented IR, and this was higher in girls than in boys (7. 0% vs. 3.4%, $p < 0.05$). Table S1 shows the age and anthropometric, biochemical, and lifestyle data of the total sample classified by group of age. Table S2 shows the age and anthropometric, biochemical and lifestyle data of the total sample classified by sex and HOMA-IR. There was a relationship between HOMA-IR and age, zBMI, fat mass, and activity coefficient.

Table 1. Anthropometric, physical activity and blood biochemical parameters in the schoolchildren studied according to sex.

	Total (n = 854)	Girls (n = 441)	Boys (n = 413)	p-Value
Age (years)	10.1 ± 0.9	10.2 ± 0.9	10.1 ± 1.0	0.262
8–10 years [%(n)]	63.35 (541)	61.7 (272)	65.1 (269)	0.295
11–13 years [%(n)]	36.65 (313)	38.3 (169)	34.9 (144)	
Madrid [%(n)]	55.6 (475)	59.0 (260)	52.1 (215)	0.237
Barcelona [%(n)]	7.9 (67)	7.3 (32)	8.5 (35)	
Sevilla [%(n)]	12.4 (106)	11.6 (51)	13.3 (55)	
A Coruña [%(n)]	13.4 (114)	11.3 (50)	15.5 (64)	
Valencia [%(n)]	10.8 (92)	10.9 (48)	10.7 (44)	
Body composition				
Weight (kg)	39.4 ± 9.3	39.6 ± 9.2	39.2 ± 9.5	0.432
Height (m) #	1.43 ± 0.09	1.44 ± 0.09	1.43 ± 0.08	0.008
BMI (kg/m^2)	19.0 ± 3.1	18.9 ± 3.0	19.1 ± 3.3	0.774
Z-BMI #	0.69 ± 1.13	0.57 ± 1.04	0.81 ± 1.21	0.001
Nutritional status by BMI				
Underweight [%(n)]	0.82 (7)	0.68 (3)	0.97 (4)	<0.001
Normal weight [%(n)]	59.4 (507)	63.3 (279)	55.21 (228)	
Overweight [%(n)]	27.4 (234)	29.0 (128)	25.7 (106)	
Obesity [%(n)]	12.4 (106)	7.0 (31)	18.2 (75)	
Body fat (%) #	27.6 ± 5.7	29.3 ± 4.7	25.8 ± 6.1	<0.001
Nutritional status by body fat percentage				
Low fat [%(n)]	0.35 (3)	0.23 (1)	0.49 (2)	<0.001
Normal fat [%(n)]	36.6 (312)	40.4 (178)	32.6 (134)	
Excessive body fat [%(n)]	31.6 (269)	33.1 (146)	29.9 (123)	
Obesity [%(n)]	31.5 (268)	26.3 (116)	37.0 (152)	
Physical activity				
Activity coefficient	1.52 ± 0.11	1.53 ± 0.11	1.53 ± 0.11	0.006
Biochemical data				
Glucose (mg/dL)	84.4 ± 9.7	83.4 ± 10.0	85.5 ± 9.2	0.134
Insulin (mcU/mL)	6.3 ± 4.4	7.1 ± 4.8	5.5 ± 3.7	<0.001
QUICKI	0.38 ± 0.05	0.38 ± 0.05	0.39 ± 0.04	<0.001
HOMA-IR	1.33 ± 0.97	1.48 ± 1.08	1.17 ± 0.80	<0.001
IR [%(n)]	5.27 (45)	7.03 (31)	3.39 (14)	<0.001

BMI, body mass index; IR, insulin resistance. # Variable follow a normal distribution. For comparison of means, the Mann-Whitney U test was used if the distribution of the variables was not homogeneous, the Student's t-test for homogeneous distributions. The χ^2 test was used to determine the significance of the differences between proportions, p-values < 0.05 were considered statistically significant.

3.2. Diet Quality

Dietary data, diet quality data, and scores for each of the sub-items are shown in Table 2, and Table S3 shows the dietary intake of those foods and nutrients included in the different diet quality indices. Boys had higher caloric intake than girls, and the underestimation of their diet was more significant than in girls (7.65 ± 18.95% vs. −11.27 ± 23.28%). They also scored significantly lower on all three calculated quality indices (HEI-2015, DASH, and aDASH). According to the HEI-2015 sub-items, boys scored lower on servings of total fruit, whole fruit, green vegetables, and legumes. Likewise, boys scored lower on the (PUFAs+MUFAs)/SFAs ratio and sodium level than girls. Girls scored higher on red meat, sugar-sweetened beverages, sodium (according to DASH and aDASH), and vegetables and fruits (according to aDASH) than boys. Table S4 shows diet quality as a function of age and sex, with younger children having better scores in diet quality index according to allboth indices. The diets of the younger schoolchildren had a better ratio of unsaturated to saturated fatty acids and consumed more dairy products. While older children consumed more red meat, more sugary drinks, and more sodium per 1000 kcal.

Table 2. Diet quality in the schoolchildren studied according to sex.

	Total (n = 854)	Girls (n = 441)	Boys (n = 413)	p-Value
Energy intake (EI) (kcal) #	2105 ± 350	2066 ± 338	2145 ± 358	<0.001
Energy expenditure (EE) (kcal)	2125 ± 377	1899 ± 265	2365 ± 325	<0.001
EI/EE (%) #	101.7 ± 22.9	110.7 ± 22.8	92.2 ± 18.7	<0.001
Proteins (% EI)	15.6 ± 2.3	15.6 ± 2.3	15.5 ± 2.4	0.356
Carbohydrates (% EI)	41.0 ± 5.1	40.9 ± 5.0	41.1 ± 5.2	0.399
Lipids (% EI) #	41.8 ± 4.8	41.9 ± 4.7	41.8 ± 4.9	0.454
HEI-2015 (total score) #	59.2 ± 8.5	60.3 ± 8.5	57.9 ± 8.4	<0.001
Total Fruits (score)	3.7 ± 1.5	3.8 ± 1.4	3.6 ± 1.5	0.013
Whole Fruits (score)	3.9 ± 1.5	4.0 ± 1.5	3.8 ± 1.6	0.016
Total Vegetables (score)	3.1 ± 1.4	3.2 ± 1.4	2.9 ± 1.3	<0.001
Greens and Beans (score)	3.8 ± 1.7	3.94 ± 1.6	3.7 ± 1.8	0.087
Whole Grains (score)	1.1 ± 1.8	1.1 ± 1.7	1.2 ± 1.8	0.307
Dairy (score)	7.0 ± 2.1	6.9 ± 2.1	7.1 ± 2.0	0.224
Total Protein Foods (score)	4.9 ± 0.4	4.9 ± 0.3	4.9 ± 0.4	0.469
Seafood and Plant Proteins (score)	1.9 ± 1.4	2.0 ± 1.4	1.8 ± 1.3	0.170
(PUFAs + MUFAs)/SFAs (score)	3.2 ± 2.2	3.4 ± 2.3	2.9 ± 2.0	0.001
Refined Grains (score)	6.9 ± 2.4	7.0 ± 2.4	6.8 ± 2.4	0.206
Sodium (score)	8.6 ± 2.0	8.8 ± 1.8	8.4 ± 2.1	0.018
Added Sugars (score)	8.70 ± 1.57	8.76 ± 1.55	8.63 ± 1.58	0.145
Saturated Fats (score)	2.4 ± 2.1	2.5 ± 2.2	2.2 ± 2.1	0.026
DASH (total score)	23.4 ± 3.7	23.9 ± 3.7	22.8 ± 3.7	<0.001
Red meat (score)	3.0 ± 1.4	3.12 ± 1.4	2.8 ± 1.4	<0.001
Sugar drinks (score)	1.79 ± 0.41	1.83 ± 0.38	1.75 ± 0.43	0.006
Sodium (score)	2.8 ± 1.5	3.0 ± 1.4	2.6 ± 1.5	<0.001
Whole Grains (score)	3.6 ± 0.8	3.6 ± 0.8	3.6 ± 0.8	0.502
Low-fat dairy (score)	3.2 ± 1.2	3.2 ± 1.2	3.2 ± 1.2	0.885
Vegetables (score)	3.0 ± 1.4	3.1 ± 1.5	2.9 ± 1.4	0.139
Seeds, nuts and legumes (score)	3.0 ± 1.4	3.0 ± 1.4	2.9 ± 1.5	0.174
Fruits (score)	3.0 ± 1.4	3.1 ± 1.4	3.0 ± 1.4	0.252
aDASH (total score)	23.3 ± 3.9	23.9 ± 3.8	22.7 ± 4.0	<0.001
Red meat (score)	3.0 ± 1.4	3.1 ± 1.4	2.8 ± 1.4	0.001
Sugar drinks (score)	1.80 ± 0.40	1.83 ± 0.38	1.77 ± 0.42	0.018
Sodium (score)	2.9 ± 1.5	3.0 ± 1.5	2.7 ± 1.5	0.028
Whole Grains (score)	3.6 ± 0.8	3.6 ± 0.8	3.6 ± 0.8	0.472
Low-fat dairy (score)	3.2 ± 1.2	3.2 ± 1.2	3.2 ± 1.2	0.873
Vegetables (score)	3.0 ± 1.4	3.1 ± 1.4	2.8 ± 1.4	0.001
Seeds, nuts and legumes (score)	3.0 ± 1.4	3.0 ± 1.4	2.9 ± 1.5	0.139
Fruits (score)	3.0 ± 1.4	3.1 ± 1.4	2.9 ± 1.4	0.060

HEI-2015. Healthy eating index; DASH. Dietary Approaches to Stop Hypertension; aDASH. Alternate Dietary Approaches to Stop Hypertension; MUFAs, monounsaturated fatty acids; PUFAs, polyunsaturated fatty acids; SFAs, saturated fatty acids. # Variable follow a normal distribution. For comparison of means, the Mann-Whitney U test was used if the distribution of the variables was not homogeneous, the Student's t-test for homogeneous distributions, p-values < 0.05 were considered statistically significant.

3.3. *Risk of IR*

Energy expenditure was higher in girls and boys with IR versus their non-IR counterparts. There were no differences in energy or the scores of the different diet quality indices between schoolchildren with and without IR (Table 3). Girls with IR scored lower in vegetables, green vegetables, and legumes according to the HEI-2015 parameters and lower in whole grains according to the DASH score than girls without IR. Boys with IR scored lower on sugar-sweetened beverages according to the DASH and aDASH scores than did boys without IR. Table S5 shows the consumption of foods and nutrients related to the analyzed, diet quality indices subitems by sex and IR, but there were no significant differences to highlight. Table S6 shows the quality of the diet according to the HOMA-IR in the two age groups. In the younger girls group, those with IR consumed more protein and more protein products than their peers without IR, however there were no differences in the score for protein products. In the older age group, boys with IR consumed more skimmed dairy products, which make them score better on this item in the DASH and aDASH.

Table 3. Diet quality in the schoolchildren studied according to HOMA-IR and sex.

	Total			Girls			Boys		
	HOMA-IR ≤ 3.16 (*n* = 809)	**HOMA-IR > 3.16 (*n* = 45)**	*p*	**HOMA-IR ≤ 3.16 (*n* = 410)**	**HOMA-IR > 3.16 (*n* = 31)**	*p*	**HOMA-IR ≤ 3.16 (*n* = 399)**	**HOMA-IR > 3.16 (*n* = 14)**	*p*
Energy intake (kcal) #	2110 ± 350	2002 ± 330	0.022	2071 ± 336	1998 ± 361	0.123	2150 ± 360	2010 ± 261	0.076
Energy expenditure (kcal)	2121 ± 372	2194 ± 445	0.429	1892 ± 266	1998 ± 237 #	0.016	2356 ± 315	2629 ± 495	0.029
EI/EE (%)	102.1 ± 22.8	94.2 ± 22.7	0.150	111.4 ± 22.7	101.4 ± 22.0	0.013	92.6 ± 18.6	78.4 ± 15.2 #	0.002
Proteins (%)	15.6 ± 2.3	15.8 ± 2.4	0.467	15.6 ± 2.3	15.6 ± 2.1	0.983	15.5 ± 2.3	16.1 ± 2.8	0.286
Carbohydrates (%)	41.0 ± 5.1	40.3 ± 5.0	0.263	40.9 ± 5.0	40.6 ± 4.4 #	0.365	41.2 ± 5.2	39.9 ± 6.3	0.237
Lipids (%) #	41.8 ± 4.8	42.1 ± 5.1	0.336	41.8 ± 4.7	42.1 ± 4.4	0.402	41.8 ± 4.8	42.3 ± 6.6	0.357
HEI-2015 (total score)	59.2 ± 8.5	57.5 ± 8.6	0.094	60.5 ± 8.4	57.8 ± 9.4 #	0.044	57.9 ± 8.5	56.8 ± 6.7 #	0.319
Total Fruits (score)	3.7 ± 1.4	3.6 ± 1.6	0.930	3.8 ± 1.4	3.8 ± 1.6	0.748	3.6 ± 1.5	3.2 ± 1.5	0.402
Whole Fruits (score)	3.9 ± 1.5	3.8 ± 1.7	0.855	4.0 ± 1.5	4.0 ± 1.6	0.783	3.8 ± 1.6	3.6 ± 1.7	0.703
Total Vegetables (score)	3.1 ± 1.3	2.9 ± 1.5	0.317	3.3 ± 1.3	2.7 ± 1.5	0.043	2.9 ± 1.3	3.2 ± 1.4	0.409
Greens and Beans (score)	3.8 ± 1.7	3.6 ± 1.8	0.171	4.0 ± 1.6	3.3 ± 2.0	0.039	3.7 ± 1.8	4.1 ± 1.3	0.716
Whole Grains (score)	1.1 ± 1.8	0.6 ± 0.8	0.262	1.1 ± 1.8	0.5 ± 0.7	0.122	1.2 ± 1.8	0.8 ± 1.0	0.680
Dairy (score)	7.0 ± 2.1	7.2 ± 1.9	0.609	6.9 ± 2.1	7.3 ± 2.0	0.354	7.1 ± 2.0	7.0 ± 1.8	0.790
Total Protein Foods (score)	4.9 ± 0.4	4.9 ± 0.3	0.822	4.9 ± 0.3	4.9 ± 0.2	0.866	4.9 ± 0.4	4.8 ± 0.5	0.954
Seafood and Plant Proteins (score)	1.9 ± 1.4	2.1 ± 1.5	0.559	2.0 ± 1.4	2.1 ± 1.4	0.767	1.8 ± 1.3	2.1 ± 1.7	0.685
(PUFAs + MUFAs)/SFAs (score)	3.2 ± 2.2	2.8 ± 2.3	0.247	3.5 ± 2.3	2.9 ± 2.4 #	0.088	2.9 ± 2.0	2.7 ± 2.3	0.722
Refined Grains (score)	6.9 ± 2.4	6.7 ± 2.4	0.633	7.0 ± 2.4	6.5 ± 2.5 #	0.111	6.8 ± 2.4	7.3 ± 2.3	0.367
Sodium (score)—S	8.6 ± 2.0	8.4 ± 2.2	0.716	8.8 ± 1.8	8.9 ± 1.9	0.488	8.4 ± 2.1	7.3 ± 2.4	0.052
Added Sugars (score)	8.7 ± 1.6	8.8 ± 1.6	0.595	8.7 ± 1.6	8.9 ± 1.4	0.735	8.6 ± 1.6	8.5 ± 1.9	0.802
Saturated Fats (score)	2.4 ± 2.2	2.1 ± 1.8	0.635	2.6 ± 2.2	2.2 ± 2.0 #	0.158	2.2 ± 2.1	2.1 ± 1.6	0.810
DASH (total score)	23.3 ± 3.7	23.6 ± 3.6	0.605	23.9 ± 3.6	23.6 ± 3.8	0.684	22.7 ± 3.7	23.6 ± 3.3	0.381
Red meat (score)	2.9 ± 1.4	3.2 ± 1.5	0.316	3.1 ± 1.4	3.2 ± 1.4	0.768	2.8 ± 1.4	3.1 ± 1.6	0.433
Sugar drinks (score)—S	1.8 ± 0.4	1.7 ± 0.4	0.292	1.8 ± 0.4	1.8 ± 0.4	0.920	1.8 ± 0.4	1.5 ± 0.5	0.024
Sodium (score)—S	2.8 ± 1.5	3.2 ± 1.6	0.098	3.0 ± 1.4	3.5 ± 1.5	0.073	2.6 ± 1.5	2.6 ± 1.7	0.886
Whole Grains (score)	3.6 ± 0.8	3.3 ± 0.6	0.035	3.6 ± 0.8	3.3 ± 0.6	0.042	3.6 ± 0.8	3.4 ± 0.6	0.484
Low-fat dairy (score)—I	3.2 ± 1.2	3.6 ± 1.1	0.034	3.2 ± 1.2	3.5 ± 1.2	0.129	3.2 ± 1.2	3.7 ± 1.1	0.115
Vegetables (score)	3.0 ± 1.4	2.8 ± 1.6	0.424	3.1 ± 1.4	2.7 ± 1.6	0.142	2.9 ± 1.4	3.1 ± 1.5	0.538
Seeds, nuts and legumes (score)	3.0 ± 1.4	3.1 ± 1.5	0.656	3.1 ± 1.4	2.8 ± 1.6	0.419	2.9 ± 1.5	3.6 ± 1.2	0.081
Fruits (score)	3.0 ± 1.4	2.8 ± 1.3	0.212	3.1 ± 1.4	2.8 ± 1.4	0.337	3.0 ± 1.4	2.6 ± 1.3	0.324
aDASH (total score)	23.3 ± 3.9	23.5 ± 3.8	0.996	23.9 ± 3.8	23.7 ± 3.8	0.480	22.7 ± 4.0	23.0 ± 3.8	0.760
Red meat (score)	2.9 ± 1.4	3.1 ± 1.5	0.499	3.1 ± 1.4	3.2 ± 1.4	0.827	2.8 ± 1.4	2.9 ± 1.6	0.726
Sugar drinks (score)—S, R	1.8 ± 0.4	1.8 ± 0.4	0.408	1.8 ± 0.4	1.9 ± 0.3	0.571	1.8 ± 0.4	1.5 ± 0.5	0.015
Sodium (score)—S	2.8 ± 1.5	2.8 ± 1.5	0.969	2.9 ± 1.4	3.1 ± 1.5	0.477	2.7 ± 1.5	2.2 ± 1.4	0.176
Whole Grains (score)	3.6 ± 0.8	3.4 ± 0.6	0.073	3.6 ± 0.8	3.3 ± 0.6	0.097	3.6 ± 0.8	3.4 ± 0.6	0.508
Low-fat dairy (score)—I	3.2 ± 1.2	3.6 ± 1.1	0.028	3.2 ± 1.2	3.5 ± 1.2	0.095	3.2 ± 1.2	3.6 ± 1.1	0.148
Vegetables (score)	3.0 ± 1.4	2.9 ± 1.5	0.663	3.2 ± 1.4	2.8 ± 1.5	0.162	2.8 ± 1.4	3.1 ± 1.5	0.387
Seeds, nuts and legumes (score)	3.0 ± 1.4	3.1 ± 1.5	0.517	3.1 ± 1.4	2.9 ± 1.6	0.571	2.9 ± 1.5	3.6 ± 1.2	0.079
Fruits (score)	3.0 ± 1.4	2.8 ± 1.4	0.431	3.1 ± 1.4	3.0 ± 1.4	0.612	2.9 ± 1.4	2.6 ± 1.3	0.405

HEI-2015. Healthy eating index; DASH. Dietary Approaches to Stop Hypertension; aDASH. Alternate Dietary Approaches to Stop Hypertension MUFAs, monounsaturated fatty acids; PUFAs, polyunsaturated fatty acids; SFAs, saturated fatty acids. Two-way ANOVA analysis: S: differences according to sex; I: differences according to insulin resistance (IR) score; R: interaction between sex and IR. # Variable follow a normal distribution. For comparison of means, the Mann-Whitney U test was used if the distribution of the variables was not homogeneous, the Student's *t*-test for homogeneous distributions, *p*-values < 0.05 were considered statistically significant.

Table 4 presents the logistic regression analyses with tertiles of the different diet quality indices (the limits of each tertile are shown in Table S7) and the risk of IR. The results are shown in crude (Model 1) and adjusted for age, sex, z-BMI, and activity coefficient (Model 2).

For HEI-2015, in girls, having a score higher than the 33rd percentile (≥56.8 points) was associated with a lower risk of having IR (adjusted model 2: OR = odds ratio (95% confidence interval): 0.43 (0.19–0.96), $p = 0.020$), both in the crude and adjusted models. No association was found in boys. In the DASH and aDASH, no risk of IR was found according to tertiles, neither in the total sample nor by sex.

a quality diet. Therefore, the diet quality indices calculated based on our data could be affected.

Due to the contradictory results in different publications, further studies on the relationship between IR and diet quality indices would be desirable. Moreover, very few studies have addressed this problem in the pediatric population. Therefore, there is a need to focus on this age group.

Finally, in our study, insulin resistance is related to age, i.e., older children have more IR, which is consistent with the insulin peak that is described in adolescence [32]. In addition, older schoolchildren score worse on diet quality indices, and consume more red meat, sugar-sweetened beverages, and sodium. This is consistent with a study conducted with the HEI-2010 where younger children had the highest overall diet quality [77].

One of the main strengths of our study was that it included a large sample size. In addition, there is little literature on IR in children in this age range, and this is one of the first studies to relate IR in children with indicators of a healthy diet. In addition, the three-day log was used to calculate the diet quality indices in our study, which is one of the most appropriate methods. One of the main limitations of the study is that being a large community-based study, Tanner stratification by a clinician was not possible and information on the pubertal status of the participating children was not included. As only the tricipital fold could be obtained in most of the participants, the method for estimating fat mass has limitations and is not the gold standard. Although the physical activity questionnaire has been used in other studies, it is not a validated questionnaire. Finally, the sample studied is a convenience sample, so the conclusions of our study should be confirmed in future research.

5. Conclusions

We can conclude that greater adherence to a healthy dietary pattern, as assessed by a higher HEI-2015 score, is associated with a reduced risk of IR in schoolchildren, especially in girls. Regarding adherence to the DASH diet, no relationship with IR was observed in our study. Given the above results, promoting a healthy diet based on the HEI-2015 index in this age group could help reduce IR and, therefore, prevent the development of type 2 diabetes and metabolic syndrome.

Supplementary Materials: The following supporting information can be downloaded at: https://www.mdpi.com/article/10.3390/nu14204232/s1, Table S1. Anthropometric, physical activity and blood biochemical parameters in the schoolchildren studied according to sex and age. Table S2. Anthropometric, physical activity and blood biochemical parameters in the schoolchildren studied according to sex and HOMA-IR. Table S3. Parameters of different index in the schoolchildren studied according to sex. Table S4. Diet quality in the schoolchildren studied according to sex and age. Table S5. Parameters of different index in the schoolchildren studied according to HOMA-IR and sex. Table S6. Diet quality in the schoolchildren studied according to HOMA-IR and age. Table S7. Tertiles of the different indices according to sex.

Author Contributions: A.M.L.-S. and R.M.O.: Project administration, funding acquisition. M.D.S.-G. and A.A.: Data curation. M.D.S.-G.: Formal analysis, writing—original draft. V.L.-K.: Visualization. All authors: Writing—review & editing. All authors have read and agreed to the published version of the manuscript.

Funding: This work was supported by the FISS project PI060318, INCERHPAN-UCM contract 210/2008, and a predoctoral contract financed by Complutense University of Madrid and Banco Santander (CT63/19-CT64/19).

Institutional Review Board Statement: This study was conducted in accordance with the guidelines of the Declaration of Helsinki. This study was approved by the Human Research Review Committee of the Faculty of Pharmacy of the Complutense University of Madrid (PI060318 approved at 17 March 2006).

Informed Consent Statement: Informed consent was obtained from all subjects involved in the study.

Data Availability Statement: The data presented in this study are available on request from the corresponding author.

Conflicts of Interest: The authors declare no conflict of interest.

References

1. Liang, Y.; Hou, D.; Zhao, X.; Wang, L.; Hu, Y.; Liu, J.; Cheng, H.; Yang, P.; Shan, X.; Yan, Y.; et al. Childhood obesity affects adult metabolic syndrome and diabetes. *Endocrine* **2015**, *50*, 87–92. [CrossRef]
2. Temneanu, O.; Trandafir, L.; Purcarea, M.R. Type 2 diabetes mellitus in children and adolescents: A relatively new clinical problem within pediatric practice. *J. Med. Life* **2016**, *9*, 235–239. [PubMed]
3. Thumann, B.F.; Michels, N.; Felső, R.; Hunsberger, M.; Kaprio, J.; Moreno, L.A.; Siani, A.; Tornaritis, M.; Veidebaum, T.; De Henauw, S.; et al. Associations between sleep duration and insulin resistance in European children and adolescents considering the mediating role of abdominal obesity. *PLoS ONE* **2020**, *15*, e0235049. [CrossRef] [PubMed]
4. Zupo, R.; Sardone, R.; Donghia, R.; Castellana, F.; Lampignano, L.; Bortone, I.; Misciagna, G.; De Pergola, G.; Panza, F.; Lozupone, M.; et al. Traditional Dietary Patterns and Risk of Mortality in a Longitudinal Cohort of the Salus in Apulia Study. *Nutrients* **2020**, *12*, 1070. [CrossRef] [PubMed]
5. Filippou, C.D.; Tsioufis, C.P.; Thomopoulos, C.G.; Mihas, C.C.; Dimitriadis, K.S.; Sotiropoulou, L.I.; Chrysochoou, C.A.; Nihoyannopoulos, P.I.; Tousoulis, D.M. Dietary Approaches to Stop Hypertension (DASH) Diet and Blood Pressure Reduction in Adults with and without Hypertension: A Systematic Review and Meta-Analysis of Randomized Controlled Trials. *Adv. Nutr. Int. Rev. J.* **2020**, *11*, 1150–1160. [CrossRef]
6. Joyce, B.T.; Wu, D.; Hou, L.; Dai, Q.; Castaneda, S.F.; Gallo, L.C.; Talavera, G.A.; Sotres-Alvarez, D.; Van Horn, L.; Beasley, J.; et al. DASH diet and prevalent metabolic syndrome in the Hispanic Community Health Study/Study of Latinos. *Prev. Med. Rep.* **2019**, *15*, 100950. [CrossRef]
7. Corsino, L.; Sotres-Alvarez, D.; Butera, N.M.; Siega-Riz, A.M.; Palacios, C.; Pérez, C.M.; Albrecht, S.S.; Giacinto, R.A.E.; Perera, M.J.; Van Horn, L.; et al. Association of the DASH dietary pattern with insulin resistance and diabetes in US Hispanic/Latino adults: Results from the Hispanic Community Health Study/Study of Latinos (HCHS/SOL). *BMJ Open Diabetes Res. Care* **2017**, *5*, e000402. [CrossRef]
8. Conlin, P.R.; Chow, D.; Miller, E.R.; Svetkey, L.P.; Lin, P.-H.; Harsha, D.W.; Moore, T.J.; Sacks, F.M.; Appel, L.J. The effect of dietary patterns on blood pressure control in hypertensive patients: Results from the dietary approaches to stop hypertension (DASH) trial. *Am. J. Hypertens.* **2000**, *13*, 949–955. [CrossRef]
9. Monfort-Pires, M.; Folchetti, L.D.; Previdelli, A.N.; Siqueira-Catania, A.; De Barros, C.R.; Ferreira, S.R.G. Healthy Eating Index is associated with certain markers of inflammation and insulin resistance but not with lipid profile in individuals at cardiometabolic risk. *Appl. Physiol. Nutr. Metab.* **2014**, *39*, 497–502. [CrossRef]
10. Sotos-Prieto, M.; Bhupathiraju, S.; Falcon, L.; Gao, X.; Tucker, K.; Mattei, J. Association between a Healthy Lifestyle Score and inflammatory markers among Puerto Rican adults. *Nutr. Metab. Cardiovasc. Dis.* **2015**, *26*, 178–184. [CrossRef]
11. Krijger, J.A.; Nicolaou, M.; Nguyen, A.N.; Voortman, T.; Hutten, B.A.; Vrijkotte, T.G. Diet quality at age 5–6 and cardiovascular outcomes in preadolescents. *Clin. Nutr. ESPEN* **2021**, *43*, 506–513. [CrossRef] [PubMed]
12. Hooshmand, F.; Asghari, G.; Yuzbashian, E.; Mahdavi, M.; Mirmiran, P.; Azizi, F. Modified Healthy Eating Index and Incidence of Metabolic Syndrome in Children and Adolescents: Tehran Lipid and Glucose Study. *J. Pediatr.* **2018**, *197*, 134–139.e2. [CrossRef] [PubMed]
13. Sal, S.; Bektas, M. Effectiveness of Obesity Prevention Program Developed for Secondary School Students. *Am. J. Health Educ.* **2022**, *53*, 45–55. [CrossRef]
14. Bricarello, L.P.; Souza, A.D.M.; Alves, M.D.A.; Retondario, A.; Fernandes, R.; Trindade, E.B.S.D.M.; Zeni, L.A.Z.R.; Vasconcelos, F.D.A.G.D. Association between DASH diet (Dietary Approaches to Stop Hypertension) and hypertension in adolescents: A cross-sectional school-based study. *Clin. Nutr. ESPEN* **2020**, *36*, 69–75. [CrossRef] [PubMed]
15. Öztürk, Y.E.; Bozbulut, R.; Döğer, E.; Bideci, A.; Köksal, E. The relationship between diet quality and insulin resistance in obese children: Adaptation of the Healthy Lifestyle-Diet Index in Turkey. *J. Pediatr. Endocrinol. Metab.* **2018**, *31*, 391–398. [CrossRef]
16. Rahimi, H.; Yuzbashian, E.; Zareie, R.; Asghari, G.; Djazayery, A.; Movahedi, A.; Mirmiran, P. Dietary approaches to stop hypertension (DASH) score and obesity phenotypes in children and adolescents. *Nutr. J.* **2020**, *19*, 112. [CrossRef] [PubMed]
17. Martínez-González, M.A.; García-Arellano, A.; Toledo, E.; Salas-Salvadó, J.; Buil-Cosiales, P.; Corella, D.; Covas, M.I.; Schröder, H.; Arós, F.; Gómez-Gracia, E.; et al. A 14-Item Mediterranean Diet Assessment Tool and Obesity Indexes among High-Risk Subjects: The PREDIMED Trial. *PLoS ONE* **2012**, *7*, e43134.
18. Trichopoulou, A.; Kouris-Blazos, A.; Wahlqvist, M.L.; Gnardellis, C.; Lagiou, P.; Polychronopoulos, E.; Vassilakou, T.; Lipworth, L.; Trichopoulos, D. Diet and overall survival in elderly people. *BMJ* **1995**, *311*, 1457–1460. [CrossRef]
19. Trichopoulou, A.; Orfanos, P.; Norat, T.; Bueno-De-Mesquita, B.; Ocké, M.C.; Peeters, P.H.; Van Der Schouw, Y.T.; Boeing, H.; Hoffmann, K.; Boffetta, P.; et al. Modified Mediterranean diet and survival: EPIC-elderly prospective cohort study. *BMJ* **2005**, *330*, 991. [CrossRef]
20. De Batlle, J.; Garcia-Aymerich, J.; Barraza-Villarreal, A.; Antó, J.M.; Romieu, I. Mediterranean diet is associated with reduced asthma and rhinitis in Mexican children. *Allergy* **2008**, *63*, 1310–1316. [CrossRef]

21. Aparicio-Ugarriza, R.; Cuenca-García, M.; Gonzalez-Gross, M.; Julián, C.; Bel-Serrat, S.; Moreno, L.A.; Breidenassel, C.; Kersting, M.; Arouca, A.B.; Michels, N.; et al. Relative validation of the adapted Mediterranean Diet Score for Adolescents by comparison with nutritional biomarkers and nutrient and food intakes: The Healthy Lifestyle in Europe by Nutrition in Adolescence (HELENA) study. *Public Health Nutr.* **2019**, *22*, 2381–2397. [CrossRef] [PubMed]
22. Levitan, E.B.; Lewis, C.E.; Tinker, L.F.; Eaton, C.B.; Ahmed, A.; Manson, J.E.; Snetselaar, L.G.; Martin, L.W.; Trevisan, M.; Howard, B.V.; et al. Mediterranean and DASH Diet Scores and Mortality in Women With Heart Failure: The Women's Health Initiative. *Circ. Hear. Fail.* **2013**, *6*, 1116–1123. [CrossRef] [PubMed]
23. Fung, T.T.; Chiuve, S.E.; McCullough, M.L.; Rexrode, K.M.; Logroscino, G.; Hu, F.B. Adherence to a DASH-Style Diet and Risk of Coronary Heart Disease and Stroke in Women. *Arch. Intern. Med.* **2008**, *168*, 713–720. [CrossRef] [PubMed]
24. Krebs-Smith, S.M.; Pannucci, T.E.; Subar, A.F.; Kirkpatrick, S.I.; Lerman, J.L.; Tooze, J.A.; Wilson, M.M.; Reedy, J. Update of the Healthy Eating Index: HEI-2015. *J. Acad. Nutr. Diet.* **2018**, *118*, 1591–1602. [CrossRef] [PubMed]
25. Kennedy, E.T.; Ohls, J.; Carlson, S.; Fleming, K.; Kennedy, E.T.; Ohls, J.; Carlson, S.; Fleming, K.; Kennedy, E.T.; Ohls, J.; et al. The Healthy Eating Index: Design and Applications. *J. Am. Diet. Assoc.* **1995**, *95*, 1103–1108. [CrossRef]
26. Askari, M.; Daneshzad, E.; Naghshi, S.; Bellissimo, N.; Suitor, K.; Azadbakht, L. Healthy eating index and anthropometric status in young children: A cross-sectional study. *Clin. Nutr. ESPEN* **2021**, *45*, 306–311. [CrossRef]
27. Rodríguez, L.A.; Mundo-Rosas, V.; Méndez-Gómez-Humarán, I.; Pérez-Escamilla, R.; Shamah-Levy, T. Dietary quality and household food insecurity among Mexican children and adolescents. *Matern. Child Nutr.* **2016**, *13*, e12372. [CrossRef]
28. Pérez-Gimeno, G.; Rupérez, A.I.; Vázquez-Cobela, R.; Herráiz-Gastesi, G.; Gil-Campos, M.; Aguilera, C.M.; Moreno, L.A.; Trabazo, M.R.L.; Bueno-Lozano, G. Energy Dense Salty Food Consumption Frequency Is Associated with Diastolic Hypertension in Spanish Children. *Nutrients* **2020**, *12*, 1027. [CrossRef] [PubMed]
29. Glenn, A.J.; Hernández-Alonso, P.; Kendall, C.W.; Martínez-González, M.; Corella, D.; Fitó, M.; Martínez, J.; Alonso-Gómez, M.; Wärnberg, J.; Vioque, J.; et al. Longitudinal changes in adherence to the portfolio and DASH dietary patterns and cardiometabolic risk factors in the PREDIMED-Plus study. *Clin. Nutr.* **2021**, *40*, 2825–2836. [CrossRef]
30. Royo-Bordonada, M.A.; Garcés, C.; Gorgojo, L.; Martín-Moreno, J.M.; Lasunción, M.A.; Rodríguez-Artalejo, F.; Fernández, O.; De Oya, M. Saturated fat in the diet of Spanish children: Relationship with anthropometric, alimentary, nutritional and lipid profiles. *Public Health Nutr.* **2006**, *9*, 429–435. [CrossRef]
31. Van der Aa, M.P.; Fazeli Farsani, S.; Knibbe, C.A.J.; de Boer, A.; van der Vorst, M.M.J. Population-Based Studies on the Epidemiology of Insulin Resistance in Children. *J. Diabetes Res.* **2015**, *2015*, 362375. [CrossRef]
32. Moran, A.; Jacobs, D.R., Jr.; Steinberger, J.; Steffen, L.M.; Pankow, J.S.; Hong, C.P.; Sinaiko, A.R. Changes in insulin resistance and cardiovascular risk during adolescence: Establishment of differential risk in males and females. *Circulation* **2008**, *117*, 2361–2368. [CrossRef]
33. Chiarelli, F.; Marcovecchio, M.L. Insulin resistance and obesity in childhood. *Eur. J. Endocrinol.* **2008**, *159* (Suppl. S1), S67–S74. [CrossRef] [PubMed]
34. Rodríguez-Rodríguez, E.; Salas-González, M.D.; Ortega, R.M.; López-Sobaler, A.M. Leukocytes and Neutrophil–Lymphocyte Ratio as Indicators of Insulin Resistance in Overweight/Obese School-Children. *Front. Nutr.* **2022**, *8*, 1318. [CrossRef] [PubMed]
35. World Health Organization. *Physical Status: The Use and Interpretation of Anthropometry. Report of the WHO Expert Committee*; World Health Organization: Geneva, Switzerland, 1995; p. 543.
36. WHO. WHO/Europe | Nutrition—Body Mass Index—BMI. World Health Organization. 2021. Available online: https://www.euro.who.int/en/health-topics/disease-prevention/nutrition/a-healthy-lifestyle/body-mass-index-bmi (accessed on 5 September 2022).
37. Onis, M.; Onyango, A.; Borghi, E.; Siyam, A.; Nishida, C.; Siekmann, J. Development of a WHO growth reference for school-aged children and adolescents. *Bull World Health Organ.* **2007**, *85*, 660–667. [CrossRef] [PubMed]
38. Stewart, A.; Marfell-Jones, M.; Olds, T.; de Ridder, H. *International Protocol for Anthropometric Assessment*; Hutt, L., Ed.; International Society for the Advancement of Kinanthropometry: Glasgow, UK, 2011.
39. Dezenberg, C.; Nagy, T.; Gower, B.; Johnson, R.; Goran, M. Predicting body composition from anthropometry in pre-adolescent children. *Int. J. Obes.* **1999**, *23*, 253–259. [CrossRef]
40. McCarthy, H.D.; Cole, T.J.; Fry, T.; Jebb, S.A.; Prentice, A.M. Body fat reference curves for children. *Int. J. Obes.* **2006**, *30*, 598–602. [CrossRef]
41. Ortega, R.M.; Requejo, A.M.; López-Sobaler, A.M. Activity questionnaire. In *Nutriguía. Manual of Clinical Nutrition in Primary Care*; Requejo, R.M., Ortega, R., Eds.; Complutense Madrid: Madrid, Spain, 2006; p. 468.
42. Peral-Suárez, Á.; Cuadrado-Soto, E.; Perea, J.M.; Navia, B.; López-Sobaler, A.M.; Ortega, R.M. Physical activity practice and sports preferences in a group of Spanish schoolchildren depending on sex and parental care: A gender perspective. *BMC Pediatr.* **2020**, *20*, 337. [CrossRef] [PubMed]
43. Rodríguez-Rodríguez, E.; Ortega, R.M.; Carvajales, P.A.; González-Rodríguez, L.G. Relationship between 24 h urinary potassium and diet quality in the adult Spanish population. *Public Health Nutr.* **2014**, *18*, 850–859. [CrossRef] [PubMed]
44. Ortega, R.M.; Rodríguez-Rodríguez, E.; Aparicio, A.; Jiménez, A.I.; López-Sobaler, A.M.; González-Rodríguez, L.G.; Andrés, P. Poor zinc status is associated with increased risk of insulin resistance in Spanish children. *Br. J. Nutr.* **2011**, *107*, 398–404. [CrossRef]

45. Ortega, R.M.; Requejo, A.M.; Quintas, E.; Sánchez-Quiles, B.; López-Sobaler, A.M.; Andrés, P. Estimated energy balance in female university students: Differences with respect to body mass index and concern about body weight. *Int. J. Obes.* **1996**, *20*, 1127–1129.
46. WHO. Energy and Protein Requirements: Report of a Joint FAO/WHO/UNU Expert Consultation. 1985. Available online: https://apps.who.int/iris/handle/10665/40157 (accessed on 5 September 2022).
47. Institute of Medicine. *Dietary Reference Intakes for Energy, Carbohydrate, Fiber, Fat, Fatty Acids, Cholesterol, Protein, and Amino Acids*; National Academies Press: Washington, DC, USA, 2005; p. 10490. [CrossRef]
48. Ortega, R.M.; Requejo, A.M.; López-Sobaler, A.M. Questionnaires for dietetic studies and the assessment of nutritional status. In *Nutriguía. Manual of Clinical Nutrition in Primary Care*; Requejo, R.M., Ortega, R., Eds.; Complutense Madrid: Madrid, Spain, 2006; pp. 456–459.
49. Ortega, R.M.; López-Sobaler, A.M.; Andrés, P.; Requejo, A.M.; Aparicio, A.M.L. *DIAL Software for Assessing Diets and Food Calculations*, for Windows, version 3.0.0.5; Department of Nutrition (UCM) & Alceingeniería, S.A.: Madrid, Spain, 2013.
50. Ortega, R.; López-Sobaler, A.; Andrés, P.; Aparicio, A. Food nutritional composition. In *A Tool for the Design and Evaluation of Food and Diets*; Departament of Nutrition and Food Science: Singapore; Comlutense University of Madrid: Madrid, Spain, 2021.
51. Ortega, R.M.; Perez-Rodrigo, C.; Lopez-Sobaler, A.M. Dietary assessment methods: Dietary records. *Nutr. Hosp.* **2015**, *31*, 38–45. [CrossRef]
52. Neese, J.W.; Duncan, P.B.D. *Development and Evaluation of a Hexokinase/Glucose-6-Phosphate Dehydrogenase Procedure for Use as a National Glucose Reference Method*; U.S. Public Health Service, Center for Disease Control, Bureau of Laboratories, Clinical Chemistry Division: Atlanta, GA, USA, 1976.
53. El Kenz, H.; Bergmann, P. Evaluation of immunochemiluminometric assays for the measurement of insulin and C-peptide using the ADVIA Centaur. *Clin. Lab.* **2004**, *50*, 171–174. [PubMed]
54. Hřebícek, J.; Janout, V.; Malincíková, J.; Horáková, D.; Cízek, L. Detection of Insulin Resistance by Simple Quantitative Insulin Sensitivity Check Index QUICKI for Epidemiological Assessment and Prevention. *J. Clin. Endocrinol. Metab.* **2002**, *87*, 144. [CrossRef] [PubMed]
55. Tripathy, D.; Carlsson, M.; Almgren, P.; Isomaa, B.; Taskinen, M.R.; Tuomi, T.; Groop, L.C. Insulin secretion and insulin sensitivity in relation to glucose tolerance: Lessons from the Botnia Study. *Diabetes* **2000**, *49*, 975–980. [CrossRef]
56. Albareda, M.; Rodríguez-Espinosa, J.; Murugo, M.; de Leiva, A.; Corcoy, R. Assessment of insulin sensitivity and beta-cell function from measurements in the fasting state and during an oral glucose tolerance test. *Diabetologia* **2000**, *43*, 1507–1511. [CrossRef]
57. Keskin, M.; Kurtoglu, S.; Kendirci, M.; Atabek, M.E.; Yazici, C. Homeostasis Model Assessment Is More Reliable Than the Fasting Glucose/Insulin Ratio and Quantitative Insulin Sensitivity Check Index for Assessing Insulin Resistance Among Obese Children and Adolescents. *Pediatrics* **2005**, *115*, e500–e503. [CrossRef] [PubMed]
58. Manios, Y.; Moschonis, G.; Kourlaba, G.; Bouloubasi, Z.; Grammatikaki, E.; Spyridaki, A.; Hatzis, C.; Kafatos, A.; Fragiadakis, G.A. Prevalence and independent predictors of insulin resistance in children from Crete, Greece: The Children Study. *Diabet. Med.* **2007**, *25*, 65–72. [CrossRef]
59. Kostovski, M.; Simeonovski, V.; Mironska, K.; Tasic, V.; Gucev, Z. Metabolic Profiles in Obese Children and Adolescents with Insulin Resistance. *Open Access Maced. J. Med Sci.* **2018**, *6*, 511–518. [CrossRef]
60. Mat, S.H.C.; Yaacob, N.M.; Hussain, S. Rate of Weight Gain and its Association with Homeostatic Model Assessment-Insulin Resistance (HOMA-IR) among Obese Children attending Paediatric Endocrine Clinic, Hospital Universiti Sains Malaysia. *J. ASEAN Fed. Endocr. Soc.* **2021**, *36*, 149–155. [CrossRef] [PubMed]
61. Wei, J.; Luo, X.; Zhou, S.; He, X.; Zheng, J.; Sun, X.; Cui, W. Associations between iron status and insulin resistance in Chinese children and adolescents: Findings from the China Health and Nutrition Survey. *Asia Pac. J. Clin. Nutr.* **2019**, *28*, 819–825. [CrossRef] [PubMed]
62. Mastrangelo, A.; Martos-Moreno, G.; García, A.; Barrios, V.; Rupérez, F.J.; Chowen, J.A.; Barbas, C.; Argente, J. Insulin resistance in prepubertal obese children correlates with sex-dependent early onset metabolomic alterations. *Int. J. Obes.* **2016**, *40*, 1494–1502. [CrossRef] [PubMed]
63. Jeffery, S.C.; Hosking, J.; Jeffery, A.N.; Murphy, M.J.; Voss, L.D.; Wilkin, T.J.; Pinkney, J. Insulin resistance is higher in prepubertal girls but switches to become higher in boys at age 16: A Cohort Study (EarlyBird 57). *Pediatr. Diabetes* **2017**, *19*, 223–230. [CrossRef] [PubMed]
64. Tovar, A.; Risica, P.M.; Ramirez, A.; Mena, N.; Lofgren, I.E.; Stowers, K.C.; Gans, K.M. Exploring the Provider-Level Socio-Demographic Determinants of Diet Quality of Preschool-Aged Children Attending Family Childcare Homes. *Nutrients* **2020**, *12*, 1368. [CrossRef] [PubMed]
65. Thomson, J.L.; Landry, A.; Tussing-Humphreys, L.M.; Goodman, M.H. Diet quality of children in the United States by body mass index and sociodemographic characteristics. *Obes. Sci. Pract.* **2019**, *6*, 84–98. [CrossRef] [PubMed]
66. Landry, M.J.; Asigbee, F.M.; Vandyousefi, S.; Khazaee, E.; Ghaddar, R.; Boisseau, J.B.; House, B.T.; Davis, J.N. Diet Quality Is an Indicator of Disease Risk Factors in Hispanic College Freshmen. *J. Acad. Nutr. Diet.* **2019**, *119*, 760–768. [CrossRef] [PubMed]
67. Frazier-Wood, A.C.; Kim, J.; Davis, J.S.; Jung, S.Y.; Chang, S. In cross-sectional observations, dietary quality is not associated with CVD risk in women; in men the positive association is accounted for by BMI. *Br. J. Nutr.* **2015**, *113*, 1244–1253. [CrossRef] [PubMed]

68. Mattei, J.; Sotos-Prieto, M.; Bigornia, S.J.; Noel, S.E.; Tucker, K.L. The Mediterranean Diet Score Is More Strongly Associated with Favorable Cardiometabolic Risk Factors over 2 Years Than Other Diet Quality Indexes in Puerto Rican Adults. *J. Nutr.* **2017**, *147*, 661–669. [CrossRef]
69. Dong, Y.; Chen, L.; Gutin, B.; Zhu, H. Total, insoluble, and soluble dietary fiber intake and insulin resistance and blood pressure in adolescents. *Eur. J. Clin. Nutr.* **2018**, *73*, 1172–1178. [CrossRef] [PubMed]
70. DiNicolantonio, J.J.; O'Keefe, J.H. Good Fats versus Bad Fats: A Comparison of Fatty Acids in the Promotion of Insulin Resistance, Inflammation, and Obesity. *Mo. Med.* **2017**, *114*, 303–307.
71. Malik, V.S.; Popkin, B.M.; Bray, G.A.; Després, J.-P.; Willett, W.C.; Hu, F.B. Sugar-Sweetened Beverages and Risk of Metabolic Syndrome and Type 2 Diabetes: A meta-analysis. *Diabetes Care* **2010**, *33*, 2477–2483. [CrossRef] [PubMed]
72. Guillermo, C.; Boushey, C.J.; Franke, A.A.; Monroe, K.R.; Lim, U.; Wilkens, L.R.; Le Marchand, L.; Maskarinec, G. Diet Quality and Biomarker Profiles Related to Chronic Disease Prevention: The Multiethnic Cohort Study. *J. Am. Coll. Nutr.* **2019**, *39*, 216–223. [CrossRef] [PubMed]
73. Jacobs, S.; Boushey, C.J.; Franke, A.A.; Shvetsov, Y.B.; Monroe, K.R.; Haiman, C.A.; Kolonel, L.N.; Le Marchand, L.; Maskarinec, G. A priori-defined diet quality indices, biomarkers and risk for type 2 diabetes in five ethnic groups: The Multiethnic Cohort. *Br. J. Nutr.* **2017**, *118*, 312–320. [CrossRef] [PubMed]
74. Asemi, Z.; Esmaillzadeh, A. DASH Diet, Insulin Resistance, and Serum hs-CRP in Polycystic Ovary Syndrome: A Randomized Controlled Clinical Trial. *Horm. Metab. Res.* **2014**, *47*, 232–238. [CrossRef]
75. Shirani, F.; Salehi-Abargouei, A.; Azadbakht, L. Effects of Dietary Approaches to Stop Hypertension (DASH) diet on some risk for developing type 2 diabetes: A systematic review and meta-analysis on controlled clinical trials. *Nutrition* **2013**, *29*, 939–947. [CrossRef]
76. Chiavaroli, L.; Viguiliouk, E.; Nishi, S.K.; Mejia, S.B.; Rahelić, D.; Kahleova, H.; Salas-Salvadó, J.; Kendall, C.W.C.; Sievenpiper, J.L. DASH Dietary Pattern and Cardiometabolic Outcomes: An Umbrella Review of Systematic Reviews and Meta-Analyses. *Nutrients* **2019**, *11*, 338. [CrossRef]
77. Banfield, E.C.; Liu, Y.; Davis, J.S.; Chang, S.; Frazier-Wood, A.C. Poor Adherence to US Dietary Guidelines for Children and Adolescents in the National Health and Nutrition Examination Survey Population. *J. Acad. Nutr. Diet.* **2015**, *116*, 21–27. [CrossRef]

nutrients

MDPI

Article

Extended Inter-Meal Interval Negatively Impacted the Glycemic and Insulinemic Responses after Both Lunch and Dinner in Healthy Subjects

Xuejiao Lu [1], Zhihong Fan [1,2,*], Anshu Liu [1], Rui Liu [1], Xinling Lou [1] and Jiahui Hu [1]

[1] College of Food Science and Nutritional Engineering, China Agricultural University, Beijing 100083, China
[2] Key Laboratory of Precision Nutrition and Food Quality, Department of Nutrition and Health, China Agricultural University, Beijing 100083, China
* Correspondence: daisyfan@cau.edu.cn; Tel.: +86-10-62737717

Abstract: This study aimed to investigate the glycemic and insulinemic effects of lunch timing based on a fixed feeding window, and the effects of apple preload on postprandial glucose and insulin responses after nutrient-balanced lunch and the subsequent high-fat dinner in healthy participants. Twenty-six participants completed four randomized, crossover experimental trials: (1) early standardized lunch at 12:00 (12S); (2) apple preload to 12S (12A+S); (3) late standardized lunch at 14:00 (14S); and (4) apple preload to 14S (14A+S); wherein twenty participants' blood samples were collected for insulin analysis following the lunch trails. In each experimental trial, each participant equipped with a continuous glucose monitor (CGM) was provided with a standardized breakfast and a high-fat dinner to be consumed at 8:00 and 18:00, respectively. The late lunch (14S) resulted in significantly elevated glucose peak, delayed insulin peak time, decreased insulin sensitivity, and increased insulin resistance following the lunch; also decreased glycemic response following the subsequent dinner and larger blood glucose fluctuation over the 24-h period compared with the 12S. The 14A+S significantly reduced the glucose peak, the insulin peak time and the glycemic variability following the lunch, also the 24-h glycemic variability compared with the 14S. The insulin sensitivity was significantly improved in the 12A+S, compared with that of the 12S. In conclusion, the present study found that an extra 2-h inter-meal fasting before and after lunch resulted in elevated glycemic response in both macronutrient-balanced meal and high-fat meal in healthy subjects. The negative impact of a late lunch could be partly reversed by the apple preload, without a trade-off of insulin secretion.

Keywords: meal timing; inter-meal interval; apple preload; glucose; insulin

1. Introduction

Obesity and hyperglycemia are regarded as among the most concerning public health issues in most parts of the world [1,2]. Recently, the timing of food ingestion, in addition to the composition of food, has been proposed as a crucial determinant of the postprandial metabolic parameters associated with glycemic control and body weight management [3,4].

Evidence from epidemiological studies is accumulating that the late eating is associated with elevated risk of obesity, diabetes and cardiovascular diseases [5–8]. Several trials indicated that the glycemic homeostasis and insulin sensitivity tend to deteriorate in the later part of the day in healthy subjects [9–11]. Low glycemic meal could elicit greater glucose response and insulin response when consumed at 22:00 than it could when consumed at 18:00 [12]. Previous studies suggested that early dinner would be beneficial to keep a mild postprandial glycemic response and a stable blood glucose level at night [13,14], while a shortened feeding window in time-restriction feeding would help to lose weight and improve glycemic control [15].

With respect to the effect of the meal timing on metabolic consequences, most trials focused on the dinner time or the feeding window between breakfast and dinner [3,16]. It is well accepted that the timing of dinner plays a significant role in blood glucose regulation. However, the timing of lunch, which sets the inter-meal fasting period after breakfast and before dinner, was yet to be fully explored. Limited studies referring the postprandial glycemic response of lunch gave inconsistent results regarding the postprandial glycemic response after lunch and the subsequent meal [9,17,18]. Given that the time setting of the meals were different among studies, it is not easy to precisely compare the results in different studies, as the confounders of unequal meal intervals and feeding windows cannot be ruled out.

Several epidemiological studies have suggested that late lunch patterns were related to less effective weight-loss [19–21], increased insulin resistance [19], and the risk of polycystic ovary syndrome [22]. A few randomized controlled trials investigating different lunch timing demonstrated that late lunch resulted in decreased glucose tolerance, disturbed stress hormone rhythm [23], negative body composition change, and reduced microbiota diversity in healthy women [24]. However, there is paucity of data regarding the impact of the inter-meal interval on the postprandial glucose and insulin response and the glycemic excursion over the 24-h period, based on the same feeding window in real life settings.

It has been already proved that preload can be a convenient dietary strategy to improve postprandial glucose homeostasis [25,26]. Previous studies reported that the fruits and dried fruits such as apple [27–29], when consumed 30 min prior to a rice meal, significantly mitigated the postprandial glycemic response in extents of 31.4% to 50% cut of the incremental area under the curve (iAUC). However, it had not yet been confirmed whether the fruit preload could exert a remarkable glycemic stabilizing effect in nutrient-balanced mixed meals with different mealtime settings.

In the present study, we set the early lunch at 12:00 and the late lunch at 14:00, with a fixed feeding window of 10 h (8:00 to 18:00). The glycemic and insulinemic effects of lunch timing, and the effects of the apple preload on postprandial glucose and insulin responses after nutrient-balanced lunch and the subsequent high-fat dinner, were investigated in healthy participants. We supposed that the inter-meal interval would pose an impact on postprandial glycemic response based on a fixed feeding window. The first research hypothesis was that an extended before-meal fasting might resulted in exaggerated postprandial blood glucose excursion in the context of lunch and dinner, irrespective of the macronutrient composition of the meal. The second hypothesis was that the apple preload would partly reverse the negative glycemic and insulinemic responses elicited by a prolonged inter-meal fasting.

2. Materials and Methods

2.1. Participants and Ethics

Participants of both sexes volunteered via social media advertisements and were included if the following criteria were satisfied: generally healthy university students with a normal body mass index (BMI) between 18.5 and 24.0 kg/m^2 [30]; a regular sleep–wake cycle; consuming 3 meals a day with lunch no later than 13:00 and dinner no later than 19:00; and a regular menstrual cycle (if female). The following exclusion criteria were applied to select the potential participants: allergies or intolerance to any of the test foods; a change in body weight of more than 5 kg within the past six months; any metabolic diseases covering diabetes, hypertension, metabolic syndrome; gastrointestinal disorders such as gastroesophageal reflux disease; heavy drinking; active smoking; use of medications or supplements known to affect circadian rhythms or metabolism; sleeping and eating disorders; engagement in competitive or endurance sports.

All participants interested in the study completed a questionnaire on daily lifestyle habits (dietary, sleep, physical activity) and health status by a face-to-face interview for an initial pre-screening. After that, eligible participants attended the lab for a morning, fasted, baseline tested visit prior to the first trial condition where a 2-h oral glucose tolerance test (OGTT) was administered. Before the test, body mass and fat mass were measured using a body fat scale (HBF-371, OMRON, Yangzhou, China) where the visceral fat index and basal metabolic rate (BMR) were obtained in the meantime. Resting blood pressure was measured in duplicate using an electronic blood pressure monitor (HEM-7200, OMRON, Dalian, China). Waist and hip circumferences were measured using a tape measure in duplicate to the nearest 0.1 cm.

The study protocol was conducted according to the principles laid down in the Helsinki Declaration, granted by the Ethics Committee of China Agricultural University (ethics number CAUHR-20220202), and registered at the Chinese Clinical Trial Registry (ChiCTR2200057791). All eligible individuals provided the written informed consent.

2.2. Study Design and Procedures

Each participant attended four separate experimental conditions in an unblinded, randomized, crossover design, with each session being separated by a wash-out period of at least three days. As shown in Figure 1, each test session spanned four consecutive days from around 18:00 on Day −1 until 8:00 on Day 2. Participants were advised to maintain a regular sleep routine and not to perform any strenuous exercise during each study session.

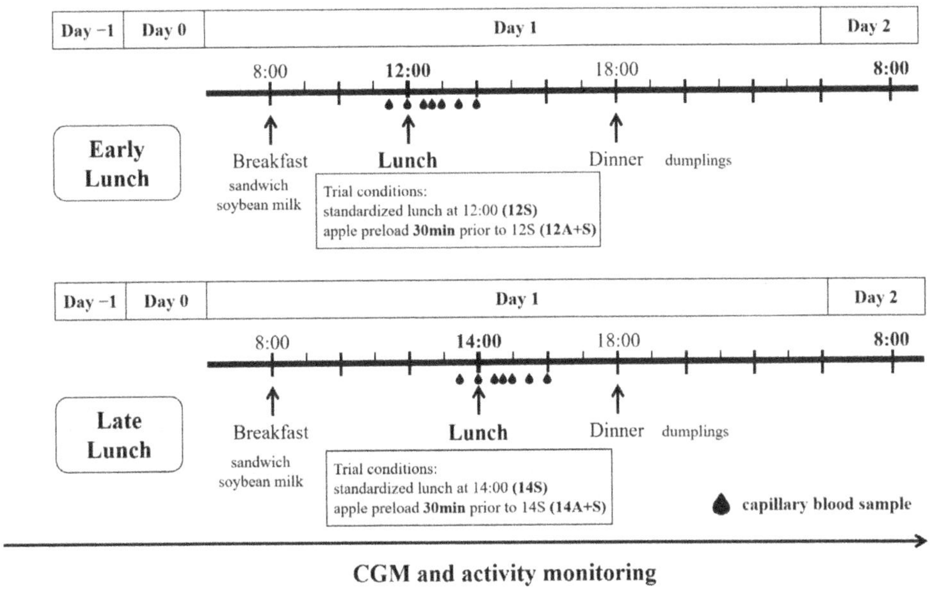

Figure 1. Summary of the study experimental design. CGM, continuous glucose monitor. All participants underwent four lunch trial conditions in a randomized order, including standardized lunch at 12:00 (12S); apple preload 30 min prior to 12S (12A+S); standardized lunch at 14:00 (14S); apple preload 30 min prior to 14S (14A+S).

The procedures were identical during each test session, except for the type and timing of the served lunch meals on the test day (Day 1). On Day −1, participants were fitted with a continuous glucose monitor (CGM) (Abbott, Shanghai, China) and a smart bracelet (Xiaomi, Beijing, China) to monitor the physical activity during each test session. On Day 0, participants were instructed to consume the breakfast, lunch and dinner in a 'free-living' environment at 8:00, 12:00, and 18:00, respectively, and abstain from fruits, coffee, tea or alcohol. On Day 1 (trial day), participants arrived at the laboratory around 7:45 following an overnight fast lasting 12 h and were provided with a standardized breakfast at 8:00. Then, participants were not allowed to consume anything other than water until the lunch meal, which was one of the four test meals: (1) standardized lunch at 12:00 (12S); (2) apple preload 30 min prior to 12S (12A+S); (3) standardized lunch at 14:00 (14S); and (4) apple preload 30 min prior to 14S (14A+S). In the non-preload meal, water was given 30 min before the meal instead of apple. At 18:00, the participants were provided with the traditional Chinese food—dumplings—as the subsequent standardized dinner. The contents and energy of breakfast and lunch were identical for each participant. For the dinner, the participants consumed the dumplings ad libitum during the first trial, and the quantity ingested was replicated in the subsequent trials for each participant.

2.3. Test Meal Components

The standardized breakfast containing 383 kcal had a macronutrient composition of approximately 58% carbohydrate, 21% fat and 21% protein, consisting of a sandwich (toast, sliced ham, tomato, lettuce) and a cup of soy milk. The macronutrient contents and energy composition of the test lunch meals are shown in Table 1. The standardized lunch (S) consisted of white rice, lactose-free low-fat milk, egg and a vegetable salad (romaine lettuce, cherry tomato, broccoli with roasted sesame dressing and sesame oil), and 119.2 g water for weight balance. The standardized lunch with apple preload (A+S) included red fuji apple containing 15.0 g of available carbohydrate consumed 30 min prior to a standardized lunch with reduced amount of rice. The two test lunch meals were tightly matched in energy content and macronutrient distribution as 60% from carbohydrate, 25% from fat and 15% from protein. The energy of dumplings for a high-fat dinner were distributed as follows: 54% from fat, 33% from carbohydrate and 13% from protein, with 37 kcal per dumpling. The energy content and macronutrient composition of the served food items were calculated from China Food Composition Tables, manufacturer data and determination experiments. All the test meals were prepared and weighed by the study staff on the day of each session, immediately served to the volunteers and consumed within 15 min to avoid possible retrogradation of starch.

Table 1. The composition, macronutrient and energy contents of the lunch test meals [1].

Test Meals	Carbohydrate (g)	Protein (g)	Fat (g)	Energy (kcal)	Detail Content
S [2]	97.2	24.2	18.0	648	Lactose-free low-fat milk 200 g, roasted sesame dressing 25 mL, egg yolk 10 g, egg white 50 g, romaine Lettuce 25 g, cherry tomato 75 g, broccoli 50 g, sesame oil 2 g, uncooked rice 100 g, water119.2 g
A+S [3]	97.2	24.1	18.1	648	Lactose-free low-fat milk 200 g, roasted sesame dressing 25 mL, egg yolk 10 g, egg white 50 g, romaine Lettuce 25 g, cherry tomato 75 g, broccoli 50 g, sesame oil 1 g, uncooked rice 80.2 g, apple 140 g

[1] The nutritional contents of the lunch test meals were obtained from China Food Composition Tables, manufacturers and determination experiments.[2] S: standardized lunch.[3] A+S: apple preload 30 min prior to standardized lunch.

2.4. Continuous Glucose Monitoring

The continuous glucose monitor (CGM) was performed on Day −1 at approximately 18:00 and the sensor was removed on Day 2 of the study at approximately 9:00. The data reported in this paper represented interstitial glucose readings recorded every 15 min and occasional missing values were imputed by averaging adjacent values.

2.5. Blood Collection and Analysis

During lunch meals, blood glucose concentrations from finger prick samples were measured by a handheld, commercial glucometer (LifeScan Inc., Milpitas, CA, USA) at −40 and −30 min, after which the participants consumed water or apple, and further blood was drawn at −15, 0, 15, 30, 45, 60 and 90 min. In addition, a 150 μL capillary blood sample from the fingertip was collected into EDTA K2-treated centrifuge tubes (WanDGL Ltd., Jinan, China) at −30, 0, 30, 45, 60 and 90 min for insulin measurement. Within 30 min of blood collection the blood samples were centrifuged at 1000×g for 15 min with 60 μL supernatant plasma dispensed into 0.5 mL Eppendorf tubes and stored at −80 °C until the assay. Plasma insulin concentrations were determined using an ELISA-based test kit (JunLB Ltd., Beijing, China).

2.6. Data Processing and Statistical Analysis

The postprandial glycemic data analysis was based on the change values relative to the fasting concentration. The postprandial insulin data were based on the percent change in insulin relative to the fasting insulin concentration to eliminate inter-personal variability. The incremental area under the curve (iAUC) was calculated for postprandial interstitial/capillary glucose and insulin responses using the trapezoid method above the baseline concentration. The incremental peak values (ΔPeak) for postprandial glucose and insulin were calculated. To assess the postprandial glucose variability, the following indexes were assessed: the largest amplitudes of glucose excursion (LAGE); the standard deviation of blood glucose (SD); the coefficient of variation in blood glucose (CV); continuous overlapping net glycemic action (CONGA-1), defined as the SD of the glycemic changes recorded between a specific point and a point one hour earlier; and J-index, calculated as $0.324 \times (\text{mean glucose} + \text{SD glucose})^2$. To estimate insulin sensitivity, the insulin sensitivity index was calculated as 10,000/square root of (fasting glucose × fasting insulin × mean glucose × mean insulin) [31]. An index of postprandial insulin resistance (HOMA-PP) was calculated for each lunch trial using the following equation [32]: HOMA-PP ($\times 10^3$) = iAUC glucose × iAUC insulin/22.5, which has been validated against the minimal model and the intravenous glucose tolerance test [33,34].

The analysis of 24 h glycemic response monitored by CGM, covering from 8:00 on Day 1 to 8:00 on Day 2, was based on the absolute glucose concentrations. Total area under the curve (tAUC), the mean glucose concentrations (Mean), the max glucose concentrations (Peak), LAGE and SD were calculated for 24 h glycemic response. Additionally, the percentage of the glucose changes greater than 2.5 for each participant (GC > 2.5) was calculated to reflect the 24 h glucose excursion, as well as the percentage of GC > 5.0. We also defined $\Delta P_{L\text{-}D}$ as the difference between the postprandial glucose peak values after lunch and dinner to represent the glucose fluctuation over a day.

A power calculation was conducted with the PASS 13 *Power Analysis and Sample Size* software (NCSS, Kaysville, UT, USA), based on a previous study [27]. A sample size of $n = 11$ was required to provide 80% power to detect a change of 167.8 mmol·min/L in iAUC ($p < 0.05$), assuming that the standard deviation (SD) is lower than 55.15 mmol·min/L.

All the statistical analysis was performed using the SPSS version 23.0 (SPSS Inc. Chicago, IL, USA). Two-way repeated measures ANOVA was chosen to assess the effects of treatment, time, and the interaction of treatment and time. One-way analysis of variance ANOVA and Duncan's multiple range test were performed to analyze the differences in the above-mentioned parameters. The variables are presented as the mean ± standard deviation (SD) or the mean value with standard error (SE), with $p < 0.05$ considered statistically significant.

3. Results

3.1. Baseline Characteristics of Participants

A total of 26 participants were enrolled in the study and completed four treatments with CGM, wherein 20 participants' capillary blood was collected for glucose and insulin analysis, as the remaining 6 participants had difficulty in collecting 150 μL capillary blood from the fingertip. Baseline characteristics of study participants are presented in Table 2.

Table 2. Baseline characteristics of study participants.

Characteristics	Mean ± SD (Male/Female)
Number of participants (male/female)	26(12/14)
Age, years	20.8 ± 0.9
Body composition	
BMI, kg/m^2	21.4 ± 2.1/20.8 ± 1.7
Waist: hip ratio	0.7 ± 0.0/0.8 ± 0.0
Waist: height ratio	0.4 ± 0.0/0.4 ± 0.0
Fat mass, %	15.6 ± 4.1/24.1 ± 4.0
Visceral fat index	4.8 ± 2.0/2.3 ± 1.2
Basal metabolic rate (BMR), kcal/day	1391.6 ± 203.4
Systolic blood pressure, mmHg	114.0 ± 12.3
Diastolic blood pressure, mmHg	67.0 ± 9.0
Habitual meal timing	
Breakfast	8:02 ± 0:42
Lunch	11:42 ± 0:25
Dinner	17:41 ± 0:22

The total daily energy intake on Day 1 was 1646 ± 137 kcal, with a distribution spread of 23% at breakfast, 39% at lunch and 38% at dinner meal. None of the participants reported any physical or gastrointestinal discomfort in each trial.

3.2. Postprandial Interstitial Glycemic Responses Following the Lunch Test Meals

The postprandial interstitial glycemic responses following the lunch test meals are shown in Figure 2. The postprandial interstitial glycemic responses of the 12A+S were remarkably lower than that of 12S, manifesting a significant lower glucose level from 30 min to 150 min and at 225 min. The 14A+S led to a significant lower glucose level at 30, 45 and 120 min than that of 14S. With respect to the different lunch timings, the 14S elicited a lower fasting glucose at 0 min, a higher peak value at 45 min, and lower glucose values at 225 and 240 min than the 12S did ($p < 0.05$). The postprandial glucose response pattern of the early lunch at 12:00 was characterized by three small peaks instead of one sharp peak in the case of the late lunch at 14:00 during the whole 270 min.

Table 3 shows the postprandial interstitial glycemic parameters for the lunch test meals. The apple preload treatments (A+S) elicited significantly lower $\Delta Peak_{270}$, $LAGE_{270}$ and other glycemic variability indices than their standardized lunch (S) counterparts, regardless of the meal timing. In addition, the degree of improvement in 12A+S trial conditions was superior to 14A+S for all the parameters. The $iAUC_{0\text{-}270}$ of 12A+S was significantly improved and achieved a 33.7% reduction compared with that of the 12S. There were no differences in the postprandial interstitial glycemic parameters except for the $\Delta Peak_{270}$ between 12S and 14S. In addition, the numbers of participants who had peak blood glucose concentrations exceeding 10 mmol/L were eight in the 14S test while only two in the 12S test.

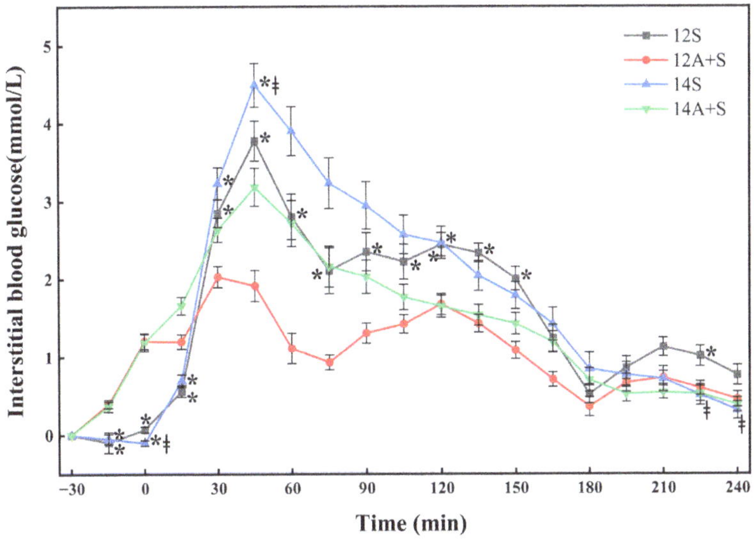

Figure 2. Postprandial interstitial glycemic responses following the lunch test meals. 12S, standardized lunch at 12:00; 12A+S, apple preload 30 min prior to 12S; 14S, standardized lunch at 14:00; 14A+S, apple preload 30 min prior to 14S. −30 min = the commencement of the water or apple; 0 min = the time when the standardized mixed meal was given. Data are mean ± SE (n = 26). * Apple preload treatments (A+S) different from their standardized lunch (S) counterparts ($p < 0.05$), ‡ 14S different from 12S ($p < 0.05$).

Table 3. Postprandial interstitial glycemic parameters for the lunch test meals (mean ± SE, n = 26).

Test Meals	$iAUC_{0-270}$ (mmol·min/L)	$\Delta Peak_{270}$ (mmol/L)	$LAGE_{270}$ (mmol/L)	SD	CV (%)	CONGA-1	J-Index
12S	432.6 ± 29.7 ^{a}	3.9 ± 0.2 ^{a}	4.2 ± 0.3 ^{ab}	1.3 ± 0.1 ^{ab}	21.1 ± 1.1 ^{ab}	2.0 ± 0.1 ^{ab}	17.1 ± 0.8 ^{ab}
12A+S	287.0 ± 18.3 ^{b}	2.4 ± 0.1 ^{b}	2.5 ± 0.1 ^{c}	0.7 ± 0.0 ^{c}	12.7 ± 0.6 ^{c}	1.1 ± 0.1 ^{c}	13.0 ± 0.5 ^{c}
14S	482.1 ± 34.6 ^{a}	4.7 ± 0.3 ^{c}	5.0 ± 0.2 ^{a}	1.6 ± 0.1 ^{a}	24.6 ± 1.1 ^{a}	2.5 ± 0.1 ^{a}	20.5 ± 1.2 ^{a}
14A+S	392.0 ± 26.9 ^{a}	3.4 ± 0.2 ^{a}	3.5 ± 0.2 ^{b}	1.0 ± 0.1 ^{b}	17.4 ± 1.1 ^{b}	1.6 ± 0.1 ^{b}	15.9 ± 0.8 ^{b}

^{a,b,c} Different superscript letters denote that mean values within a column are significantly different ($p < 0.05$). 12S, standardized lunch at 12:00; 12A+S, apple preload 30 min prior to 12S; 14S, standardized lunch at 14:00; 14A+S, apple preload 30 min prior to 14S.

3.3. Postprandial Capillary Glucose and Insulin Responses Following the Lunch Test Meals

Figure 3 shows the postprandial capillary glucose and insulin responses and parameters in four lunch trial conditions. The outcomes of capillary glucose were generally consistent with those obtained by CGM, except that a delayed peak value occurred at 45 min in the postprandial glycemic response curve and a larger glucose $iAUC_{0-120}$ in 14S ($p < 0.05$). The insulin data showed that the 14S induced higher insulin $iAUC_{0-120}$ and insulin $\Delta Peak_{120}$ than 12S did ($p < 0.05$). Apple preload restored the peak glucose concentration and the glucose iAUC after the late lunch but failed to reverse the insulin iAUC to that of the early lunch's level.

As shown in Figure 4, there was a significant difference between the 14S and the 14A+S in terms of the glucose peak time. Furthermore, an obvious delay in the insulin peak time was observed in the 14S. The insulin sensitivity index of the 14S was 30.7% lower than that of the 12S, while the 12A+S achieved a 20% increase in the insulin sensitivity index compared with 12S ($p < 0.05$). In addition, the HOMA-PP of 14S was 3.05-fold higher than that of the 12S ($p < 0.05$). The delayed insulin peak time of the 14S was restored, but the impaired insulin sensitivity and insulin resistance was not recovered by the apple preload treatment (14A+S).

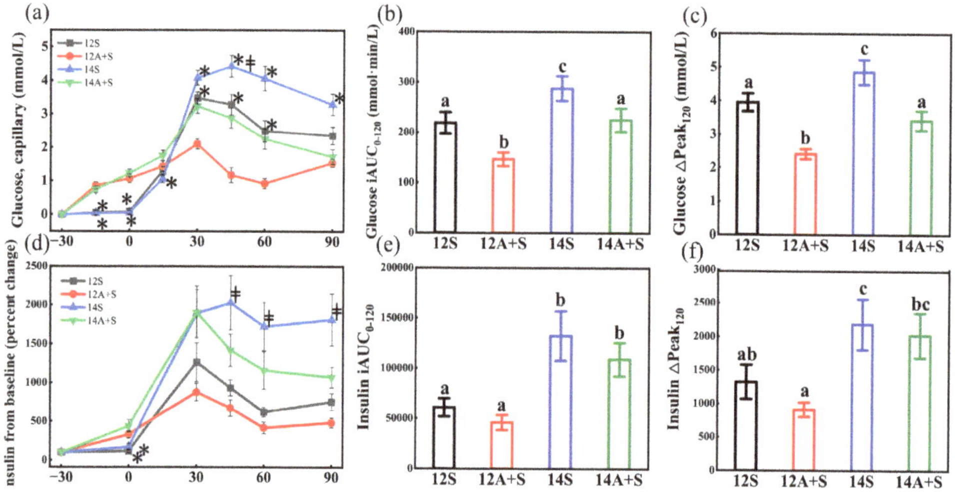

Figure 3. Postprandial capillary glucose responses and parameters (**a**–**c**), postprandial capillary insulin responses and parameters (**d**–**f**) in four lunch trial conditions. 12S, standardized lunch at 12:00; 12A+S, apple preload 30 min prior to 12S; 14S, standardized lunch at 14:00; 14A+S, apple preload 30 min prior to 14S. −30 min = the commencement of the water or apple; 0 min = the time when the standardized mixed meal was given. Data are mean ± SE (n = 20). * Apple preload treatments (A+S) different from their standardized lunch (S) counterparts ($p < 0.05$), ‡ 14S different from 12S ($p < 0.05$). Significant differences ($p < 0.05$) are represented by different letters on the bars in figure (**b**,**c**,**e**,**f**).

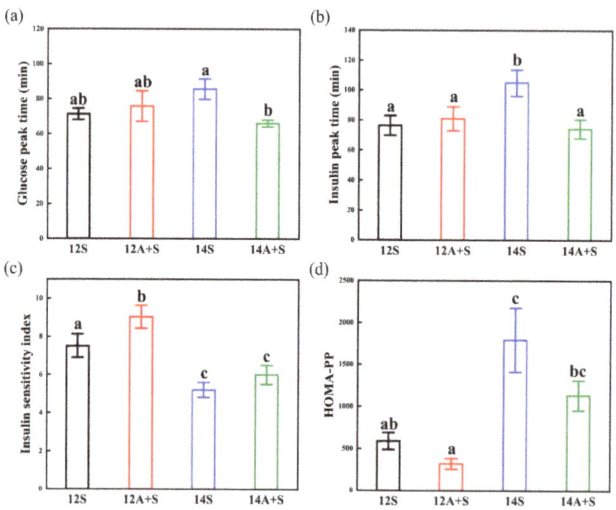

Figure 4. Glucose peak time (**a**), insulin peak time (**b**), insulin sensitivity index (**c**) and HOMA-PP (**d**) in four lunch test meals. 12S, standardized lunch at 12:00; 12A+S, apple preload 30 min prior to 12S; 14S, standardized lunch at 14:00; 14A+S, apple preload 30 min prior to 14S. Data are mean ± SE (n = 20). Significant differences ($p < 0.05$) are represented by different letters on the bars.

3.4. Postprandial Interstitial Glycemic Responses Following the Subsequent Meals

Figure 5 shows the postprandial interstitial glycemic response and parameters after a high-fat dinner, the subsequent meal of the test lunch. There were no differences in the postprandial interstitial glycemic responses and parameters between A+S and S, while the meal timing made a difference indicated by a higher glucose value from 45 to 180 min and increased ΔPeak, LAGE, SD in the subsequent meal of 12S ($p < 0.05$). The difference in the postprandial glucose $iAUC_{0\text{-}180}$ after dinner between S and its A+S counterparts did not reach the significant level, though about half of the subjects showed increased glycemic $iAUC_{0\text{-}180}$ after the dinner following the apple preload lunch meals.

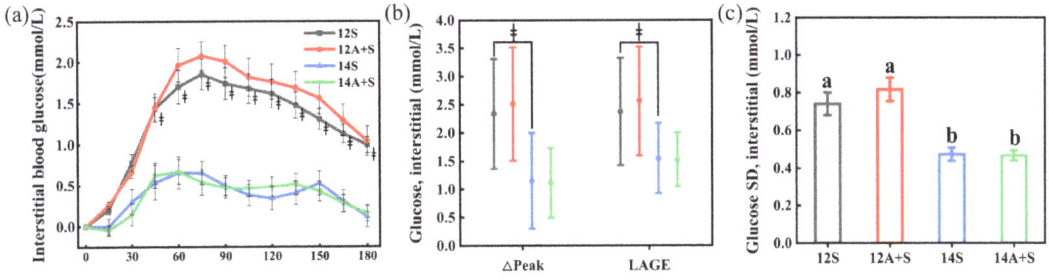

Figure 5. The postprandial interstitial glycemic response (**a**), the ΔPeak and LAGE (**b**), and the SD (**c**) at a subsequent high-fat dinner. 12S, standardized lunch at 12:00; 12A+S, apple preload 30 min prior to 12S; 14S, standardized lunch at 14:00; 14A+S, apple preload 30 min prior to 14S. 0 min = the commencement of the dinner. Data are mean ± SD in figure (**b**), mean ± SE in other figures. ‡ 14S different from 12S ($p < 0.05$). Significant differences ($p < 0.05$) are represented by different letters on the bars in figure (**c**).

3.5. 24 h Interstitial Glucose Trace

Figure 6 shows the interstitial glucose trace for 24 h in four lunch test trials, and the 24 h interstitial glucose parameters are shown in Table 4. There was no significant difference in the 24 h Mean and 24 h tAUC among the four lunch trial conditions. The apple preload intervention (A+S) elicited significantly lower 24 h Peak, LAGE, SD and $\Delta P_{L\text{-}D}$ than their non-preload counterparts did, regardless of the lunch meal timing. The treatment 12A+S attained a smallest percentage of the glucose changes greater than 5.0 mmol/L (GC > 5.0), while the 14A+S achieved a significant decrease in the percentage of GC > 2.5 and GC > 5.0 compared with the 14S. The 24 h Peak, LAGE, the percentage of GC > 5.0 and $\Delta P_{L\text{-}D}$ of the 14S were significantly higher than those of the 12S.

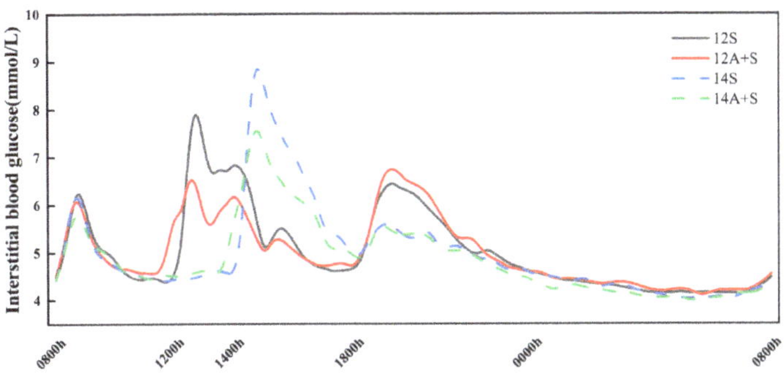

Figure 6. The interstitial glucose trace for 24 h in lunch testing trial conditions. 12S, standardized lunch at 12:00; 12A+S, apple preload 30 min prior to 12S; 14S, standardized lunch at 14:00; 14A+S, apple preload 30 min prior to 14S.

Table 4. The 24 h interstitial glucose parameters for four lunch trail conditions (mean ± SE, n = 26).

Test Meals	24 h Mean (mmol/L)	24 h tAUC (mmol·h/L)	24 h Peak (mmol/L)	24 h LAGE (mmol/L)	24 h SD	GC > 2.5 (%)	GC > 5.0 (%)	$\Delta P_{L\text{-}D}$
12S	5.0 ± 0.1	120.8 ± 1.5	8.4 ± 0.2 ^{a}	4.7 ± 0.3 ^{a}	1.0 ± 0.0 ^{ac}	6.1 ± 0.8 ^{a}	0.2 ± 0.1 ^{a}	1.4 ± 0.2 ^{a}
12A+S	5.0 ± 0.1	120.2 ± 1.8	7.4 ± 0.2 ^{b}	3.6 ± 0.2 ^{b}	0.9 ± 0.0 ^{b}	4.0 ± 1.0 ^{ab}	0.0 ± 0.0 ^{a}	0.3 ± 0.2 ^{b}
14S	5.0 ± 0.1	119.5 ± 1.7	9.4 ± 0.3 ^{c}	5.7 ± 0.3 ^{c}	1.1 ± 0.1 ^{c}	5.9 ± 0.7 ^{a}	1.3 ± 0.4 ^{b}	3.3 ± 0.2 ^{c}
14A+S	4.9 ± 0.1	116.6 ± 2.0	8.0 ± 0.2 ^{ab}	4.1 ± 0.2 ^{ab}	0.9 ± 0.0 ^{ab}	3.6 ± 0.6 ^{b}	0.2 ± 0.1 ^{a}	2.1 ± 0.3 ^{d}

^{a,b,c,d} Different superscript letters denote that mean values within a column are significantly different ($p < 0.05$). 12S, standardized lunch at 12:00; 12A+S, apple preload 30 min prior to 12S; 14S, standardized lunch at 14:00; 14A+S, apple preload 30 min prior to 14S.

4. Discussion

The present study found that the postprandial glycemic response following a late lunch (14S), which had an extra 2-h interval between the breakfast and the lunch, was characterized by significantly elevated glucose peak, delayed insulin peak time, decreased insulin sensitivity index, increased insulin resistance following the lunch, and also decreased glycemic response following the subsequent dinner and larger blood glucose fluctuation over the 24-h period when compared with an early lunch (12S). The apple preload prior to a mixed meal with balanced macronutrient composition could partly reverse the negative impact of a late lunch by lowering the glucose peak, restoring the insulin peak time, reducing the glycemic variability, and improving the 24-h glycemic variability. The apple preload treatment improved the insulin sensitivity when applied at an early lunch, but failed to recover the insulin sensitivity at a late lunch.

Several crossover trials have shown the detrimental glucose metabolism consequences caused by late dinner in healthy participants [12–14,35,36], but only one study investigated the effect of a late lunch on postprandial blood glucose levels [23]. In the present study, the early and late lunch were set at 12:00 and 14:00, respectively, with a 2-h difference between meals, while the previous study was set at 13:00 and 16:30, respectively, with a 3.5-h difference between the two treatments [23]. Though there was only a 2-h difference in breakfast–lunch intervals, we observed a significant higher peak glucose and insulin value, decreased insulin sensitivity index, and greater postprandial insulin resistance in the late lunch trial condition (14S) compared with that of the early lunch.

The early-phase insulin response was supposed to play a critical role in determining postprandial hyperglycemia [37,38]. In the present study, compared with the 12S, the 14S showed delayed insulin peak synchronized with a delayed and elevated glucose peak, suggesting that the late lunch might induce a weak early-phase insulin secretion due to an insufficient β-cell response, which in turn posed a need for a sustained insulin supply to bring the glucose concentration back towards the baseline.

The larger postprandial glucose fluctuation in the 14S could be attributed to the prolonged meal interval between breakfast and lunch resulting from late eating, which might change the energy substrate utilization as fasting carbohydrate oxidation would decrease under a late eating condition [23]. Previous studies observed that breakfast skipping increased fat oxidation and decreased carbohydrate oxidation before lunch [39,40], while the increased preprandial lipid oxidation could be an independent predictor of postprandial glycemic response [41]. A prolonged fasting period might be considered as a stress that leads to enhanced lipolysis in both healthy subjects [42] and the type 2 diabetes [43]. An elevated plasma free fatty acid level, however, was associated with increased insulin resistance and hepatic glucose production [44,45].

The circadian rhythm could be another possible factor in the postprandial hyperglycemia in the condition of a late lunch, as the endogenous rhythms promote a steady increase in glucose concentration through the 24-h day independent of the behavioral cycle [46,47]. The misalignment between the circadian cycle and the fasting/feeding and sleep/wake cycle has been shown to elevate both the glucose and insulin level due to the exaggerated postprandial glucose response in the later part of day [48]. However, in the late-night feeding trials, the intervals between the lunch and dinner were different (5 h vs. 9 h [12], 5 h vs. 8 h [13], 6 h vs. 9 h (with a snack in the afternoon) [14], 7 h vs. 10.5 h [35] and 6 h vs. 11 h [36]) in the previous studies. The interval between meals might play a larger role in effecting the postprandial glycemic response, in addition to the circadian factor. The outcome of the present study supported our research hypothesis, as a 2-h longer interval between breakfast and lunch resulted in a poorer glycemic response after lunch, while a 2-h shorter lunch–dinner interval induced a milder glycemic response after dinner.

The window for food consumption in this study was 10 h (between 8:00 and 18:00). It is reported that a mild time-restricted eating (TRE), especially those focused on the early part of the day [49], could improve blood glucose control and insulin sensitivity [14,15]. The present study suggests a possibility that part of the benefit of the TRE might be attributed to the shortened meal interval in addition to the circadian rhythm factor, since the timing of the intermediate meal influenced the 24-h glucose fluctuation, with a fixed eating schedule of first and last meal. Given that most studies on TRE only specified the eating window between the first and last meal [50,51], the effect of the interval between meals deserves investigation in future studies.

Previous studies have confirmed the hypoglycemic effect of the apple preload based on a rice meal [27,29]. However, this study is the first to show that the apple preload could ameliorate the postprandial glycemic response following a mixed meal, whether at a routine lunch time or at a late lunch setting, based on a nutrient-balanced meal composition. The apple preload effectively reversed some of the negative effects of a later lunch, achieved a smaller blood glucose peak, a non-delayed glucose peak time and insulin peak time, reduced iAUC of blood glucose, and milder glycemic excursion compared with its non-preload counterpart.

The insulin data indicated that the apple preload did not elicit increased absolute insulin secretion. On the contrary, it was found that the insulin $iAUC_{0\text{-}120}$ of 12A+S and 14A+S achieved a 24.5% and 17.6% decrease, respectively, compared with their non-preload counterparts. The insulin levels were lower in 12A+S at 60–120 min and 14A+S at 75–120 min, compared with their non-preload counterparts. What is more, the 12A+S showed improved postprandial insulin sensitivity index compared with the 12S. These results suggested that the hypoglycemic effect of the apple preload might relate to an enhanced insulin sensitivity rather than elevated insulin production.

The mechanism of the apple preload's action on the improvement of postprandial glycemic response is yet to be elucidated. As our previous study found that the co-ingestion of apple and rice did not elicit the same effect as apple preload 30 min prior to the rice did [27], it seems that the effect of the preload treatment depends on a precise timing [52]. The sugar component of apple may play an important role, as the fructose [53] and apple sugar solution [27] could attenuate the postprandial glycemic response in a preload setting. Though the fructose does not stimulate a rapid insulin response, the ingestion of sugars such as glucose and sucrose can be a stimulus for the cephalic phase insulin response (CPIR) that helps to improve glucose control [54,55]. The action of CPIR could not be ruled out though the insulin data within 10 min was not collected. It is possible that a synergy of the induction of CPIR and the catalytic effect of fructose on liver glucose metabolism [56,57] enabled the remarkable preload effect without up-regulating the insulin, as a greater fructose/glucose ratio in the moiety of sugar components was found to favor the hypoglycemic effect of fruits and dried fruits [58]. Another possibility is the modification of the pattern of entero-insular axis peptides by preload apple carbohydrate. It is reported that slow-digest carbohydrates could increase glucagon-like peptide-1(GLP-1) while decreasing gastric inhibitory polypeptide (GIP) [59]. The carbohydrate in apple is of low glycemic type with glycemic index of 36 [60], while apple polyphenols were shown to be associated with the inhibition capacity of α-glucosidase [61].

To the best of our knowledge, this is the first study to explore the effect of between-meal interval on postprandial glycemic effect of two meal settings with two macronutrient composition meals. It is also the first report with respect to the effect of a preload on the hyperglycemia caused by delayed meals. The glucose and insulin responses during the postprandial period based on different lunch timing were investigated, as well as the glycemic response following the next meal. The glycemic effect of the test meals was double checked with CGM and capillary blood tests. In this study, the mealtime on Day 0 was stipulated and the breakfast on Day 1 was provided for the participants in order to minimize any variations in baseline metabolite concentrations prior to the lunch trial and diminish possible carryover effects derived from the participant's previous meal. Given that the irregular mealtime is a common problem in working people of contemporary society, the study on the glycemic response based on meal intervals is relevant.

However, the limitations of our study must be considered. Firstly, this is a one-day experiment with healthy participants, so the effect of delayed lunch and prolonged inter-meal fasting is still to be confirmed by long-term intervention studies and in the prediabetic and the diabetic groups. Secondly, the CPIR after meal was not determined. Thirdly, the possible change in the level of hormones such as cortisol and GLP-1, as well as the possible change in digestion process, were not investigated. Finally, only the apple preload was tested in the trial. The possible preload effect of other food could not be extrapolated from the present study.

5. Conclusions

In conclusion, the present study found that an extra 2-h inter-meal fasting before and after lunch could result in elevated glycemic response in both macronutrient-balanced meals and high-fat meals in healthy participants. The negative impact of a late lunch could be partly reversed by an apple load prior to lunch, without a trade-off of insulin secretion. The result of the present study suggests that the between-meal interval may be a potential key determinant in glycemic stability in addition to the diurnal rhythm. The possible metabolic consequences of late lunch times and the mechanism need to be investigated, especially in people of impaired glucose tolerance and the diabetes patients.

Author Contributions: Conceptualization, Z.F.; methodology, Z.F., X.L. (Xuejiao Lu) and A.L.; formal analysis, X.L. (Xuejiao Lu); investigation, X.L. (Xuejiao Lu), R.L., A.L., X.L. (Xinling Lou) and J.H.; writing—original draft preparation, X.L. (Xuejiao Lu); writing—review and editing, Z.F. All authors have read and agreed to the published version of the manuscript.

Funding: This research received no external funding.

Institutional Review Board Statement: The study protocol was conducted according to the principles laid down in the Helsinki Declaration and approved by the Ethics Committee of China Agricultural University (ethics number CAUHR-20220202).

Informed Consent Statement: Informed consent was obtained from all subjects involved in the study. Written informed consent has been obtained from the patients to publish this paper.

Data Availability Statement: The data presented in this study are available on request from the corresponding author.

Acknowledgments: We sincerely thank all the volunteers participated in the blood glucose tests for their time and corporation.

Conflicts of Interest: The authors declare no conflict of interest.

References

1. Ford, N.D.; Patel, S.A.; Narayan, M.V. Obesity in Low- and Middle-Income Countries: Burden, Drivers, and Emerging Challenges. *Annu. Rev. Public Health* **2017**, *38*, 145–164. [CrossRef] [PubMed]
2. Sun, H.; Saeedi, P.; Karuranga, S.; Pinkepank, M.; Ogurtsova, K.; Duncan, B.B.; Stein, C.; Basit, A.; Chan, J.C.N.; Mbanya, J.C.; et al. IDF Diabetes Atlas: Global, regional and country-level diabetes prevalence estimates for 2021 and projections for 2045. *Diabetes Res. Clin. Pract.* **2022**, *183*, 109119. [CrossRef]
3. Papakonstantinou, E.; Oikonomou, C.; Nychas, G.; Dimitriadis, G.D. Effects of Diet, Lifestyle, Chrononutrition and Alternative Dietary Interventions on Postprandial Glycemia and Insulin Resistance. *Nutrients* **2022**, *14*, 823. [CrossRef]

4. Davis, R.; Rogers, M.; Coates, A.M.; Leung, G.K.W.; Bonham, M.P. The Impact of Meal Timing on Risk of Weight Gain and Development of Obesity: A Review of the Current Evidence and Opportunities for Dietary Intervention. *Curr. Diabetes Rep.* **2022**, *22*, 147–155. [CrossRef]
5. Dashti, H.S.; Gomez-Abellan, P.; Qian, J.; Esteban, A.; Morales, E.; Scheer, F.; Garaulet, M. Late eating is associated with cardiometabolic risk traits, obesogenic behaviors, and impaired weight loss. *Am. J. Clin. Nutr.* **2020**, *113*, 154–161. [CrossRef]
6. Martínez-Lozano, N.; Tvarijonaviciute, A.; Ríos, R.; Barón, I.; Scheer, F.A.J.L.; Garaulet, M. Late Eating Is Associated with Obesity, Inflammatory Markers and Circadian-Related Disturbances in School-Aged Children. *Nutrients* **2020**, *12*, 2881. [CrossRef] [PubMed]
7. McHill, A.W.; Phillips, A.J.; Czeisler, C.A.; Keating, L.; Yee, K.; Barger, L.K.; Garaulet, M.; Scheer, F.A.; Klerman, E.B. Later circadian timing of food intake is associated with increased body fat. *Am. J. Clin. Nutr.* **2017**, *106*, 1213–1219. [CrossRef] [PubMed]
8. Thomas, E.A.; Zaman, A.; Cornier, M.; Catenacci, V.A.; Tussey, E.J.; Grau, L.; Arbet, J.; Broussard, J.L.; Rynders, C.A. Later Meal and Sleep Timing Predicts Higher Percent Body Fat. *Nutrients* **2021**, *13*, 73. [CrossRef] [PubMed]
9. Saad, A.; Dalla Man, C.; Nandy, D.K.; Levine, J.A.; Bharucha, A.E.; Rizza, R.A.; Basu, R.; Carter, R.E.; Cobelli, C.; Kudva, Y.C.; et al. Diurnal Pattern to Insulin Secretion and Insulin Action in Healthy Individuals. *Diabetes* **2012**, *61*, 2691–2700. [CrossRef]
10. Leung, G.K.W.; Huggins, C.E.; Bonham, M.P. Effect of meal timing on postprandial glucose responses to a low glycemic index meal: A crossover trial in healthy volunteers. *Clin. Nutr.* **2019**, *38*, 465–471. [CrossRef]
11. Van Cauter, E.; Shapiro, E.T.; Tillil, H.; Polonsky, K.S. Circadian modulation of glucose and insulin responses to meals: Relationship to cortisol rhythm. *Am. J. Physiol.* **1992**, *262*, E467–E475. [CrossRef] [PubMed]
12. Gu, C.; Brereton, N.; Schweitzer, A.; Cotter, M.; Duan, D.; Børsheim, E.; Wolfe, R.R.; Pham, L.V.; Polotsky, V.Y.; Jun, J.C. Metabolic Effects of Late Dinner in Healthy Volunteers—A Randomized Crossover Clinical Trial. *J. Clin. Endocrinol. Metab.* **2020**, *105*, 2789–2802. [CrossRef]
13. Kajiyama, S.; Imai, S.; Hashimoto, Y.; Yamane, C.; Miyawaki, T.; Matsumoto, S.; Ozasa, N.; Tanaka, M.; Kajiyama, S.; Fukui, M. Divided consumption of late-night-dinner improves glucose excursions in young healthy women: A randomized cross-over clinical trial. *Diabetes Res. Clin. Pract.* **2018**, *136*, 78–84. [CrossRef]
14. Nakamura, K.; Tajiri, E.; Hatamoto, Y.; Ando, T.; Shimoda, S.; Yoshimura, E. Eating Dinner Early Improves 24-h Blood Glucose Levels and Boosts Lipid Metabolism after Breakfast the Next Day: A Randomized Cross-Over Trial. *Nutrients* **2021**, *13*, 2424. [CrossRef] [PubMed]
15. Che, T.; Yan, C.; Tian, D.; Zhang, X.; Liu, X.; Wu, Z. Time-restricted feeding improves blood glucose and insulin sensitivity in overweight patients with type 2 diabetes: A randomised controlled trial. *Nutr. Metab.* **2021**, *18*, 1–10. [CrossRef]
16. Lopez-Minguez, J.; Gómez-Abellán, P.; Garaulet, M. Timing of Breakfast, Lunch, and Dinner. Effects on Obesity and Metabolic Risk. *Nutrients* **2019**, *11*, 2624. [CrossRef]
17. Service, F.J.; Hall, L.D.; Westland, R.E.; O'Brien, P.C.; Go, V.L.; Haymond, M.W.; Rizza, R.A. Effects of size, time of day and sequence of meal ingestion on carbohydrate tolerance in normal subjects. *Diabetologia* **1983**, *25*, 316–321. [CrossRef] [PubMed]
18. Zhao, W.; Liu, Z.; Fan, Z.; Wu, Y.; Lou, X.; Liu, A.; Lu, X. Diurnal differences in glycemic responses, insulin responses and cognition after rice-based meals. *Asia Pac. J. Clin. Nutr.* **2022**, *31*, 57–65. [PubMed]
19. Garaulet, M.; Gomez-Abellan, P.; Alburquerque-Bejar, J.J.; Lee, Y.C.; Ordovas, J.M.; Scheer, F.A. Timing of food intake predicts weight loss effectiveness. *Int. J. Obes. Lond* **2013**, *37*, 604–611. [CrossRef]
20. Ruiz-Lozano, T.; Vidal, J.; de Hollanda, A.; Scheer, F.A.J.L.; Garaulet, M.; Izquierdo-Pulido, M. Timing of food intake is associated with weight loss evolution in severe obese patients after bariatric surgery. *Clin. Nutr.* **2016**, *35*, 1308–1314. [CrossRef]
21. Garaulet, M.; Vera, B.; Bonnet-Rubio, G.; Gómez-Abellán, P.; Lee, Y.; Ordovás, J.M. Lunch eating predicts weight-loss effectiveness in carriers of the common allele at PERILIPIN1: The ONTIME (Obesity, Nutrigenetics, Timing, Mediterranean) study. *Am. J. Clin. Nutr.* **2016**, *104*, 1160–1166. [CrossRef] [PubMed]
22. Kulshreshtha, B.; Sharma, N.; Pant, S.; Sharma, L.; Pahuja, B.; Singh, P. PCOS patients differ in meal timings rather than total caloric or macronutrient intake in comparison to weight matched controls. *Eur. J. Obstet. Gyn. R. B.* **2022**, *270*, 11–16. [CrossRef]
23. Bandin, C.; Scheer, F.A.; Luque, A.J.; Avila-Gandia, V.; Zamora, S.; Madrid, J.A.; Gomez-Abellan, P.; Garaulet, M. Meal timing affects glucose tolerance, substrate oxidation and circadian-related variables: A randomized, crossover trial. *Int. J. Obes. Lond* **2015**, *39*, 828–833. [CrossRef] [PubMed]
24. Collado, M.C.; Engen, P.A.; Bandín, C.; Cabrera Rubio, R.; Voigt, R.M.; Green, S.J.; Naqib, A.; Keshavarzian, A.; Scheer, F.A.J.L.; Garaulet, M. Timing of food intake impacts daily rhythms of human salivary microbiota: A randomized, crossover study. *FASEB J.* **2018**, *32*, 2060–2072. [CrossRef] [PubMed]
25. Nesti, L.; Mengozzi, A.; Tricò, D. Impact of Nutrient Type and Sequence on Glucose Tolerance: Physiological Insights and Therapeutic Implications. *Front. Endocrinol.* **2019**, *10*, 144. [CrossRef]
26. Wee, M.S.M.; Henry, C.J. Reducing the glycemic impact of carbohydrates on foods and meals: Strategies for the food industry and consumers with special focus on Asia. *Compr. Rev. Food Sci. F.* **2020**, *19*, 670–702. [CrossRef] [PubMed]
27. Lu, J.; Zhao, W.; Wang, L.; Fan, Z.; Zhu, R.; Wu, Y.; Zhou, Y. Apple Preload Halved the Postprandial Glycaemic Response of Rice Meal in Healthy Subjects. *Nutrients* **2019**, *11*, 2912. [CrossRef] [PubMed]
28. Zhao, W.; Wang, L.; Fan, Z.; Lu, J.; Zhu, R.; Wu, Y.; Lu, X. Co-ingested vinegar-soaked or preloaded dried apple mitigated acute postprandial glycemia of rice meal in healthy subjects under equicarbohydrate conditions. *Nutr. Res.* **2020**, *83*, 108–118. [CrossRef]

29. Lu, X.; Lu, J.; Fan, Z.; Liu, A.; Zhao, W.; Wu, Y.; Zhu, R. Both Isocarbohydrate and Hypercarbohydrate Fruit Preloads Curbed Postprandial Glycemic Excursion in Healthy Subjects. *Nutrients* **2021**, *13*, 2470. [CrossRef]
30. Bei-Fan, Z.; Cooperative, M.A.G.W. Predictive values of body mass index and waist circumference for risk factors of certain related diseases in Chinese adults: Study on optimal cut-off points of body mass index and waist circumference in Chinese adults. *Asia Pac. J. Clin. Nutr.* **2002**, *11*, S685–S693. [CrossRef]
31. Matsuda, M.; DeFronzo, R.A. Insulin sensitivity indices obtained from oral glucose tolerance testing—Comparison with the euglycemic insulin clamp. *Diabetes Care.* **1999**, *22*, 1462–1470. [CrossRef] [PubMed]
32. Brynes, A.E.; Edwards, C.M.; Ghatei, M.A.; Dornhorst, A.; Morgan, L.M.; Bloom, S.R.; Frost, G.S. A randomised four-intervention crossover study investigating the effect of carbohydrates on daytime profiles of insulin, glucose, non-esterified fatty acids and triacylglycerols in middle-aged men. *Brit. J. Nutr.* **2003**, *89*, 207–218. [CrossRef] [PubMed]
33. Caumo, A.; Bergman, R.N.; Cobelli, C. Insulin sensitivity from meal tolerance tests in normal subjects: A minimal model index. *J. Clin. Endocrinol. Metab.* **2000**, *85*, 4396–4402. [CrossRef] [PubMed]
34. Aloulou, I.; Brun, J.; Mercier, J. Evaluation of insulin sensitivity and glucose effectiveness during a standardized breakfast test: Comparison with the minimal model analysis of an intravenous glucose tolerance test. *Metab. Clin. Exp.* **2006**, *55*, 676–690. [CrossRef] [PubMed]
35. Sato, M.; Nakamura, K.; Ogata, H.; Miyashita, A.; Nagasaka, S.; Omi, N.; Yamaguchi, S.; Hibi, M.; Umeda, T.; Nakaji, S.; et al. Acute effect of late evening meal on diurnal variation of blood glucose and energy metabolism. *Obes. Res. Clin. Pract.* **2011**, *5*, e220–e228. [CrossRef]
36. Tsuchida, Y.; Hata, S.; Sone, Y. Effects of a late supper on digestion and the absorption of dietary carbohydrates in the following morning. *J. Physiol. Anthropol.* **2013**, *32*, 9. [CrossRef]
37. Del Prato, S. Loss of early insulin secretion leads to postprandial hyperglycaemia. *Diabetologia* **2003**, *46*, M2–M8. [CrossRef]
38. Tsujino, D.; Nishimura, R.; Taki, K.; Miyashita, Y.; Morimoto, A.; Tajima, N. Daily glucose profiles in Japanese people with normal glucose tolerance as assessed by continuous glucose monitoring. *Diabetes Technol. Ther.* **2009**, *11*, 457–460. [CrossRef]
39. Nas, A.; Mirza, N.; Hägele, F.; Kahlhöfer, J.; Keller, J.; Rising, R.; Kufer, T.A.; Bosy-Westphal, A. Impact of breakfast skipping compared with dinner skipping on regulation of energy balance and metabolic risk. *Am. J. Clin. Nutr.* **2017**, n151332. [CrossRef]
40. Kobayashi, F.; Ogata, H.; Omi, N.; Nagasaka, S.; Yamaguchi, S.; Hibi, M.; Tokuyama, K. Effect of breakfast skipping on diurnal variation of energy metabolism and blood glucose. *Obes. Res. Clin. Pract.* **2014**, *8*, e249–e257. [CrossRef]
41. Ando, T.; Nakae, S.; Usui, C.; Yoshimura, E.; Nishi, N.; Takimoto, H.; Tanaka, S. Effect of diurnal variations in the carbohydrate and fat composition of meals on postprandial glycemic response in healthy adults: A novel insight for the second-meal phenomenon. *Am. J. Clin. Nutr.* **2018**, *108*, 332–342. [CrossRef] [PubMed]
42. Jensen, M.D.; Haymond, M.W.; Gerich, J.E.; Cryer, P.E.; Miles, J.M. Lipolysis during fasting. Decreased suppression by insulin and increased stimulation by epinephrine. *J. Clin. Invest.* **1987**, *79*, 207–213. [CrossRef] [PubMed]
43. Imai, S.; Kajiyama, S.; Hashimoto, Y.; Yamane, C.; Miyawaki, T.; Ozasa, N.; Tanaka, M.; Fukui, M. Divided consumption of late-night-dinner improves glycemic excursions in patients with type 2 diabetes: A randomized cross-over clinical trial. *Diabetes Res. Clin. Pract.* **2017**, *129*, 206–212. [CrossRef] [PubMed]
44. Jiao, P.; Ma, J.; Feng, B.; Zhang, H.; Alan-Diehl, J.; Eugene-Chin, Y.; Yan, W.; Xu, H. FFA-Induced Adipocyte Inflammation and Insulin Resistance: Involvement of ER Stress and IKKβ Pathways. *Obesity* **2011**, *19*, 483–491. [CrossRef]
45. Boden, G. Role of fatty acids in the pathogenesis of insulin resistance and NIDDM. *Diabetes* **1997**, *46*, 3–10. [CrossRef]
46. Morgan, L.; Arendt, J.; Owens, D.; Folkard, S.; Hampton, S.; Deacon, S.; English, J.; Ribeiro, D.; Taylor, K. Effects of the endogenous clock and sleep time on melatonin, insulin, glucose and lipid metabolism. *J. Endocrinol.* **1998**, *157*, 443–451. [CrossRef]
47. Shea, S.A.; Hilton, M.F.; Orlova, C.; Ayers, R.T.; Mantzoros, C.S. Independent Circadian and Sleep/Wake Regulation of Adipokines and Glucose in Humans. *J. Clin. Endocrinol. Metab.* **2005**, *90*, 2537–2544. [CrossRef] [PubMed]
48. Scheer, F.A.; Hilton, M.F.; Mantzoros, C.S.; Shea, S.A. Adverse metabolic and cardiovascular consequences of circadian misalignment. *Proc. Natl. Acad. Sci. USA* **2009**, *106*, 4453–4458. [CrossRef]
49. Xie, Z.; Sun, Y.; Ye, Y.; Hu, D.; Zhang, H.; He, Z.; Zhao, H.; Yang, H.; Mao, Y. Randomized controlled trial for time-restricted eating in healthy volunteers without obesity. *Nat. Commun.* **2022**, *13*, 1003. [CrossRef]
50. Adafer, R.; Messaadi, W.; Meddahi, M.; Patey, A.; Haderbache, A.; Bayen, S.; Messaadi, N. Food Timing, Circadian Rhythm and Chrononutrition: A Systematic Review of Time-Restricted Eating's Effects on Human Health. *Nutrients* **2020**, *12*, 3770. [CrossRef] [PubMed]
51. Pellegrini, M.; Cioffi, I.; Evangelista, A.; Ponzo, V.; Goitre, I.; Ciccone, G.; Ghigo, E.; Bo, S. Effects of time-restricted feeding on body weight and metabolism. A systematic review and meta-analysis. *Rev. Endocr. Metab. Disord.* **2020**, *21*, 17–33. [CrossRef]
52. Almiron-Roig, E.; Palla, L.; Guest, K.; Ricchiuti, C.; Vint, N.; Jebb, S.A.; Drewnowski, A. Factors that determine energy compensation: A systematic review of preload studies. *Nutr. Rev.* **2013**, *71*, 458–473. [CrossRef]
53. Heacock, P.M.; Hertzler, S.R.; Wolf, B.W. Fructose prefeeding reduces the glycemic response to a high-glycemic index, starchy food in humans. *J. Nutr.* **2002**, *132*, 2601–2604. [CrossRef]
54. Pullicin, A.J.; Glendinning, J.I.; Lim, J. Cephalic phase insulin release: A review of its mechanistic basis and variability in humans. *Physiol. Behav.* **2021**, *239*, 113514. [CrossRef]
55. Dhillon, J.; Lee, J.Y.; Mattes, R.D. The cephalic phase insulin response to nutritive and low-calorie sweeteners in solid and beverage form. *Physiol. Behav.* **2017**, *181*, 100–109. [CrossRef] [PubMed]

56. Petersen, K.F.; Laurent, D.; Yu, C.; Cline, G.W.; Shulman, G.I. Stimulating effects of low-dose fructose on insulin-stimulated hepatic glycogen synthesis in humans. *Diabetes* **2001**, *50*, 1263–1268. [CrossRef] [PubMed]
57. Geidl-Flueck, B.; Gerber, P. Insights into the Hexose Liver Metabolism—Glucose versus Fructose. *Nutrients* **2017**, *9*, 1026. [CrossRef] [PubMed]
58. Zhu, R.; Fan, Z.; Dong, Y.; Liu, M.; Wang, L.; Pan, H. Postprandial Glycaemic Responses of Dried Fruit-Containing Meals in Healthy Adults: Results from a Randomised Trial. *Nutrients* **2018**, *10*, 694. [CrossRef]
59. Angarita Dávila, L.; Bermúdez, V.; Aparicio, D.; Céspedes, V.; Escobar, M.; Durán-Agüero, S.; Cisternas, S.; de Assis Costa, J.; Rojas-Gómez, D.; Reyna, N.; et al. Effect of Oral Nutritional Supplements with Sucromalt and Isomaltulose versus Standard Formula on Glycaemic Index, Entero-Insular Axis Peptides and Subjective Appetite in Patients with Type 2 Diabetes: A Randomised Cross-Over Study. *Nutrients* **2019**, *11*, 1477. [CrossRef]
60. Atkinson, F.S.; Foster-Powell, K.; Brand-Miller, J.C. International Tables of Glycemic Index and Glycemic Load Values: 2008. *Diabetes Care.* **2008**, *31*, 2281–2283. [CrossRef]
61. De Oliveira Raphaelli, C.; Dos Santos Pereira, E.; Camargo, T.M.; Vinholes, J.; Rombaldi, C.V.; Vizzotto, M.; Nora, L. Apple Phenolic Extracts Strongly Inhibit α-Glucosidase Activity. *Plant Food. Hum. Nutr.* **2019**, *74*, 430–435. [CrossRef] [PubMed]

nutrients

MDPI

Review

Obesity, Gut Microbiota, and Metabolome: From Pathophysiology to Nutritional Interventions

Zivana Puljiz [1,†], Marko Kumric [2,†], Josip Vrdoljak [2], Dinko Martinovic [2], Tina Ticinovic Kurir [2,3], Marin Ozren Krnic [2], Hrvoje Urlic [2], Zeljko Puljiz [4,5], Jurica Zucko [1], Petra Dumanic [6], Ivana Mikolasevic [7] and Josko Bozic [2,*]

1 Laboratory for Bioinformatics, Faculty of Food Technology and Biotechnology, University of Zagreb, 10000 Zagreb, Croatia; jzucko@pbf.hr (J.Z.)
2 Department of Pathophysiology, University of Split School of Medicine, 21000 Split, Croatia; marko.kumric@mefst.hr (M.K.); dinko.martinovic@mefst.hr (D.M.); tticinov@mefst.hr (T.T.K.)
3 Department of Endocrinology, Diabetes and Metabolic Diseases, University Hospital of Split, 21000 Split, Croatia
4 Department of Internal Medicine, University of Split School of Medicine, 21000 Split, Croatia
5 Department of Gastroenterology and Hepatology, University Hospital of Split, 21000 Split, Croatia
6 Medical Laboratory Diagnostic Division, University Hospital of Split, 21000 Split, Croatia
7 Department of Gastroenterology and Hepatology, University Hospital Centre Rijeka, School of Medicine, University of Rijeka, 51000 Rijeka, Croatia
* Correspondence: josko.bozic@mefst.hr
† These authors contributed equally to this work.

Abstract: Obesity is a disorder identified by an inappropriate increase in weight in relation to height and is considered by many international health institutions to be a major pandemic of the 21st century. The gut microbial ecosystem impacts obesity in multiple ways that yield downstream metabolic consequences, such as affecting systemic inflammation, immune response, and energy harvest, but also the gut–host interface. Metabolomics, a systematized study of low-molecular-weight molecules that take part in metabolic pathways, represents a serviceable method for elucidation of the crosstalk between hosts' metabolism and gut microbiota. In the present review, we confer about clinical and preclinical studies exploring the association of obesity and related metabolic disorders with various gut microbiome profiles, and the effects of several dietary interventions on gut microbiome composition and the metabolome. It is well established that various nutritional interventions may serve as an efficient therapeutic approach to support weight loss in obese individuals, yet no agreement exists in regard to the most effective dietary protocol, both in the short and long term. However, metabolite profiling and the gut microbiota composition might represent an opportunity to methodically establish predictors for obesity control that are relatively simple to measure in comparison to traditional approaches, and it may also present a tool to determine the optimal nutritional intervention to ameliorate obesity in an individual. Nevertheless, a lack of adequately powered randomized trials impedes the application of observations to clinical practice.

Keywords: obesity; gut microbiota; metabolome; Mediterranean diet; Roux-en-Y gastric bypass; ketogenic diet

1. Introduction

Obesity is a disorder identified by an inappropriate increase in weight in relation to height, mainly owing to fat tissue accumulation, and is widely considered a pandemic of the 21st century. The available data indicate that the etiological triad of genome, diet, and gut microbiota plays a crucial role in the pathophysiology of obesity. The gut microbial ecosystem impacts obesity in multiple ways that yield downstream metabolic consequences, such as affecting systemic inflammation, immune response, and energy harvest, but also gut–host interface (gastrointestinal mucus layer, epithelial permeability, and inflammation of the digestive system) [1]. The gut microbiota constitution can govern the efficacy of energy harvest from foodstuffs, and changes in dietary style have been related to changes in the composition of gut microbial populations. Our ability to examine the microbiota constitution was significantly improved with the advent of a metagenomic approach, which has already granted the production of the human gut microbiome gene catalogue and enabled stratification of individuals according to their respective gut genomic profile into different enterotypes [2–4]. Furthermore, metabolomics, a systematized study of low-molecular-weight molecules that take part in metabolic pathways, represents a serviceable method for elucidation of the crosstalk between hosts' metabolism and gut microbiota.

Metabolomics is a promising tool to upgrade current clinical assessments based on single metabolites by identifying metabolic signatures (biomarkers) that represent global biochemical changes in disease and predict responses to side effects of drug treatment (pharmacometabolomics) [5]. The gut microbiota composition can be affected by various external factors, with dietary habits being recognized as the most important one. In many pathological states, such as celiac disease, irritable bowel syndrome (IBS), and also some neurological disorders (epilepsy), specific dietary regimens such as a Mediterranean diet, ketogenic diet, and gluten-free diet are recognized as therapeutic. It seems that all of the above noted dietary patterns can affect the gut microbiota composition, especially when implemented over longer time periods [6]. Accordingly, it is worth mentioning that nutritional intervention may serve as an efficient therapeutic approach to support weight loss in obese individuals, yet no agreement exists in regard to the most effective dietary protocol, both in the short and long term [6].

In the present review, we provide an overview of the current evidence concerning the association between gut microbiota and obesity. As identification of the gut microbiota–metabolites is warranted in order to clarify the interactions between the host and the intestinal flora, we investigated changes in the metabolome, which is a necessary step to appropriately model metabolic interactions inside the intestinal ecosystem. Hence, in this review, we primarily discussed clinical and preclinical studies that explored the association of obesity and related metabolic disorders with various gut microbiome profiles, and the effects of weight loss interventions with different dietary approaches on gut microbiome and the metabolome.

2. Gut Microbiota and Obesity

Gut microbiota represents one of the most complex ecosystems in nature, harboring large bacterial populations in the intestine and colon, most of which are anaerobic bacteria (95% of total organisms) [7]. Gut microbiota in humans is mainly composed of five distinct phyla: *Actinobacteria*, *Bacteroidetes*, *Verrucomicrobia*, *Firmicutes*, and *Proteobacteria*, with *Bacteroidetes* and *Firmicutes* representing as much as 90% of all gut bacteria [8–11]. Nevertheless, a large between-subjects diversity in gut microbiome is present [9]. Three bacterial clusters have been described so far: *Bacteroides* (enterotype 1), *Prevotella* (enterotype 2), and *Ruminococcus* (enterotype 3) [8]. The *Prevotella* enterotype is related to a high-carbohydrate diet, whereas the *Bacteroides* enterotype has been mainly correlated with a diet rich in animal fat and protein [9]. This simplified classification, although not generally accepted [12], may contribute to a more comprehensive understanding of the complex relations operating between gut microbiota and obesity.

Dysbiosis, a disturbance in microbiota composition, has been linked to three distinct mechanisms that can all take place at the same time: (1) the loss of "beneficial" bacteria, (2) an excessive growth of potentially hazardous bacteria, and (3) a reduction in the microbial diversity [13,14]. A growing amount of evidence suggests that changed microbiota can have an impact on host physiology through a number of different channels, including improved energy harvest, changes in the immune system, metabolic signaling, and inflammatory pathways [15–17]. For instance, evidence suggests that microbiota may be a significant contributor that favors weight gain, fat storage, and insulin resistance [18].

The gut bacteria are involved in energy homeostasis via formation of short chain fatty acid (SCFAs) and extraction of energy through fermentation processes [19]. The gut microbiota may also enable increased intestinal absorption by increasing villous vascularization and increased triglyceride storage via modulation of fasting-induced adipose factor, a lipoprotein lipase (LPL) activity inhibitor [20]. Interestingly, a study of a mice model demonstrated that the obese phenotype may be transmissible by transplanting the gut microbiota [21]. Subsequently, a focus was placed on the identification of bacterial strains that are implicated in the pathophysiology of obesity.

In general, obese individuals are characterized by decreased bacterial diversity [22–24], as well as gene richness [25,26]. In fact, recent studies in European populations have demonstrated that people with less complex microbiomes have greater obesity in general, higher levels of inflammatory markers, more pronounced insulin resistance, and dyslipidemia [27]. Likewise, obese and overweight individuals with low diversity microbiota demonstrate an increase in microbial richness upon the introduction of an energy-restricted diet [28]. Unlike the original reports suggested that obesity is associated with a higher ratio of *Firmicutes*-to-*Bacteroidetes* [29], more recent studies failed to demonstrate such an association, suggesting that the ratio is not a significant factor in human obesity [30–35]. In fact, the mere fact that most studies showed decreased bacterial diversity in obesity is also suggestive that separate changes in gut microbiome composition at family, genus, or species level are more relevant for pathophysiology of obesity than the aforementioned ratio. Nonetheless, many researchers have studied ways in which diet can modulate the *Firmicutes*-to-*Bacteroidetes* ratio. In accordance with the previously noted findings, early studies demonstrated that the *Firmicutes*-to-*Bacteroidetes* ratio is increased in obese individuals but decreases upon weight loss after bariatric surgery and/or calorie-restriction diets, whereas other studies mostly failed to demonstrate the same [36,37]. It is however important to note that all these studies were largely underpowered to demonstrate any difference and, thus, perhaps the strongest evidence implying a lack of link between the *Firmicutes*-to-*Bacteroidetes* ratio and obesity came from a meta-analysis that showed no difference in the abundance of *Firmicutes* and *Bacteroidetes*, nor the *Firmicutes*-to-*Bacteroidetes* ratio [38]. Overall, although the mere ratio of these two phyla is obviously not a representative indicator of obesity, the studies that covered this topic revealed some important findings with regard to the microbiota–obesity interconnection. Specifically, it challenged the concept of a unique taxonomic signature related to obesity, thus steering future research into identifying taxonomic markers for stratification of patients into subgroups.

Apart from *Bacteroidetes* and *Firmicutes*, other studies have associated obesity with distinct bacteria including family *Christensenellaceae* and the genera *Akkermansia*, *Bifidobacteria*, *Methanobacteriales*, and *Lactobacillus* [39]. Specifically, the *Christensenellaceae* family was recently associated with weight loss and several gene expression pathways in subcutaneous adipose tissue, such as protein–amino acid N-glycation, whereas its relative quantity was inversely related to the host BMI [39,40]. Moreover, supplementation with *A. muciniphila* was shown to improve metabolic parameters in obese patients [41]. Furthermore, unlike the abundance of *Lactobacillus reuteri* and *L. gasseri* that positively correlated with obesity, *L. paracasei* negatively correlated with obesity, indicating that obesity related bacteria are species specific, and that bacteria in the same genus may yield opposing effects [42]. The role of bacteria in obesity seems to be strain specific, i.e., both beneficial and harmful bacteria can exist within the same taxon. Hence, it is rather challenging to categorize obesity-related bacterial communities according to their taxonomic relationships.

The relationship between obesity and gut microbiota is a two-way street. Namely, gut microbiota may promote obesity through the direct interaction between the microbiota and cells in the GI tract or indirect interaction between produced metabolites and remote organs. The underlying mechanisms may include the effects of gut microbiota on fat storage, appetite, absorption of energy, circadian rhythm, and chronic inflammation, all of which can promote obesity [43]. Finally, results from interventions that affect gut microbiota (fecal microbiota transplant, use of pre-, pro-, and synbiotics) have pointed out that such changes might herald favorable outcomes in obese individuals [44].

A very important, yet commonly neglected, aspect concerning the research of gut microbiome are technical challenges in defining the gut microbiome composition. This issue is well illustrated in a meta-analysis by Walters et al. [45]. Namely, the results of their meta-analysis, in which only studies involving high-throughput sequencing of the 16S rRNA gene were analyzed, showed that the gut microbiota composition is not clustered by subject's body mass index (BMI) but rather by study, suggesting that the per-study effect exceeds the biological effect. Moreover, additional bias can arise as a consequence of various differences in data analysis and sample processing, such as the method of DNA extraction, the method of sequencing, the choice of primers, and the bioinformatic analysis (taxonomy database and assignment algorithm used) [46–48].

3. Metabolites and Obesity: The Role of Gut Microbiota

Although obesity is an established metabolic disorder, the pathophysiological mechanisms linking weight gain with metabolic profiles are still elusive. Gut microbiome dysbiosis seen in obese individuals may have an impact on the metabolism and excretion of microbiota byproducts, and may therefore affect hosts' physiology. Recent data imply that the intestinal microbiota of obese individuals is characterized by alterations in any circulating metabolites and is associated with fasting levels of a number of metabolites, such as lipids and lipid-like metabolites, amino acids and their byproducts, bile acids derivatives, and metabolites derived from the degradation of polyphenols, choline, carnitine, and purines [49,50]. Metabolite profiling might represent an opportunity to methodically establish predictors for obesity control that are relatively simple to measure in comparison to traditional approaches. In addition, these predictors may also constitute a tool to evaluate responses/effects to different intervention approaches.

A meta-analysis by Moore et al. recognized 37 metabolites that are associated with BMI, identifying 18 of them, including butyrylcarnitine, a marker of whole-body fatty acid oxidation, and histidine for the first time [51]. The meta-analysis was performed on studies that included both American and Chinese populations and it included 947 patients. Perhaps the biggest drawback of this meta-analysis was the fact that it included cross-sectional studies exclusively. On the other hand, Zhao et al. performed a 5-year follow-up cohort study that included 300 Mexican American healthy women [52]. At baseline, the authors observed 7 metabolites that are associated with BMI in both cohorts. The metabolites were as follows: asparagine, glycine, glutamic acid, kynurenic acid, methyl succinate, urate, and serine. At the end of the follow-up period, the investigators identified 6 metabolites (acetylcholine, acetyl glycine, hippuric acid, leucine, urate, and xanthine) whose baseline levels heralded significant weight gain during a 5-year follow-up in both cohorts. It is worth mentioning that 4 of these were the same as metabolites from the aforementioned meta-analysis. The only metabolite for which an association with BMI had not been established previous to Zhao et al.'s study was methylsuccinate, a metabolite involved in isoleucine catabolism [53].

Accordingly, Geidenstam et al. showed that metabolite risk score may represent a useful tool on both the individual and population levels, and markedly so if additional metabolites are recognized to improve predictive power [54]. It is important to address that none of the anthropometric, lifestyle, and glycemic risk factors that are standardly measured may fully explain the association between changes in BMI (ΔBMI) and metabolite risk score. The proportion of ΔBMI variance explained by the metabolite risk score was similar to that for the best predictive risk factors (glycemic measures). The above-noted metabolite risk score is comprised of eight metabolites, including malate, niacinamide, xanthine, tyrosine, uridine, and three lipids (TAG 56:6, SM24:0, and TAG 56:2), and could point to independent processes related to weight gain. Regarding the lipids, Geidenstam et al. showed that TAG 56:6 positively correlates with ΔBMI, whereas TAG 56:2 and SM 24:0 show the opposite trend. Such a pattern is in corroboration with previous findings showing that lipids with varying saturation levels have the opposite connection with obesity-related phenotypes [54]. Furthermore, the authors observed that SM 24:0 and TAG 56:2 are associated with decreased weight gain in both of the analyzed cohorts (Mexico City Diabetes Study and Framingham Heart Study), and increased diabetes risk in the former cohort (i.e., they have negative effect sizes in the metabolite risk score that is protective against diabetes), while TAG 56:6 had the opposite direction of effect. Overall, these results suggest that metabolite risk score can be employed as an independent predictor for studying weight gain.

Studies that have examined branched chain amino acid (BCAA) supplementation in both animal and human models suggest that circulating AAs can contribute to insulin resistance, conceivably by affecting insulin signaling in skeletal muscles [55,56]. On a cellular level, the underlying pathophysiological mechanisms seem to include activation of the mTOR, JUN, and IRS1 signaling pathways [57,58]. On the other hand, several authors have shown an improvement of glucose homeostasis in animals fed with a leucine-rich diet [59]. In line with this, it is noteworthy to mention that multiple amino acids, particularly BCAAs, are modulators of insulin secretion [60]. In particular, from a panel of more than 60 metabolites, aromatic amino acids and BCAAs appear to be the best predictors for the future development of diabetes mellitus. A single measurement of these amino acids in fasting state has demonstrated incremental benefit over standard risk factors (such as fasting glucose, dietary patterns, and BMI) [61]. Obese adults, with or without diabetes mellitus, were shown to have hyperaminoacidemia. Specifically, increases in the concentrations of the BCAAs, alongside phenylalanine and tyrosine, have been described [62]. Dissimilar to other essential amino acids, BCAAs are degraded in skeletal muscle, and it has been shown that BCAA circulating levels are elevated post-prandially [63]. Such elevations in BCAAs may influence glucose homeostasis since oxidation of BCAAs spares glucose utilization in skeletal muscle [64]. McCormack et al. demonstrated that an increase in circulating BCAAs is significantly associated with obesity in both children and adolescent populations and is capable of predicting the development of insulin resistance [65]. In fact, increased concentrations of BCAAs are also present in young obese individuals, and their metabolomic profiles are in accordance with increased catabolism of BCAAs, whereas elevations in BCAAs are positively associated with insulin resistance measured after 18 months, independent of the initial BMI [65]. The background of obesity-related increases in the concentrations of BCAAs remains elusive. It is plausible that obese people have more BCAAs in their diet, which could in the context of a high-fat diet characteristic of diet-induced obesity also portend to insulin resistance development [66]. Conversely, insulin resistance may in turn lead to the failure of insulin's physiologic capacity to suppress BCAA levels [67]. Shaham et al. have shown that some obese individuals display differential sensitivity to insulin-induced suppression of proteolysis compared to lipolysis by using metabolite profiling technology [68]. Interestingly, they showed that elevated BCAA concentration in serum may appear long before other indices of insulin resistance become abnormal [69]. In a cross-sectional study that included 182 non-diabetic individuals, Seibert et al. performed a measurement of 24 plasma amino acids and an insulin suppression test. The study showed that in 14/24 amino acids, concentrations were significantly higher in men than women, whereas glycine was found to be lower in men. The majority of these amino acids positively correlated with steady-state plasma glucose (SSPG), while glycine concentration demonstrated a negative correlation. Leucine, isoleucine, glutamic acid, and tyrosine concentrations showed the strongest correlation with SSPG. In comparison to women, men were more prone to demonstrate a rather unfavorable amino acid profile, with a higher concentration of amino acids associated with higher insulin resistance and less glycine [70].

In recent years, a possible connection between serum AAs, gut microbiome, and obesity was explored. In a pivotal study, Calvani et al. demonstrated a plausible link between gut microflora metabolism and obesity phenotype using a nuclear magnetic resonance (NMR)-based metabolomic approach [71]. The study included 15 morbidly obese male patients (having BMI > 40 kg/m^2) and 10 heathy controls. The analysis of NMR spectra of urine samples showed a clear distinction between morbidly obese subjects and lean counterparts in this regard. The main metabolites that were involved in discrimination between the two were xanthine, hippuric acid, trigonelline, and 2-hydroxyisobutyrate. All the aforementioned metabolites were higher in lean subjects, with the exception of 2-hydroxyisobutyrate, which was more abundant in morbidly obese subjects. A very important concept that had also arisen from the findings of this study is the change in urinary metabolic phenotype following bariatric surgery. Namely, hippuric acid and trigonelline levels significantly increased following surgery, whereas 2-hydroxyisobutyrate and xanthine concentrations approximated to the lean subject values. Multiple studies concerning this problem were conducted henceforth, yet most of them were also conducted on relatively small samples, thus failing to reach final conclusions. Nevertheless, some larger studies are worth mentioning in this regard. Liu et al. demonstrated that *Bacteroides thetaiotaomicron*, a glutamate-fermenting commensal, is less abundant in obese individuals in comparison to lean counterparts, and that the amount inversely correlates with serum glutamate concentration [72]. Moreover, it was shown that bariatric surgery led to reversal of these alterations, with regards to both *B. thetaiotaomicron* and glutamate serum levels. On the other hand, gavage with *B. thetaiotaomicron* led to a reduction in serum glutamate concentration, and attenuated diet-induced body-weight gain in a mice model. Multiple important conclusions were reached in an extensive analysis by Org et al., in which the association between gut microbiota and the metabolome was explored using 16S ribosomal RNA gene sequencing and NMR spectroscopy in 531 Finnish adult male subjects [73]. Firstly, a number of connections were established between the gut microbiota composition and serum levels of amino acids, lipids, glucose, and fatty acids. It is important to note that the association existed both with regards to the diversity and richness of microbiota. For instance, acetate levels correlated with microbial diversity, and both acetate and glutamine positively correlated with microbial richness. In addition, associations between multiple bacterial genera and serum concentrations of the above-noted molecules were also established. Furthermore, significant associations between trimethylamine N-oxide (TMAO), a gut microbiota-dependent metabolite, and both coronary artery disease and stroke were found. Finally, the altered composition of microbiota and a significant microbiota–metabolite relationship dependent on glucose tolerance and BMI were detected.

4. Effect of Nutritional Interventions on Gut Microbiota and Metabolites

An important note to consider when discussing the effect of nutritional interventions on gut microbiota is that gut microbiome can respond very rapidly to dietary changes, in a way that even short-term (days) animal or plant-based diets may alter microbial community structure and overwhelm inter-individual differences in microbial gene expression. In a seminal study, David et al. demonstrated that changes in gut microbiome are much more prominent with animal-based than plant-based diets. In fact, it is even hypothesized that there is an evolutionary basis that might explain such a rapid switch between carnivorous and herbivorous functional profiles of gut microbiome [74]. The authors suggested that the intake of meat in ancestral humans was relatively volatile, and largely depended on season and hunting success, whereas more abundant plant food offered a sort of a fallback source for nutrients [75]. Therefore, it seems that evolution enabled gut microbiota to rapidly adjust to the imposed changes in diet and subsequently enhance dietary flexibility in humans. On the other hand, dietary interventions seem to elicit transient changes in microbiota composition and it remains elusive as to what duration is warranted to induce a permanent change to the core microbial profile [76]. Finally, it is worth noting that according to recent reports, the composition of one's gut microbiome is more closely related to specific dietary preferences rather than the nutritional interventions, and, thus, these factors might make it more difficult to see the overall microbiome responses to different diets [77]. An overview of the effects of dietary interventions on gut microbiota and the metabolome is summarized in Figures 1 and 2.

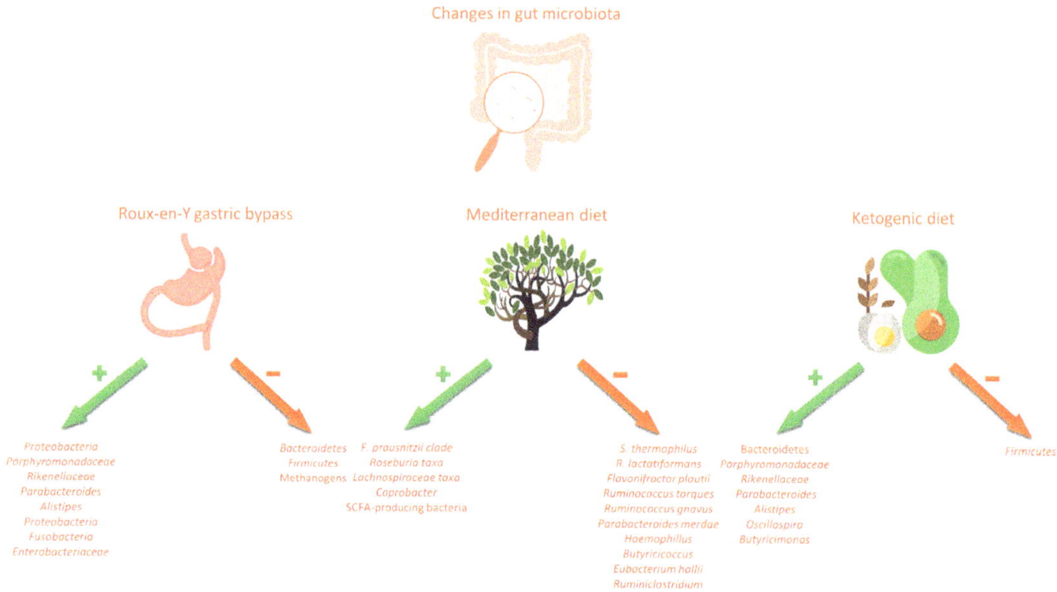

Figure 1. Changes in gut microbiota following dietary interventions (Roux-en-Y gastric bypass, Mediterranean diet, ketogenic diet).

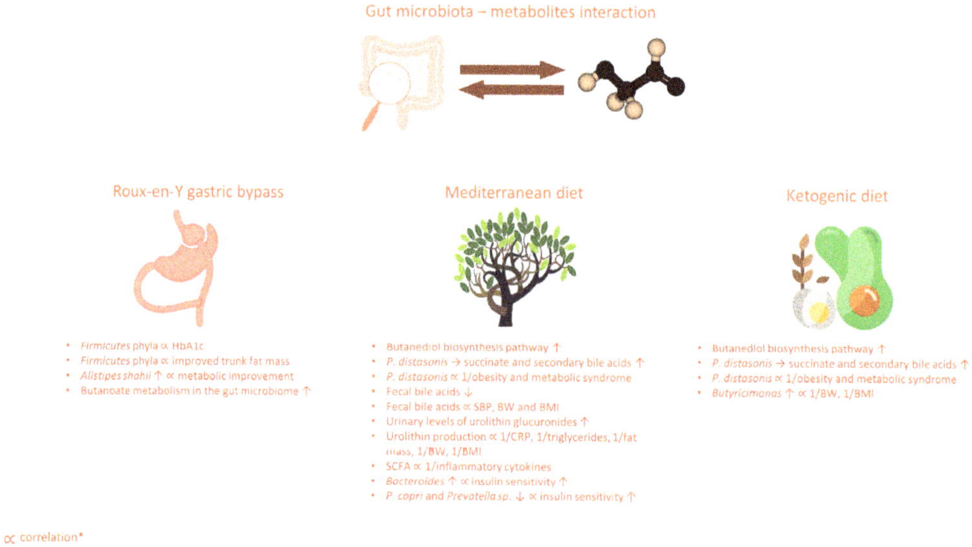

Figure 2. Metabolic changes following dietary interventions (Roux-en-Y gastric bypass, Mediterranean diet, ketogenic diet). * negative correlation was represented as "∝ 1/parameter". Arrows indicate increase or decrease in bacteria abundance. Abbreviations: BMI: body mass index; BW: body weight; CRP: C-reactive protein; HbA1c: hemoglobin A1c; SBP: systolic blood pressure.

4.1. Ketogenic Diet

A ketogenic diet (KD) represents a dietary intervention based on a very low-carbohydrate, high-fat diet which was disputed for a long time. Multiple studies aimed to elucidate the effect of a KD on the composition of gut microbiome in the setting of various pathological states, especially various neurological disorders in which a KD is often recommended by physicians [78].

Preclinical studies consistently demonstrate that this dietary pattern reduces the microbial diversity of the gut, most likely as a result of decreased polysaccharide levels leading to a reduction in bacteria that utilize energy from the gut [79–81]. For instance, Olson et al. demonstrated such a reduction in mouse models of epilepsy, but also found that the relative abundance of *Akkermansia muciniphila* and *Parabacteroides* increased [81]. Moreover, the authors even recognized these bacteria as mediators of seizure protection, as it was demonstrated in their study that a KD reduces seizure incidence, an effect abrogated by antibiotic treatment. Notably, Olson et al. also found a link between reduced seizure susceptibility and glutamate in hippocampus, thus further elucidating the role of the gut–brain axis in this regard. Newell et al. reported somewhat different results in a mice model of autism spectrum disorder; although the results were consistent in terms of gut bacteria total abundance, the authors also reported a reduction in the abundance of *A. muciniphila* [80]. Ma et al. also reported interesting preclinical results in which they showed that shaping of gut microbiota by a KD may mitigate the risk of neurodegeneration and offer benefits to neurovascular health [79]. Specifically, aside from decreased diversity, an increased abundance was observed for multiple beneficial bacteria, including the above-noted *A. muciniphila* but also *Lactobacillus* and SCFA-producing bacteria, whereas bacterial taxa that mediate a pro-inflammatory response (*Desulfovibrio* and *Turicibacter*) were down-regulated. Finally, a recent study used metabolomics to elucidate the molecular effects of a KD in models of healthy and tumor xenograft mice models [82]. Metabolomic profiling on plasma samples revealed distinct metabolic fingerprints in the group of mice bearing breast cancer and, importantly, such fingerprints have dissipated following a KD, which resulted in recovery to the metabolic status of healthy mice. Thus, the authors concluded that a KD has a significant molecular effect on tumor growth inhibition beyond a mere constraint of energy supply to tumor cells.

On the other hand, a KD was extensively studied in the setting of neurological disorders. A study in children with refractory epilepsy showed that 6 months of a KD led to a reduction in gut microbiota diversity, but, more importantly, that changes in gut microbiota could predict the efficacy of a KD in terms of a reduction in seizure occurrence. Additionally, a KD caused an increase in *Bacteroidetes* abundance and decrease in *Firmicutes* abundance, which can be interpreted in light of their ratio that we previously discussed [83]. Furthermore, it was demonstrated in multiple sclerosis patients that a KD caused a reduction in terms of both abundance and diversity of gut bacteria at first, yet that a prolonged KD not only restores bacterial diversity, but also leads to more abundant gut microbiota in comparison to the baseline [84]. Notably, it was shown at baseline that the biofermentative function of the colon is markedly impaired in patients with multiple sclerosis. Overall, although the data imply that a KD offers beneficial effects on neurological disorders, at least in part via mechanisms related to gut microbiota, it has to be addressed that adequately powered randomized controlled trials (RCTs) are needed to confirm these notions, given that most data in this regard are based on small-scale studies.

Several studies aimed to elucidate the KD effects on gut microbiota and metabolic pathways in the setting of obesity. Gutiérrez-Repiso et al. compared the effect of three dietary interventions on the gut microbiota profile: KD, Mediterranean diet, and bariatric surgery [85]. In general, perhaps the most important conclusion of the study was the fact that changes in the gut microbiota profile associated with weight loss interventions are not uniform, albeit there was an overlap with regards to certain changes. Specifically, *Porphyromonadaceae* and *Rikenellaceae* families increased following both a KD and bariatric surgery. For both the aforementioned bacteria, a negative correlation with BMI was previously demonstrated, and studies on animal models associated these bacteria with weight loss using probiotics [86,87]. On the genus level, a KD led to an increase in *Parabacteroides*, and *Alistipes*, both of which were negatively associated with BMI in both adult and young populations [88–90]. In addition, *Alistipes* was previously shown to predict weight loss following dietary interventions [18]. On the other hand, the amount of *Lactobacillus* was decreased, which is in contrast to the above-noted preclinical studies. At species level, the most relevant finding was an increase in *Parabacteroides distasonis*, bacteria that negatively correlated with obesity and metabolic syndrome in previous studies, and for which a beneficial effect in reducing weight gain, hyperglycemia, and hepatic steatosis was demonstrated [91,92]. More precisely, it is presumed that *P. distasonis* improved metabolic dysfunction via production of succinate and secondary bile acids [92]. Finally, the authors applied PICRUSt to predict metagenome functional content, but found no statistically significant change in functional gut microbiota pathways that was common to the three interventions. Nevertheless, in a KD, most of the changes were on the pathways related to biosynthesis and/or degradation/utilization/assimilation, thus indicating a change in their metabolism. The butanediol biosynthesis pathway, a pathway previously implicated in body weight reduction, thermogenesis, and appetite suppression, was shown to be enriched in both KD and Mediterranean diets [93–96]. A randomized controlled pilot study explored the effects of a combined KD and synbiotics on gut microbiota [97]. The study showed that the addition of synbiotics does not provide an additional effect on the improvement of microbial diversity observed after 4 months of KD. Specifically, the genus *Oscillospira*, which was shown to have protective associations with visceral fat mass and increased energy expenditure, and *Butyricimonas*, a butyrate-producing bacterium, increased following the intervention [98–100]. Furthermore, *Butyricimonas* was positively associated with a greater weight loss and BMI reduction. Despite the addition of synbiotic supplementation not affecting microbial diversity, it seems that its addition may ameliorate inflammation in a process mediated by the gut microbiota alteration.

4.2. Mediterranean Diet

The Mediterranean diet (MD) is a dietary pattern that has its roots in the olive-growing regions of the Mediterranean Basin [101]. A high consumption of olive oil, vegetables, fruits, unprocessed cereals, nuts, legumes, fish, and other seafood, moderate consumption of old cheese and red wine, and limited consumption of dairy products and meat are its principal characteristics. Apart from demonstrating benefit in many chronic diseases, a MD's effects on individuals with obesity is well established, particularly in regard to the gut microbiota composition and the metabolome [102–104]. Moreover, since a substantial number of RCTs concerning the interrelation between a MD and gut microbiota were performed (unlike with other dietary interventions), inferences about this dietary regime are perhaps the most reliable ones.

Meslier et al. conducted an RCT which aimed to elucidate how isocaloric MD intervention affected metabolic health, gut microbiome, and the systemic metabolome over the course of 8 weeks [105]. Following the intervention, *Streptococcus thermophilus*, *Ruthenibacterium lactatiformans*, *Flavonifractor plautii*, *Ruminococcus torques*, *Ruminococcus gnavus*, and *Parabacteroides merdae* were significantly reduced, whereas five members of the *Faecalibacterium prausnitzii* clade, alongside several members of the *Roseburia* and *Lachnospiraceae* taxa, were more abundant in the treatment group compared with the control group. The effect of diet was confirmed using integration of the three meta-omics datasets, based on which the authors established a separation of the MD and control group with respect to microbiome diversity, functional modules, and metabolomic profiles. Interestingly, urinary levels of urolithin glucuronides increased following the introduction of the MD, and such changes were concordant with levels of the urolithin producers in the microbiome. Moreover, a negative correlation was established between urolithin production and CRP, triglycerides, fat mass, body weight, and BMI, thus providing a direct link between gut microbiota and functional metabolic alterations resulting from this dietary intervention. Fecal bile acids were reduced in the stools of participants in the MD group, and this may be of relevance as fecal bile acids demonstrated a positive correlation with systolic blood pressure, body weight, and BMI. Finally, the authors concluded that a MD might be helpful in improving insulin sensitivity in individuals with higher levels of multiple *Bacteroides* species, and lower levels of *P. copri* and *Prevotella* sp.

In a separate RCT, the effects of a MD on gut microbiota in obese population over the course of the year were explored [106]. The analysis showed that *Butyricicoccus, Eubacterium hallii, Haemophillus*, and *Ruminiclostridium* were reduced, whereas *Coprobacter* and uncultured bacterium increased in the energy-restricted MD group in comparison to non-energy-restricted MD. Furthermore, metabolic pathways responsible for the biosynthesis of amino acids, nucleosides, nucleotides, and carbohydrates were significantly different between the two groups of interest. As selected SCFA-producing bacteria were increasingly abundant in both subgroups, and as their number positively correlated with the MD adherence score (MedScore), the authors concluded that MD-induced changes may be modulated by SCFA-producing bacteria regardless of energy restriction. Accordingly, Pagliai et al. explored the effects of a MD on gut microbiota and SCFA production in an RCT with a 3-month follow-up [107]. Although the study suggested that 3 months of MD does not induce major changes in the gut microbiota, the observed changes in SCFA production support the role of a MD in modulating the inflammatory response. Specifically, a negative correlation was observed between SCFA and multiple inflammatory cytokines. Furthermore, the effects of a modified Mediterranean–ketogenic diet (MMKD) on gut microbiota and SCFAs have also been explored in populations with neurological disorders. It was demonstrated in a randomized crossover study that specific gut microbial signatures may herald mild cognitive impairment in Alzheimer's disease (AD) and that the MMKD may modulate both gut microbiome and metabolites associated with biomarkers of AD in cerebrospinal fluid [108]. Their results were also corroborated by two recent studies that explored the effects of an MMKD on AD and multiple sclerosis [109,110]. Finally, Barber et al. showed in a short-term (2 weeks) MD intervention that despite the relatively small difference in microbiota composition between subjects on a MD and subjects receiving a standard Western-type diet, microbial metabolism was disparate between the groups, as evidenced by urinary metabolite profiles and an abundance of microbial metabolic pathways [111]. Therefore, one can conclude that a sustained MD is needed to elicit relevant gut microbiota changes, whereas even a short-term MD might offer benefit in regard to the metabolome.

4.3. Bariatric Surgery

Often regarded as the last resort of morbid obesity treatment, bariatric surgery still represents an important weight-reduction method.

We previously noted that Gutiérrez-Repiso et al. demonstrated the expansion of *Porphyromonadaceae* and *Rikenellaceae* families, *Parabacteroides* and *Alistipes* generum, and *P. distasonis* bacterial species following Roux-en-Y gastric bypass (RYGB) surgery. These results imply that the effector arm of RYGB-induced metabolic alterations may lie in gut microbiota changes. Furthermore, RYGB also led to downregulation of most of the biosynthesis pathways, perhaps as a result of extreme caloric restriction. Pentose phosphate and sugar biosynthesis pathways were also enriched following RYGB, whereas a decrease in the nucleic acid processing pathway was demonstrated. The aforementioned pentose phosphate pathway has been related to obesity, even though the data are rather conflicting [112,113]. By applying a parallel and integrated metagenomic and metabonomic approach, Li et al. demonstrated in a rat model that RYGB surgery induces marked alterations of the main gut bacteria, increasing the concentrations of *Proteobacteria* (as much as 52-fold) while lowering the concentrations of *Bacteroidetes* and *Firmicutes* [114]. Moreover, the authors also aimed to explain their results from a mechanistic standpoint by showing that surgically induced changes in the GI anatomy, flow of nutrients, and weight are also associated with changes in urinary and fecal profiles, thus reflecting increased oligosaccharide fermentation in GI and amine generation, as well as biogenesis of p-cresol and related compounds. Furthermore, a human study further confirmed that RYGB causes profound changes in the gut microbiota composition. Specifically, a marked increase was observed in the *Gammaproteobacteria* class (especially *Enterobacteriaceae*), decrease in *Firmicutes*, and loss of methanogens. The authors argue that shortened small intestinal length, changes in acid exposure to the gastric remnant, and oxygen availability in the colon favored the growth of facultative anaerobes such as *Gammaproteobacteria* over obligate anaerobes such as *Firmicutes*. In addition, it was hypothesized that surgery itself affects food ingestion and digestion, but that it also portends more rapid transit of ingested materials to the colon. Sánchez-Alcoholado et al. compared the effects of two distinct types of bariatric surgeries on gut microbiota and metabolism: RYGB and laparoscopic sleeve gastrectomy (LSG) [115]. Overall, the authors concluded that RYGB yields overall greater gut microbiota differences than SG. Following RYGB, an increase in *Proteobacteria, Fusobacteria*, and *Enterobacteriaceae* was observed, whereas *Verrucomicrobiaceae* showed higher levels in SG. Moreover, an inverse correlation was established between *Enterobacteriaceae* and *Veillonella* with cholesterol levels, and a positive relationship was also seen between *Verrucomicrobia* and high-density lipoprotein cholesterol, thus suggesting an implication of these bacteria in the observed changes in total cholesterol levels between procedures following surgery. The difference in the observed changes may be attributed to the difference in pH between RYGB and LSG. In RYGB, the distal stomach and the small intestine are excluded from the digestive transit; hence, the stomach acidity is in part avoided and the amount of hydrochloric acid in the intestine is reduced. pH is an important determinant of the distribution of major fermentation end products and it has been also demonstrated that pH may affect bacterial growth [116,117]. For instance, butyrogenic reactions occur at mildly acidic pH [118]. Accordingly, butanoate metabolism was enriched in the gut microbiome of RYGB patients, most probably mediated by *Clostridiaceae, Clostridium*, or *Oscillospira*, all of which are butyrate producers [119]. In a separate study, Kong et al. [120] reported an increase in microbiota richness following RYGB, mostly owing to an increase in *Proteobacteria*. In addition, an association was established between gut microbiota and gene expression in white adipose tissue. Wisnewsky et al. demonstrated that RYGB and LSG produced notable changes to gut microbiota, and that the observed differences may be associated with modifications in the digestive physiology that are caused by a surgical procedure [121]. For instance, *Firmicutes* phyla in RYGB correlated with glycated hemoglobin levels and improved trunk fat mass, whereas an increase in *Alistipes shahii* after surgery was associated with metabolic improvement.

4.4. Vegan and Vegetarian Diet

Vegan and vegetarian diets are gaining popularity for their reputed health protective properties, believed to be the consequence of reduced levels of inflammation and linked to their distinct gut microbiota [122]. A general pattern observed in vegan and vegetarian populations is an increase in the abundance of bacteria involved in the fermentation of dietary fiber, such as *Ruminococcus, Lactobacillus, Clostridium, Eubacterium rectale*, and *F. prausnitzii* [123,124]. Results of observational studies assessing the effect of vegan diet on gut microbiota showed an increase of phyla *Bacteroidetes* and a higher abundance of genus *Prevotella*, while studies yielded inconclusive results regarding the diversity and richness of gut microbiota [125]. On the other hand, interventional studies assessing the effects of plant-based diets on gut microbiota support findings of an increased abundance of bacteria involved in fiber breakdown, as well as an increase in butyrate-producing bacteria such as *Ruminococcaceae, Lachnospiraceae, Coprococcus, Roseburia, Blautia, Alistipes*, and *F. prausnitzii*, while also showing inconclusive results with respect to the diversity and richness of gut microbiota [126].

The observed change in the composition of the intestinal microbiota may be caused by a variety of factors, including the types of bacteria that are consumed directly through food, the substrates that are consumed, variations in duration of transition through the gastrointestinal system, the local pH, host secretion that is affected by dietary patterns, and the regulation of gene expression in host cells and/or bacteria [127]. Evidence extracted from studies which followed the *Firmicutes*-to-*Bacteroidetes* ratio are inconsistent, as we have already discussed. Nevertheless, the observed decrease in *Firmicutes* levels in favor of *Bacteroidetes* and *Bifidobacteria* might be explained as a response to an increased intake of resistant starches. A microbiome enriched in *Firmicutes* has been associated with an increased capacity for energy harvest and obesity, which could be beneficial in preventing and treating obesity [128]. In addition, De Filippo et al. confirmed a higher abundance of *Prevotella*, a genus of the *Bacteroidetes* phyla, as a response to a vegan diet [129]. Animal and plant-based food impact the abundance of *Bacteroides*, a genus well known for pro-inflammatory effects. *Clostridium clostridioforme, B. thetaiotaomicron*, and *Faecalibacterium prausnitzii*, all considered to offer protective effects, had a higher relative abundance among vegetarians/vegans compared to omnivores. On the other hand, there are inconsistent data regarding the impact of dietary patterns on the abundance of *Clostridium* cluster XIVa. Previous data showed a lower abundance of this species in vegetarians/vegans, but a recent study confirmed *Clostridium* cluster XIVa bacteria to be a major component of gut microbiota in vegetarian women [130]. Abundance of the third major enterotype, *Ruminococcus*, was associated with long-term fruit and vegetable consumption. Species from this enterotype specialize in degrading complex carbohydrates, such as resistant starch and cellulose found in plant-based foods, which results in the production of butyrate, a molecule that offers different beneficial roles to human health including anti-inflammatory properties, lower endotoxemia, and lower arterial stiffness [131,132]. *Ruminococcus* might also play a role in the conversion of animal-derived choline to trimethylamine [133]. This might explain the importance of an animal-based diet as well as a plant-based diet to *Ruminococcus* abundance.

A diet enriched with whole grains such as barley, brown rice, or a combination of the two increased microbial diversity in a study by Martinez et al., in which participants were followed for 28 weeks [134]. The results of short-term dietary interventions which increased fiber intake showed an opposite effect, showing reduced diversity. This could be explained by a hypothesis of transitory microbial "stress", which has a mild but significant impact and causes an increase in *Enterobacteriaceae* as a result of the rapid dietary change caused by dietary intervention. Klimenko et al. found a positive association between the alpha-diversity value of microbial richness and long-term fruit and vegetable intake, as well as a negative association between alpha-diversity and BMI [135]. These results could be explained by a greater presence of butyrate-producing bacteria which can lower colonic pH, preventing the growth of pathogenic bacteria such as *Enterobacteriaceae* in individuals following a higher-fiber diet [136].

5. Future Directions and Conclusions

Based on the presented data, we may conclude that various dietary interventions can result in significant changes to both gut microbiota and metabolic pathways in obese individuals. It is worth noting that the effects do not simply depend on the type of intervention, but also on its duration and the presence of concomitant caloric restriction. Unfortunately, it is rather challenging to derive proper inferences, given that most of the data are not a result of sufficiently powered RCTs. Perhaps the strongest body of evidence is available for MD, but even for this, more data are needed to reach appropriate conclusions. The biggest task in future studies is to find out whether the response to any of these interventions can be predicted using the gut microbiota composition and/or metabolites, thus providing us with a feasible method of finding the appropriate dietary intervention. Currently, there are multiple obstacles preventing us from fulfilling this task. For a start, although it was demonstrated that gut microbial changes can occur rapidly, it remains elusive what duration, if any, is needed to induce a permanent change to the core microbial profile. Secondly, the composition of one's gut microbiome seems to be more closely related to specific dietary preferences rather than the nutritional interventions, thus further impeding the adequate designing of the studies. In line with this, nutritional intervention studies are commonly burdened by adherence issues, which poses a problem, especially in the absence of reliable adherence indicators. Finally, although there is an abundance of data in this regard, a gap still exists in our understanding of the role of gut microbiota and metabolites in the pathophysiology of obesity and metabolic syndrome. For instance, we conclude about the effects of a certain diet on a certain disorder by using serum biomarkers and, in many cases, the serum biomarkers have very limited accuracy in depicting the respective disorder.

Author Contributions: Conceptualization, Z.P. (Zivana Puljiz) and M.K.; methodology, Z.P. (Zivana Puljiz) and M.K.; writing—original draft preparation, Z.P. (Zivana Puljiz), M.K., D.M., J.V., P.D., M.O.K. and H.U.; writing—review and editing, T.T.K., Z.P. (Zeljko Puljiz), J.Z., J.V., I.M. and J.B.; visualization, Z.P. (Zivana Puljiz), M.K., D.M., P.D., M.O.K. and H.U.; supervision, T.T.K., Z.P. (Zeljko Puljiz), J.Z., I.M. and J.B.; project administration, I.M. and J.B.; funding acquisition, I.M. and J.B. All authors have read and agreed to the published version of the manuscript.

Funding: This research received no external funding.

Institutional Review Board Statement: Not applicable.

Informed Consent Statement: Not applicable.

Data Availability Statement: Not applicable.

Conflicts of Interest: The authors declare no conflict of interest.

References

1. Cox, L.M.; Blaser, M.J. Pathways in Microbe-Induced Obesity. *Cell Metab.* **2013**, *17*, 883–894. [CrossRef] [PubMed]
2. Lagier, J.C.; Dubourg, G.; Million, M.; Cadoret, F.; Bilen, M.; Fenollar, F.; Levasseur, A.; Rolain, J.M.; Fournier, P.E.; Raoult, D. Culturing the human microbiota and culturomics. *Nat. Rev. Microbiol.* **2018**, *16*, 540–550. [CrossRef] [PubMed]
3. Zhang, C.; Zhang, M.; Wang, S.; Han, R.; Cao, Y.; Hua, W.; Mao, Y.; Zhang, X.; Pang, X.; Wei, C.; et al. Interactions between gut microbiota, host genetics and diet relevant to development of metabolic syndromes in mice. *ISME J.* **2010**, *4*, 232–241. [CrossRef]
4. Zhang, C.; Zhang, M.; Pang, X.; Zhao, Y.; Wang, L.; Zhao, L. Structural resilience of the gut microbiota in adult mice under high-fat dietary perturbations. *ISME J.* **2012**, *6*, 1848–1857. [CrossRef] [PubMed]
5. Park, S.; Sadanala, K.C.; Kim, E.-K. A Metabolomic Approach to Understanding the Metabolic Link between Obesity and Diabetes. *Mol. Cells* **2015**, *38*, 587–596. [CrossRef] [PubMed]
6. Reddel, S.; Putignani, L.; Del Chierico, F. The Impact of Low-FODMAPs, Gluten-Free, and Ketogenic Diets on Gut Microbiota Modulation in Pathological Conditions. *Nutrients* **2019**, *11*, 373. [CrossRef]
7. Le Chatelier, E.; Nielsen, T.; Qin, J.; Prifti, E.; Hildebrand, F.; Falony, G.; Almeida, M.; Arumugam, M.; Batto, J.-M.; Kennedy, S.; et al. Richness of human gut microbiome correlates with metabolic markers. *Nature* **2013**, *500*, 541–546. [CrossRef]
8. Qin, J.; Li, R.; Raes, J.; Arumugam, M.; Burgdorf, K.S.; Manichanh, C.; Nielsen, T.; Pons, N.; Levenez, F.; Yamada, T.; et al. A human gut microbial gene catalogue established by metagenomic sequencing. *Nature* **2010**, *464*, 59–65. [CrossRef]
9. Heintz-Buschart, A.; Wilmes, P. Human Gut Microbiome: Function Matters. *Trends Microbiol.* **2018**, *26*, 563–574. [CrossRef]
10. Milani, C.; Duranti, S.; Bottacini, F.; Casey, E.; Turroni, F.; Mahony, J.; Belzer, C.; Delgado Palacio, S.; Arboleya Montes, S.; Mancabelli, L.; et al. The First Microbial Colonizers of the Human Gut: Composition, Activities, and Health Implications of the Infant Gut Microbiota. *Microbiol. Mol. Biol. Rev.* **2017**, *81*, e00036-17. [CrossRef]
11. Teunis, C.; Nieuwdorp, M.; Hanssen, N. Interactions between Tryptophan Metabolism, the Gut Microbiome and the Immune System as Potential Drivers of Non-Alcoholic Fatty Liver Disease (NAFLD) and Metabolic Diseases. *Metabolites* **2022**, *12*, 514. [CrossRef] [PubMed]
12. Knights, D.; Ward, T.L.; McKinlay, C.E.; Miller, H.; Gonzalez, A.; McDonald, D.; Knight, R. Rethinking "Enterotypes". *Cell Host Microbe* **2014**, *16*, 433–437. [CrossRef] [PubMed]
13. DeGruttola, A.K.; Low, D.; Mizoguchi, A.; Mizoguchi, E. Current Understanding of Dysbiosis in Disease in Human and Animal Models. *Inflamm. Bowel Dis.* **2016**, *22*, 1137–1150. [CrossRef] [PubMed]
14. Peterson, C.T.; Sharma, V.; Elmén, L.; Peterson, S.N. Immune homeostasis, dysbiosis and therapeutic modulation of the gut microbiota. *Clin. Exp. Immunol.* **2015**, *179*, 363–377. [CrossRef] [PubMed]
15. Rhee, E.P.; Cheng, S.; Larson, M.; Walford, G.A.; Lewis, G.D.; McCabe, E.; Yang, E.; Farrell, L.; Fox, C.S.; O'donnell, C.J.; et al. Lipid profiling identifies a triacylglycerol signature of insulin resistance and improves diabetes prediction in humans. *J. Clin. Investig.* **2011**, *121*, 1402–1411. [CrossRef]
16. Ralston, J.C.; Zulyniak, M.A.; Nielsen, D.E.; Clarke, S.; Badawi, A.; El-Sohemy, A.; Ma, D.W.; Mutch, D.M. Ethnic- and sex-specific associations between plasma fatty acids and markers of insulin resistance in healthy young adults. *Nutr. Metab.* **2013**, *10*, 42. [CrossRef]

17. Hodge, A.M.; English, D.R.; O'Dea, K.; Sinclair, A.J.; Makrides, M.; Gibson, R.A.; Giles, G.G. Plasma phospholipid and dietary fatty acids as predictors of type 2 diabetes: Interpreting the role of linoleic acid. *Am. J. Clin. Nutr.* **2007**, *86*, 189–197. [CrossRef]
18. Cardinelli, C.S.; Sala, P.C.; Alves, C.C.; Torrinhas, R.S.; Waitzberg, D.L. Influence of Intestinal Microbiota on Body Weight Gain: A Narrative Review of the Literature. *Obes. Surg.* **2014**, *25*, 346–353. [CrossRef]
19. Bäckhed, F.; Ding, H.; Wang, T.; Hooper, L.V.; Koh, G.Y.; Nagy, A.; Semenkovich, C.F.; Gordon, J.I. The Gut Microbiota as an Environmental Factor That Regulates Fat Storage. *Proc. Natl. Acad. Sci. USA* **2004**, *101*, 15718–15723. [CrossRef]
20. Vijay-Kumar, M.; Aitken, J.D.; Carvalho, F.A.; Cullender, T.C.; Mwangi, S.; Srinivasan, S.; Sitaraman, S.V.; Knight, R.; Ley, R.E.; Gewirtz, A.T. Metabolic Syndrome and Altered Gut Microbiota in Mice Lacking Toll-Like Receptor 5. *Science* **2010**, *328*, 228–231. [CrossRef]
21. Turnbaugh, P.J.; Hamady, M.; Yatsunenko, T.; Cantarel, B.L.; Duncan, A.; Ley, R.E.; Sogin, M.L.; Jones, W.J.; Roe, B.A.; Affourtit, J.P.; et al. A core gut microbiome in obese and lean twins. *Nature* **2009**, *457*, 480–484. [CrossRef] [PubMed]
22. Weiss, G.A.; Hennet, T. Mechanisms and consequences of intestinal dysbiosis. *Cell. Mol. Life Sci.* **2017**, *74*, 2959–2977. [CrossRef] [PubMed]
23. Sergeev, I.N.; Aljutaily, T.; Walton, G.; Huarte, E. Effects of Synbiotic Supplement on Human Gut Microbiota, Body Composition and Weight Loss in Obesity. *Nutrients* **2020**, *12*, 222. [CrossRef] [PubMed]
24. Ferrer, M.; Ruiz, A.; Lanza, F.; Haange, S.-B.; Oberbach, A.; Till, H.; Bargiela, R.; Campoy, C.; Segura, M.T.; Richter, M.; et al. Microbiota from the distal guts of lean and obese adolescents exhibit partial functional redundancy besides clear differences in community structure. *Environ. Microbiol.* **2013**, *15*, 211–226. [CrossRef]
25. Cotillard, A.; Kennedy, S.P.; Kong, L.C.; Prifti, E.; Pons, N.; Le Chatelier, E.; Almeida, M.; Quinquis, B.; Levenez, F.; Galleron, N.; et al. Dietary intervention impact on gut microbial gene richness. *Nature* **2013**, *500*, 585–588, Erratum in *Nature* **2013**, *502*, 580. [CrossRef]
26. Roland, B.C.; Lee, D.; Miller, L.S.; Vegesna, A.; Yolken, R.; Severance, E.; Prandovszky, E.; Zheng, X.E.; Mullin, G.E. Obesity increases the risk of small intestinal bacterial overgrowth (SIBO). *Neurogastroenterol. Motil.* **2018**, *30*, e13199. [CrossRef]
27. Vetrani, C.; Di Nisio, A.; Paschou, S.A.; Barrea, L.; Muscogiuri, G.; Graziadio, C.; Savastano, S.; Colao, A.; on behalf of the Obesity Programs of Nutrition, Education, Research and Assessment (OPERA) Group. From Gut Microbiota through Low-Grade Inflammation to Obesity: Key Players and Potential Targets. *Nutrients* **2022**, *14*, 2103. [CrossRef]
28. Lyra, A.; Lahtinen, S.; Tiihonen, K.; Ouwehand, A.C. Intestinal microbiota and overweight. *Benef. Microbes* **2010**, *1*, 407–421. [CrossRef]
29. Schwiertz, A.; Taras, D.; Schäfer, K.; Beijer, S.; Bos, N.A.; Donus, C.; Hardt, P.D. Microbiota and SCFA in Lean and Overweight Healthy Subjects. *Obesity* **2010**, *18*, 190–195. [CrossRef]
30. Zhang, H.; DiBaise, J.K.; Zuccolo, A.; Kudrna, D.; Braidotti, M.; Yu, Y.; Parameswaran, P.; Crowell, M.D.; Wing, R.; Rittmann, B.E.; et al. Human Gut Microbiota in Obesity and after Gastric Bypass. *Proc. Natl. Acad. Sci. USA* **2009**, *106*, 2365–2370. [CrossRef]
31. Duncan, S.H.; Lobley, G.E.; Holtrop, G.; Ince, J.; Johnstone, A.M.; Louis, P.; Flint, H.J. Human Colonic Microbiota Associated with Diet, Obesity and Weight Loss. *Int. J. Obes.* **2008**, *32*, 1720–1724. [CrossRef] [PubMed]
32. Houtman, T.A.; Eckermann, H.A.; Smidt, H.; de Weerth, C. Gut microbiota and BMI throughout childhood: The role of firmicutes, bacteroidetes, and short-chain fatty acid producers. *Sci. Rep.* **2022**, *12*, 3140. [CrossRef] [PubMed]
33. Patil, D.; Dhotre, D.; Chavan, S.G.; Sultan, A.; Jain, D.; Lanjekar, V.B.; Gangawani, J.; Shah, P.S.; Todkar, J.S.; Shah, S.; et al. Molecular analysis of gut microbiota in obesity among Indian individuals. *J. Biosci.* **2012**, *37*, 647–657. [CrossRef]
34. Tims, S.; Derom, C.; Jonkers, D.M.; Vlietinck, R.; Saris, W.H.; Kleerebezem, M.; De Vos, W.M.; Zoetendal, E.G. Microbiota conservation and BMI signatures in adult monozygotic twins. *ISME J.* **2013**, *7*, 707–717. [CrossRef]
35. Ley, R.E.; Turnbaugh, P.J.; Klein, S.; Gordon, J.I. Human Gut Microbes Associated with Obesity. *Nature* **2006**, *444*, 1022–1023. [CrossRef] [PubMed]
36. Magne, F.; Gotteland, M.; Gauthier, L.; Zazueta, A.; Pesoa, S.; Navarrete, P.; Balamurugan, R. The Firmicutes/Bacteroidetes Ratio: A Relevant Marker of Gut Dysbiosis in Obese Patients? *Nutrients* **2020**, *12*, 1474. [CrossRef]
37. Sze, M.A.; Schloss, P.D. Looking for a Signal in the Noise: Revisiting Obesity and the Microbiome. *mBio* **2016**, *7*, e01018-16. [CrossRef]
38. Wang, T.; Yu, L.; Xu, C.; Pan, K.; Mo, M.; Duan, M.; Zhang, Y.; Xiong, H. Chronic fatigue syndrome patients have alterations in their oral microbiome composition and function. *PLoS ONE* **2018**, *13*, e0203503. [CrossRef]
39. Waters, J.L.; Ley, R.E. The human gut bacteria Christensenellaceae are widespread, heritable, and associated with health. *BMC Biol.* **2019**, *17*, 83. [CrossRef]
40. Alemán, J.O.; Bokulich, N.A.; Swann, J.R.; Walker, J.M.; De Rosa, J.C.; Battaglia, T.; Costabile, A.; Pechlivanis, A.; Liang, Y.; Breslow, J.L.; et al. Fecal microbiota and bile acid interactions with systemic and adipose tissue metabolism in diet-induced weight loss of obese postmenopausal women. *J. Transl. Med.* **2018**, *16*, 244. [CrossRef]
41. Depommier, C.; Everard, A.; Druart, C.; Plovier, H.; Van Hul, M.; Vieira-Silva, S.; Falony, G.; Raes, J.; Maiter, D.; Delzenne, N.M.; et al. Supplementation with Akkermansia muciniphila in overweight and obese human volunteers: A proof-of-concept exploratory study. *Nat. Med.* **2019**, *25*, 1096–1103. [CrossRef]
42. Crovesy, L.; Ostrowski, M.; Ferreira, D.M.T.P.; Rosado, E.L.; Soares, M. Effect of Lactobacillus on body weight and body fat in overweight subjects: A systematic review of randomized controlled clinical trials. *Int. J. Obes.* **2017**, *41*, 1607–1614. [CrossRef]
43. Liu, B.-N.; Liu, X.-T.; Liang, Z.-H.; Wang, J.-H. Gut microbiota in obesity. *World J. Gastroenterol.* **2021**, *27*, 3837–3850. [CrossRef]
44. Sarmiento-Andrade, Y.; Suárez, R.; Quintero, B.; Garrochamba, K.; Chapela, S.P. Gut microbiota and obesity: New insights. *Front. Nutr.* **2022**, *9*, 1018212. [CrossRef] [PubMed]
45. Walters, W.A.; Xu, Z.; Knight, R. Meta-analyses of human gut microbes associated with obesity and IBD. *FEBS Lett.* **2014**, *588*, 4223–4233. [CrossRef]

46. Lozupone, C.A.; Stombaugh, J.; Gonzalez, A.; Ackermann, G.; Wendel, D.; Vázquez-Baeza, Y.; Jansson, J.K.; Gordon, J.I.; Knight, R. Meta-analyses of studies of the human microbiota. *Genome Res.* **2013**, *23*, 1704–1714. [CrossRef]
47. Liu, Z.; DeSantis, T.Z.; Andersen, G.; Knight, R. Accurate taxonomy assignments from 16S rRNA sequences produced by highly parallel pyrosequencers. *Nucleic Acids Res.* **2008**, *36*, e120. [CrossRef] [PubMed]
48. Nearing, J.T.; Comeau, A.M.; Langille, M.G.I. Identifying biases and their potential solutions in human microbiome studies. *Microbiome* **2021**, *9*, 113. [CrossRef] [PubMed]
49. Heianza, Y.; Sun, D.; Smith, S.R.; Bray, G.A.; Sacks, F.M.; Qi, L. Changes in Gut Microbiota–Related Metabolites and Long-term Successful Weight Loss in Response to Weight-Loss Diets: The POUNDS Lost Trial. *Diabetes Care* **2018**, *41*, 413–419. [CrossRef]
50. Sierra, A.C.; Ramos-Lopez, O.; Riezu-Boj, J.I.; Milagro, F.I.; Martinez, J.A. Diet, Gut Microbiota, and Obesity: Links with Host Genetics and Epigenetics and Potential Applications. *Adv. Nutr.* **2019**, *10*, S17–S30. [CrossRef]
51. Moore, S.C.; Matthews, C.; Sampson, J.N.; Stolzenberg-Solomon, R.Z.; Zheng, W.; Cai, Q.; Tan, Y.T.; Chow, W.-H.; Ji, B.-T.; Liu, D.K.; et al. Human metabolic correlates of body mass index. *Metabolomics* **2014**, *10*, 259–269. [CrossRef] [PubMed]
52. Zhao, H.; Shen, J.; Djukovic, D.; Daniel-MacDougall, C.; Gu, H.; Wu, X.; Chow, W.-H. Metabolomics-identified metabolites associated with body mass index and prospective weight gain among Mexican American women. *Obes. Sci. Pract.* **2016**, *2*, 309–317. [CrossRef] [PubMed]
53. Urpi-Sarda, M.; Aguilera, E.A.; Llorach, R.; Vázquez-Fresno, R.; Estruch, R.; Corella, D.; Sorli, J.; Carmona, F.; Sánchez-Pla, A.; Salas-Salvadó, J.; et al. Non-targeted metabolomic biomarkers and metabotypes of type 2 diabetes: A cross-sectional study of PREDIMED trial participants. *Diabetes Metab.* **2019**, *45*, 167–174. [CrossRef] [PubMed]
54. Geidenstam, N.; Hsu, Y.-H.H.; Astley, C.M.; Mercader, J.M.; Ridderstråle, M.; Gonzalez, M.E.; Gonzalez, C.; Hirschhorn, J.N.; Salem, R.M. Using metabolite profiling to construct and validate a metabolite risk score for predicting future weight gain. *PLoS ONE* **2019**, *14*, e0222445. [CrossRef]
55. Oakenfull, D. Physicochemical properties of dietary fiber: Overview. In *Handbook of Dietary Fiber*; Cho, S., Dreher, M.L., Eds.; Marcel Dekker Inc.: New York, NY, USA, 2001; pp. 230–241.
56. Slavin, J.L.; Jacobs, D.; Marquart, L.; Wiemer, K. The Role of Whole Grains in Disease Prevention. *J. Am. Diet. Assoc.* **2001**, *101*, 780–785. [CrossRef]
57. Newgard, C.B.; An, J.; Bain, J.R.; Muehlbauer, M.J.; Stevens, R.D.; Lien, L.F.; Haqq, A.M.; Shah, S.H.; Arlotto, M.; Slentz, C.A.; et al. A Branched-Chain Amino Acid-Related Metabolic Signature that Differentiates Obese and Lean Humans and Contributes to Insulin Resistance. *Cell Metab.* **2009**, *9*, 311–326. [CrossRef]
58. Patti, M.E.; Brambilla, E.; Luzi, L.; Landaker, E.J.; Kahn, C.R. Bidirectional modulation of insulin action by amino acids. *J. Clin. Investig.* **1998**, *101*, 1519–1529. [CrossRef]
59. Zhang, Y.; Guo, K.; LeBlanc, R.E.; Loh, D.; Schwartz, G.J.; Yu, Y.-H. Increasing Dietary Leucine Intake Reduces Diet-Induced Obesity and Improves Glucose and Cholesterol Metabolism in Mice via Multimechanisms. *Diabetes* **2007**, *56*, 1647–1654. [CrossRef]
60. Floyd, J.C.; Fajans, S.S.; Conn, J.W.; Knopf, R.F.; Rull, J. Stimulation of insulin secretion by amino acids. *J. Clin. Investig.* **1966**, *45*, 1487–1502. [CrossRef]
61. Zhou, M.; Shao, J.; Wu, C.-Y.; Shu, L.; Dong, W.; Liu, Y.; Chen, M.; Wynn, R.M.; Wang, J.; Wang, J.; et al. Targeting BCAA Catabolism to Treat Obesity-Associated Insulin Resistance. *Diabetes* **2019**, *68*, 1730–1746. [CrossRef]
62. Felig, P.; Marliss, E.; Cahill, G.F., Jr. Plasma Amino Acid Levels and Insulin Secretion in Obesity. *N. Engl. J. Med.* **1969**, *281*, 811–816. [CrossRef] [PubMed]
63. Matthews, D.E. Observations of Branched-Chain Amino Acid Administration in Humans. *J. Nutr.* **2005**, *135*, 1580S–1584S. [CrossRef] [PubMed]
64. Buse, M.G.; Biggers, J.F.; Friderici, K.H.; Buse, J.F. Oxidation of Branched Chain Amino Acids by Isolated Hearts and Diaphragms of the Rat. *J. Biol. Chem.* **1972**, *247*, 8085–8096. [CrossRef] [PubMed]
65. McCormack, S.E.; Shaham, O.; McCarthy, M.A.; Deik, A.A.; Wang, T.J.; Gerszten, R.E.; Clish, C.B.; Mootha, V.K.; Grinspoon, S.K.; Fleischman, A. Circulating branched-chain amino acid concentrations are associated with obesity and future insulin resistance in children and adolescents. *Pediatr. Obes.* **2013**, *8*, 52–61. [CrossRef]
66. Lee, J.; Vijayakumar, A.; White, P.J.; Xu, Y.; Ilkayeva, O.; Lynch, C.J.; Newgard, C.B.; Kahn, B.B. BCAA Supplementation in Mice with Diet-induced Obesity Alters the Metabolome without Impairing Glucose Homeostasis. *Endocrinology* **2021**, *162*, bqab062. [CrossRef]
67. Pozefsky, T.; Felig, P.; Tobin, J.D.; Soeldner, J.S.; Cahill, G.F. Amino acid balance across tissues of the forearm in postabsorptive man. Effects of insulin at two dose levels. *J. Clin. Investig.* **1969**, *48*, 2273–2282. [CrossRef]
68. Shaham, O.; Wei, R.; Wang, T.J.; Ricciardi, C.; Lewis, G.D.; Vasan, R.S.; Carr, S.A.; Thadhani, R.; Gerszten, R.E.; Mootha, V.K. Metabolic profiling of the human response to a glucose challenge reveals distinct axes of insulin sensitivity. *Mol. Syst. Biol.* **2008**, *4*, 214. [CrossRef]
69. Wang, T.J.; Larson, M.G.; Vasan, R.S.; Cheng, S.; Rhee, E.P.; McCabe, E.; Lewis, G.D.; Fox, C.S.; Jacques, P.F.; Fernandez, C.; et al. Metabolite profiles and the risk of developing diabetes. *Nat. Med.* **2011**, *17*, 448–453. [CrossRef]
70. Seibert, R.; Abbasi, F.; Hantash, F.M.; Caulfield, M.P.; Reaven, G.; Kim, S.H. Relationship between insulin resistance and amino acids in women and men. *Physiol. Rep.* **2015**, *3*, e12392. [CrossRef]
71. Calvani, R.; Miccheli, A.; Capuani, G.; Miccheli, A.T.; Puccetti, C.; Delfini, M.; Iaconelli, A.; Nanni, G.; Mingrone, G. Gut microbiome-derived metabolites characterize a peculiar obese urinary metabotype. *Int. J. Obes.* **2010**, *34*, 1095–1098. [CrossRef]

72. Liu, R.; Hong, J.; Xu, X.; Feng, Q.; Zhang, D.; Gu, Y.; Shi, J.; Zhao, S.; Liu, W.; Wang, X.; et al. Gut microbiome and serum metabolome alterations in obesity and after weight-loss intervention. *Nat. Med.* **2017**, *23*, 859–868. [CrossRef] [PubMed]
73. Org, E.; Blum, Y.; Kasela, S.; Mehrabian, M.; Kuusisto, J.; Kangas, A.J.; Soininen, P.; Wang, Z.; Ala-Korpela, M.; Hazen, S.L.; et al. Relationships between gut microbiota, plasma metabolites, and metabolic syndrome traits in the METSIM cohort. *Genome Biol.* **2017**, *18*, 70. [CrossRef]
74. David, L.A.; Maurice, C.F.; Carmody, R.N.; Gootenberg, D.B.; Button, J.E.; Wolfe, B.E.; Ling, A.V.; Devlin, A.S.; Varma, Y.; Fischbach, M.A.; et al. Diet rapidly and reproducibly alters the human gut microbiome. *Nature* **2014**, *505*, 559–563. [CrossRef] [PubMed]
75. Hawkes, K.; O'Connell, J.F.; Jones, N.G.B.; Oftedal, O.T.; Blumenschine, R.J. Hunting income patterns among the Hadza: Big game, common goods, foraging goals and the evolution of the human diet. *Philos. Trans. R. Soc. B Biol. Sci.* **1991**, *334*, 243–251. [CrossRef]
76. Heianza, Y.; Qi, L. Gene-Diet Interaction and Precision Nutrition in Obesity. *Int. J. Mol. Sci.* **2017**, *18*, 787. [CrossRef]
77. Johnson, A.J.; Vangay, P.; Al-Ghalith, G.A.; Hillmann, B.M.; Ward, T.L.; Shields-Cutler, R.R.; Kim, A.D.; Shmagel, A.K.; Syed, A.N.; Walter, J.; et al. Daily Sampling Reveals Personalized Diet-Microbiome Associations in Humans. *Cell Host Microbe* **2019**, *25*, 789–802.e5. [CrossRef]
78. McDonald, T.J.W.; Cervenka, M.C. Ketogenic Diets for Adult Neurological Disorders. *Neurotherapeutics* **2018**, *15*, 1018–1031. [CrossRef] [PubMed]
79. Ma, D.; Wang, A.C.; Parikh, I.; Green, S.J.; Hoffman, J.D.; Chlipala, G.; Murphy, M.P.; Sokola, B.S.; Bauer, B.; Hartz, A.M.S.; et al. Ketogenic diet enhances neurovascular function with altered gut microbiome in young healthy mice. *Sci. Rep.* **2018**, *8*, 6670. [CrossRef]
80. Newell, C.; Bomhof, M.R.; Reimer, R.A.; Hittel, D.S.; Rho, J.M.; Shearer, J. Ketogenic diet modifies the gut microbiota in a murine model of autism spectrum disorder. *Mol. Autism* **2016**, *7*, 37. [CrossRef] [PubMed]
81. Attaye, I.; van Oppenraaij, S.; Warmbrunn, M.V.; Nieuwdorp, M. The Role of the Gut Microbiota on the Beneficial Effects of Ketogenic Diets. *Nutrients* **2021**, *14*, 191. [CrossRef]
82. Licha, D.; Vidali, S.; Aminzadeh-Gohari, S.; Alka, O.; Breitkreuz, L.; Kohlbacher, O.; Reischl, R.J.; Feichtinger, R.G.; Kofler, B.; Huber, C.G. Untargeted Metabolomics Reveals Molecular Effects of Ketogenic Diet on Healthy and Tumor Xenograft Mouse Models. *Int. J. Mol. Sci.* **2019**, *20*, 3873. [CrossRef] [PubMed]
83. Xie, G.; Zhou, Q.; Qiu, C.-Z.; Dai, W.-K.; Wang, H.-P.; Li, Y.-H.; Liao, J.-X.; Lu, X.-G.; Lin, S.-F.; Ye, J.-H.; et al. Ketogenic diet poses a significant effect on imbalanced gut microbiota in infants with refractory epilepsy. *World J. Gastroenterol.* **2019**, *23*, 6164–6171. [CrossRef] [PubMed]
84. Swidsinski, A.; Dörffel, Y.; Loening-Baucke, V.; Gille, C.; Göktas, Ö.; Reißhauer, A.; Neuhaus, J.; Weylandt, K.-H.; Guschin, A.; Bock, M. Reduced Mass and Diversity of the Colonic Microbiome in Patients with Multiple Sclerosis and Their Improvement with Ketogenic Diet. *Front. Microbiol.* **2017**, *8*, 1141. [CrossRef]
85. Gutiérrez-Repiso, C.; Molina-Vega, M.; Bernal-López, M.R.; Garrido-Sánchez, L.; García-Almeida, J.M.; Sajoux, I.; Moreno-Indias, I.; Tinahones, F.J. Different Weight Loss Intervention Approaches Reveal a Lack of a Common Pattern of Gut Microbiota Changes. *J. Pers. Med.* **2021**, *11*, 109. [CrossRef] [PubMed]
86. Morales-Marroquin, E.; Hanson, B.; Greathouse, L.; De La Cruz-Munoz, N.; Messiah, S.E. Comparison of methodological approaches to human gut microbiota changes in response to metabolic and bariatric surgery: A systematic review. *Obes. Rev.* **2020**, *21*, e13025. [CrossRef]
87. Clarke, S.; Murphy, E.F.; O'Sullivan, O.; Ross, R.; O'Toole, P.; Shanahan, F.; Cotter, P.D. Targeting the Microbiota to Address Diet-Induced Obesity: A Time Dependent Challenge. *PLoS ONE* **2013**, *8*, e65790. [CrossRef]
88. Olson, C.A.; Vuong, H.E.; Yano, J.M.; Liang, Q.Y.; Nusbaum, D.J.; Hsiao, E.Y. The Gut Microbiota Mediates the Anti-Seizure Effects of the Ketogenic Diet. *Cell* **2018**, *173*, 1728–1741.e13, Erratum in *Cell* **2018**, *174*, 497. [CrossRef]
89. Deledda, A.; Palmas, V.; Heidrich, V.; Fosci, M.; Lombardo, M.; Cambarau, G.; Lai, A.; Melis, M.; Loi, E.; Loviselli, A.; et al. Dynamics of Gut Microbiota and Clinical Variables after Ketogenic and Mediterranean Diets in Drug-Naïve Patients with Type 2 Diabetes Mellitus and Obesity. *Metabolites* **2022**, *12*, 1092. [CrossRef]
90. Zeng, Q.; Li, D.; He, Y.; Li, Y.; Yang, Z.; Zhao, X.; Liu, Y.; Wang, Y.; Sun, J.; Feng, X.; et al. Discrepant gut microbiota markers for the classification of obesity-related metabolic abnormalities. *Sci. Rep.* **2019**, *9*, 13424. [CrossRef]
91. Gallardo-Becerra, L.; Cornejo-Granados, F.; Lopez, R.G.; Valdez-Lara, A.; Bikel, S.; Canizales-Quinteros, S.; López-Contreras, B.E.; Mendoza-Vargas, A.; Nielsen, H.; Ochoa-Leyva, A. Metatranscriptomic analysis to define the Secrebiome, and 16S rRNA profiling of the gut microbiome in obesity and metabolic syndrome of Mexican children. *Microb. Cell Factories* **2020**, *19*, 61. [CrossRef]
92. Wang, K.; Liao, M.; Zhou, N.; Bao, L.; Ma, K.; Zheng, Z.; Wang, Y.; Liu, C.; Wang, W.; Wang, J.; et al. Parabacteroides distasonis Alleviates Obesity and Metabolic Dysfunctions via Production of Succinate and Secondary Bile Acids. *Cell Rep.* **2019**, *26*, 222–235.e5. [CrossRef] [PubMed]
93. Booth, I.R.; Kroll, R.G. Regulation of cytoplasmic pH (pH1) in bacteria and its relationship to metabolism. *Biochem. Soc. Trans.* **1983**, *11*, 70–72. [CrossRef]
94. Davis, R.A.H.; Deemer, S.E.; Bergeron, J.M.; Little, J.T.; Warren, J.L.; Fisher, G.; Smith, D.L., Jr.; Fontaine, K.R.; Dickinson, S.L.; Allison, D.B.; et al. Dietary R, S -1,3-butanediol diacetoacetate reduces body weight and adiposity in obese mice fed a high-fat diet. *FASEB J.* **2019**, *33*, 2409–2421. [CrossRef] [PubMed]
95. Srivastava, S.; Kashiwaya, Y.; King, M.T.; Baxa, U.; Tam, J.; Niu, G.; Chen, X.; Clarke, K.; Veech, R.L. Mitochondrial biogenesis and increased uncoupling protein 1 in brown adipose tissue of mice fed a ketone ester diet. *FASEB J.* **2012**, *26*, 2351–2362. [CrossRef]

96. Stubbs, B.J.; Cox, P.J.; Evans, R.D.; Cyranka, M.; Clarke, K.; De Wet, H. A Ketone Ester Drink Lowers Human Ghrelin and Appetite. *Obesity* **2017**, *26*, 269–273. [CrossRef]
97. Gutiérrez-Repiso, C.; Hernández-García, C.; García-Almeida, J.M.; Bellido, D.; Martín-Núñez, G.M.; Sánchez-Alcoholado, L.; Alcaide-Torres, J.; Sajoux, I.; Tinahones, F.J.; Moreno-Indias, I. Effect of Synbiotic Supplementation in a Very-Low-Calorie Ketogenic Diet on Weight Loss Achievement and Gut Microbiota: A Randomized Controlled Pilot Study. *Mol. Nutr. Food Res.* **2019**, *63*, 1900167. [CrossRef] [PubMed]
98. Guo, Y.; Huang, Z.-P.; Liu, C.-Q.; Qi, L.; Sheng, Y.; Zou, D.-J. Modulation of the gut microbiome: A systematic review of the effect of bariatric surgery. *Eur. J. Endocrinol.* **2018**, *178*, 43–56. [CrossRef]
99. Palleja, A.; Kashani, A.; Allin, K.H.; Nielsen, T.; Zhang, C.; Li, Y.; Brach, T.; Liang, S.; Feng, Q.; Jørgensen, N.B.; et al. Roux-en-Y gastric bypass surgery of morbidly obese patients induces swift and persistent changes of the individual gut microbiota. *Genome Med.* **2016**, *8*, 67. [CrossRef]
100. Cerdó, T.; Ruiz, A.; Jáuregui, R.; Azaryah, H.; Torres-Espínola, F.J.; García-Valdés, L.; Segura, M.T.; Suárez, A.; Campoy, C. Maternal obesity is associated with gut microbial metabolic potential in offspring during infancy. *J. Physiol. Biochem.* **2017**, *74*, 159–169. [CrossRef]
101. Davis, C.; Bryan, J.; Hodgson, J.; Murphy, K. Definition of the Mediterranean Diet; A Literature Review. *Nutrients* **2015**, *7*, 9139–9153. [CrossRef]
102. Vrdoljak, J.; Kumric, M.; Vilovic, M.; Martinovic, D.; Tomic, I.J.; Krnic, M.; Kurir, T.T.; Bozic, J. Effects of Olive Oil and Its Components on Intestinal Inflammation and Inflammatory Bowel Disease. *Nutrients* **2022**, *14*, 757. [CrossRef] [PubMed]
103. Grahovac, M.; Kumric, M.; Vilovic, M.; Martinovic, D.; Kreso, A.; Kurir, T.T.; Vrdoljak, J.; Prizmic, K.; Božić, J. Adherence to Mediterranean diet and advanced glycation endproducts in patients with diabetes. *World J. Diabetes* **2021**, *12*, 1942–1956. [CrossRef] [PubMed]
104. Martinovic, D.; Tokic, D.; Martinovic, L.; Vilovic, M.; Vrdoljak, J.; Kumric, M.; Bukic, J.; Kurir, T.T.; Tavra, M.; Bozic, J. Adherence to Mediterranean Diet and Tendency to Orthorexia Nervosa in Professional Athletes. *Nutrients* **2022**, *14*, 237. [CrossRef] [PubMed]
105. Meslier, V.; Laiola, M.; Roager, H.M.; De Filippis, F.; Roume, H.; Quinquis, B.; Giacco, R.; Mennella, I.; Ferracane, R.; Pons, N.; et al. Mediterranean diet intervention in overweight and obese subjects lowers plasma cholesterol and causes changes in the gut microbiome and metabolome independently of energy intake. *Gut* **2020**, *69*, 1258–1268. [CrossRef] [PubMed]
106. Muralidharan, J.; Moreno-Indias, I.; Bulló, M.; Lopez, J.V.; Corella, D.; Castañer, O.; Vidal, J.; Atzeni, A.; Fernandez-García, J.C.; Torres-Collado, L.; et al. Effect on gut microbiota of a 1-y lifestyle intervention with Mediterranean diet compared with energy-reduced Mediterranean diet and physical activity promotion: PREDIMED-Plus Study. *Am. J. Clin. Nutr.* **2021**, *114*, 1148–1158. [CrossRef]
107. Pagliai, G.; Russo, E.; Niccolai, E.; Dinu, M.; Di Pilato, V.; Magrini, A.; Bartolucci, G.; Baldi, S.; Menicatti, M.; Giusti, B.; et al. Influence of a 3-month low-calorie Mediterranean diet compared to the vegetarian diet on human gut microbiota and SCFA: The CARDIVEG Study. *Eur. J. Nutr.* **2019**, *59*, 2011–2024. [CrossRef]
108. Nagpal, R.; Neth, B.J.; Wang, S.; Craft, S.; Yadav, H. Modified Mediterranean-ketogenic diet modulates gut microbiome and short-chain fatty acids in association with Alzheimer's disease markers in subjects with mild cognitive impairment. *Ebiomedicine* **2019**, *47*, 529–542. [CrossRef]
109. Di Majo, D.; Cacciabaudo, F.; Accardi, G.; Gambino, G.; Giglia, G.; Ferraro, G.; Candore, G.; Sardo, P. Ketogenic and Modified Mediterranean Diet as a Tool to Counteract Neuroinflammation in Multiple Sclerosis: Nutritional Suggestions. *Nutrients* **2022**, *14*, 2384. [CrossRef]
110. Brinkley, T.E.; Leng, I.; Register, T.C.; Neth, B.J.; Zetterberg, H.; Blennow, K.; Craft, S. Changes in Adiposity and Cerebrospinal Fluid Biomarkers Following a Modified Mediterranean Ketogenic Diet in Older Adults at Risk for Alzheimer's Disease. *Front. Neurosci.* **2022**, *16*, 906539. [CrossRef]
111. Barber, C.; Mego, M.; Sabater, C.; Vallejo, F.; Bendezu, R.A.; Masihy, M.; Guarner, F.; Espín, J.C.; Margolles, A.; Azpiroz, F. Differential Effects of Western and Mediterranean-Type Diets on Gut Microbiota: A Metagenomics and Metabolomics Approach. *Nutrients* **2021**, *13*, 2638. [CrossRef]
112. Gutiérrez-Repiso, C.; Moreno-Indias, I.; Tinahones, F.J. Shifts in gut microbiota and their metabolites induced by bariatric surgery. Impact of factors shaping gut microbiota on bariatric surgery outcomes. *Rev. Endocr. Metab. Disord.* **2021**, *22*, 1137–1156. [CrossRef] [PubMed]
113. Ge, T.; Yang, J.; Zhou, S.; Wang, Y.; Li, Y.; Tong, X. The Role of the Pentose Phosphate Pathway in Diabetes and Cancer. *Front. Endocrinol.* **2020**, *11*, 365. [CrossRef] [PubMed]
114. Li, J.; Ashrafian, H.; Bueter, M.; Kinross, J.; Sands, C.; Le Roux, C.; Bloom, S.R.; Darzi, A.; Athanasiou, T.; Marchesi, J.R.; et al. Metabolic surgery profoundly influences gut microbial-host metabolic cross-talk. *Gut* **2011**, *60*, 1214–1223. [CrossRef]
115. Sánchez-Alcoholado, L.; Gutiérrez-Repiso, C.; Gómez-Pérez, A.M.; García-Fuentes, E.; Tinahones, F.J.; Moreno-Indias, I. Gut microbiota adaptation after weight loss by Roux-en-Y gastric bypass or sleeve gastrectomy bariatric surgeries. *Surg. Obes. Relat. Dis.* **2019**, *15*, 1888–1895. [CrossRef] [PubMed]
116. O'May, G.A.; Reynolds, N.; Macfarlane, G.T. Effect of pH on an In Vitro Model of Gastric Microbiota in Enteral Nutrition Patients. *Appl. Environ. Microbiol.* **2005**, *71*, 4777–4783. [CrossRef]

117. O'May, G.A.; Reynolds, N.; Smith, A.R.; Kennedy, A.; Macfarlane, G.T. Effect of pH and Antibiotics on Microbial Overgrowth in the Stomachs and Duodena of Patients Undergoing Percutaneous Endoscopic Gastrostomy Feeding. *J. Clin. Microbiol.* **2005**, *43*, 3059–3065. [CrossRef]
118. Ilhan, Z.E.; Marcus, A.K.; Kang, D.-W.; Rittmann, B.E.; Krajmalnik-Brown, R. pH-Mediated Microbial and Metabolic Interactions in Fecal Enrichment Cultures. *Msphere* **2017**, *2*, e00047-17. [CrossRef]
119. Tremaroli, V.; Karlsson, F.; Werling, M.; Ståhlman, M.; Kovatcheva-Datchary, P.; Olbers, T.; Fändriks, L.; le Roux, C.W.; Nielsen, J.; Bäckhed, F. Roux-en-Y Gastric Bypass and Vertical Banded Gastroplasty Induce Long-Term Changes on the Human Gut Microbiome Contributing to Fat Mass Regulation. *Cell Metab.* **2015**, *22*, 228–238. [CrossRef]
120. Kong, L.-C.; Tap, J.; Aron-Wisnewsky, J.; Pelloux, V.; Basdevant, A.; Bouillot, J.-L.; Zucker, J.-D.; Doré, J.; Clément, K. Gut microbiota after gastric bypass in human obesity: Increased richness and associations of bacterial genera with adipose tissue genes. *Am. J. Clin. Nutr.* **2013**, *98*, 16–24. [CrossRef]
121. Aron-Wisnewsky, J.; Prifti, E.; Belda, E.; Ichou, F.; Kayser, B.D.; Dao, M.C.; Verger, E.O.; Hedjazi, L.; Bouillot, J.-L.; Chevallier, J.-M.; et al. Major microbiota dysbiosis in severe obesity: Fate after bariatric surgery. *Gut* **2019**, *68*, 70–82. [CrossRef]
122. Glick-Bauer, M.; Yeh, M.-C. The Health Advantage of a Vegan Diet: Exploring the Gut Microbiota Connection. *Nutrients* **2014**, *6*, 4822–4838. [CrossRef] [PubMed]
123. Xiao, W.; Zhang, Q.; Yu, L.; Tian, F.; Chen, W.; Zhai, Q. Effects of vegetarian diet-associated nutrients on gut microbiota and intestinal physiology. *Food Sci. Hum. Wellness* **2022**, *11*, 208–217. [CrossRef]
124. Sidhu, S.R.K.; Kok, C.W.; Kunasegaran, T.; Ramadas, A. Effect of Plant-Based Diets on Gut Microbiota: A Systematic Review of Interventional Studies. *Nutrients* **2023**, *15*, 1510. [CrossRef] [PubMed]
125. Losno, E.A.; Sieferle, K.; Perez-Cueto, F.J.A.; Ritz, C. Vegan Diet and the Gut Microbiota Composition in Healthy Adults. *Nutrients* **2021**, *13*, 2402. [CrossRef] [PubMed]
126. Rosario, V.D.; Fernandes, R.; Trindade, E.B.D.M. Vegetarian diets and gut microbiota: Important shifts in markers of metabolism and cardiovascular disease. *Nutr. Rev.* **2016**, *74*, 444–454. [CrossRef]
127. Salonen, A.; de Vos, W.M. Impact of Diet on Human Intestinal Microbiota and Health. *Annu. Rev. Food Sci. Technol.* **2014**, *5*, 239–262. [CrossRef]
128. Ley, R.E.; Bäckhed, F.; Turnbaugh, P.; Lozupone, C.A.; Knight, R.D.; Gordon, J.I. Obesity alters gut microbial ecology. *Proc. Natl. Acad. Sci. USA* **2005**, *102*, 11070–11075. [CrossRef]
129. De Filippo, C.; Cavalieri, D.; Di Paola, M.; Ramazzotti, M.; Poullet, J.B.; Massart, S.; Collini, S.; Pieraccini, G.; Lionetti, P. Impact of diet in shaping gut microbiota revealed by a comparative study in children from Europe and rural Africa. *Proc. Natl. Acad. Sci. USA* **2010**, *107*, 14691–14696. [CrossRef]
130. Hayashi, H.; Sakamoto, M.; Benno, Y. Fecal Microbial Diversity in a Strict Vegetarian as Determined by Molecular Analysis and Cultivation. *Microbiol. Immunol.* **2002**, *46*, 819–831. [CrossRef] [PubMed]
131. Brüssow, H.; Parkinson, S.J. You are what you eat. *Nat. Biotechnol.* **2014**, *32*, 243–245. [CrossRef]
132. Menni, C.; Lin, C.; Cecelja, M.; Mangino, M.; Matey-Hernandez, M.L.; Keehn, L.; Mohney, R.P.; Steves, C.J.; Spector, T.D.; Kuo, C.-F.; et al. Gut microbial diversity is associated with lower arterial stiffness in women. *Eur. Heart J.* **2018**, *39*, 2390–2397. [CrossRef] [PubMed]
133. Ishii, C.; Nakanishi, Y.; Murakami, S.; Nozu, R.; Ueno, M.; Hioki, K.; Aw, W.; Hirayama, A.; Soga, T.; Ito, M.; et al. A Metabologenomic Approach Reveals Changes in the Intestinal Environment of Mice Fed on American Diet. *Int. J. Mol. Sci.* **2018**, *19*, 4079. [CrossRef] [PubMed]
134. Martínez, I.; Lattimer, J.M.; Hubach, K.L.; Case, J.A.; Yang, J.; Weber, C.G.; Louk, J.A.; Rose, D.J.; Kyureghian, G.; Peterson, D.A.; et al. Gut microbiome composition is linked to whole grain-induced immunological improvements. *ISME J.* **2013**, *7*, 269–280. [CrossRef] [PubMed]
135. Klimenko, N.S.; Tyakht, A.V.; Popenko, A.S.; Vasiliev, A.S.; Altukhov, I.A.; Ischenko, D.S.; Shashkova, T.I.; Efimova, D.A.; Nikogosov, D.A.; Osipenko, D.A.; et al. Microbiome Responses to an Uncontrolled Short-Term Diet Intervention in the Frame of the Citizen Science Project. *Nutrients* **2018**, *10*, 576. [CrossRef]
136. Scott, K.P.; Duncan, S.H.; Flint, H.J. Dietary fibre and the gut microbiota. *Nutr. Bull.* **2008**, *33*, 201–211. [CrossRef]

MDPI AG
Grosspeteranlage 5
4052 Basel
Switzerland
Tel.: +41 61 683 77 34

Nutrients Editorial Office
E-mail: nutrients@mdpi.com
www.mdpi.com/journal/nutrients

www.ingramcontent.com/pod-product-compliance
Lightning Source LLC
Chambersburg PA
CBHW040753250926
51850CB00007BB/1490
9783725861279